Radio Frequency Transistors

Principles and Practical Applications

Norm Dye
Helge Granberg

Butterworth-Heinemann
Boston London Oxford Singapore Sydney Toronto Wellington

Recognizing the importance of preserving what has been written, it is the policy of Butterworth-Heinemann to have the books it publishes printed on acid-free paper, and we exert our best efforts to that end.

Library of Congress Cataloging-in-Publication Data

Dye, Norm, 1929–
Radio frequency transistors : principles and practical applications / Norm Dye, Helge Granberg.
p. cm.
Includes bibliographical references and index.
ISBN 0–7506–9059–3 : $49.95
1. Power transistors. 2. Transistor amplifiers. 3. Transistor radio transmitters. 4. Amplifiers, Radio frequency. I. Granberg, Helge, 1932– . II. Title.
TK7871.92.D96 1993
621.384' 131—dc20 92–54695
CIP

Butterworth-Heinemann
80 Montvale Avenue
Stoneham, MA 02180

Linacre House, Jordan Hill
Oxford OX2 8DP
United Kingdom

10 9 8 7 6 5 4 3 2 1

Printed in the United States of America

Editorial, design, and production services provided by HighText Publications, Inc., San Diego, California.

CONTENTS

PREFACE

This book is about radio frequency (RF) transistors. It primarily focuses on applications viewed from the perspective of a semiconductor supplier who, over the years, has been involved not only in the manufacture of RF transistors but also their use in receivers, transmitters, plasma generators, magnetic resonance imaging, etc.

Since the late 1960s, Motorola Semiconductors has been at the forefront in the development of solid state transistors for use at radio frequencies. The authors have been a part of this development since 1970. Much information has been acquired during this time, and it is our intention in writing this book to make the bulk of that information available to users of RF transistors in a concise manner and from a single source.

This book is not theoretical. It is intended to be practical as the name implies. Some mathematics is encountered during the course of the book but it is not rigorous. Formulas are not derived; however, sufficient references are cited for the reader who wishes to delve deeper into a particular subject.

This book is slanted toward power transistors and their applications because much less material is available in the literature on this subject, particularly in one location such as a book. Also, RF power is the primary experience of the authors. One chapter is devoted to low power (small signal) transistor applications in an effort to cover more completely the breadth of power levels in RF transistors.

Chapters 1 through 4 talk about RF transistor fundamentals, such as what's different about RF transistors, how they are specified, how to select a transistor, and the differences in FETs and BJTs. Also covered are topics such as classes of operation, forms of modulation, biasing and operating in a pulse mode. Chapters 5 and 6 lay the groundwork for future circuit designs by discussing such general subjects as laying out circuit boards, mounting RF devices and the importance of die temperature.

In Chapters 7, 8 and 9, the authors take the reader through various considerations in planning an amplifier design. Among the diverse topics covered are choice of circuit, stability, impedance matching (including computer aided design programs), and the power amplifier output. Chapters 10 through 12 focus on wideband techniques.

Finally, Chapter 13 describes the many factors affecting small signal (low power) amplifier design. A variety of examples illustrate the concepts in an effort to make small signal amplifier design straight forward through a step-by-step approach.

Acknowledgments

The authors wish to thank the many application engineers in the RF product operation at Motorola Semiconductors for their contributions to the book. Special recognition goes to Phuong Le for his assistance in low power applications, to Dan Moline for making available his recently introduced computer program for impedance matching with the aid of Smith Chart displays, to Bob Baeten for his assistance in computer aided design programs, to Walt Wright for answering many questions about microwaves and pulse power applications, and to Hank Pfizenmayer for his advice and expertise in filter design. Special thanks also go to Analog Instruments Co., Box 808, New Providence, NJ 07974, for their permission to reproduce the Smith Chart in several diagrams in Chapter 13. "Smith" is a registered trademark of Analog Instruments.

And special thanks go to the management of the Communications Semiconductor Products Division within Motorola Semiconductor Sector whose encouragement and support has made writing this book possible.

1

Understanding RF Data Sheet Parameters

INTRODUCTION

Data sheets are often the sole source of information about the capability and characteristics of a product. This is particularly true of unique RF semiconductor devices that are used by equipment designers all over the world. Because the circuit designer often cannot talk directly with the factory, he relies on the data sheet for his device information.[1] And for RF devices, many of the specifications are unique in themselves. Thus it is important that the user and the manufacturer of RF products speak a common language, i.e., what the semiconductor manufacturer says about his RF device is understood fully by the circuit designer.

In this chapter, a review is given of RF transistor and amplifier module parameters from maximum ratings to functional characteristics. The section is divided into five basic parts: D.C. specifications, power transistors, low power transistors, power modules, and linear modules. Comments are made about critical specifications, about how values are determined and what are their significance. A brief description of the procedures used to obtain impedance data and thermal data is set forth, the importance of test circuits is elaborated, and background information is given to help understand low noise considerations and linearity requirements.

D.C. SPECIFICATIONS

Basically, RF transistors are characterized by two types of parameters: D.C. and functional. The "D.C." specs consist (by definition) of breakdown voltages, leakage currents, h_{FE} (D.C. beta). and capacitances, while the functional specs cover gain, ruggedness, noise figure, Z_{in} and Z_{out}, S-parameters, distortion, etc. Thermal characteristics do not fall cleanly into either category since thermal resistance and power dissipation can be either D.C. or A.C. Thus, we will treat the spec of thermal resistance as a special specification and give it its own heading called "thermal characteristics." Figure 1-1 is one page of a typical RF power data sheet showing D.C. and functional specs.

ELECTRICAL CHARACTERISTICS (T_C = 25°C unless otherwise noted.)

Characteristic	Symbol	Min	Typ	Max	Unit
OFF CHARACTERISTICS					
Collector-Emitter Breakdown Voltage (I_C = 20 mAdc, I_B = 0)	$V_{(BR)CEO}$	16	–	–	Vdc
Collector-Emitter Breakdown Voltage (I_C = 20 mAdc, V_{BE} = 0)	$V_{(BR)CES}$	36	–	–	Vdc
Emitter-Base Breakdown Voltage (I_E = 5.0 mAdc, I_C = 0)	$V_{(BR)EBO}$	4.0	–	–	Vdc
Collector Cutoff Current (V_{CE} = 15 Vdc, V_{BE} = 0, T_C = 25°C)	I_{CES}	–	–	10	mAdc
ON CHARACTERISTICS					
DC Current Gain (I_C = 4.0 Adc, V_{CE} = 5.0 Vdc)	h_{FE}	20	70	150	–
DYNAMIC CHARACTERISTICS					
Output Capacitance (V_{CB} = 12.5 Vdc, I_E = 0, f = 1.0 MHz)	C_{ob}	–	90	125	pF
FUNCTIONAL TESTS					
Common-Emitter Amplifier Power Gain (V_{CC} = 12.5 Vdc, P_{out} = 45 W, I_C(Max) = 5.8 Adc, f = 470 MHz)	G_{pe}	4.8	5.4	–	dB
Input Power (V_{CC} = 12.5 Vdc, P_{out} = 45 W, f = 470 MHz)	P_{in}	–	13	15	Watts
Collector Efficiency (V_{CC} = 12.5 Vdc, P_{out} = 45 W, I_C (Max) = 5.8 Adc, f = 470 MHz)	η	55	60	–	%
Load Mismatch Stress (V_{CC} = 16 Vdc, P_{in} = Note 1, f = 470 MHz, VSWR = 20:1, All Phase Angles)	ψ*	No Degradation in Output Power			
Series Equivalent Input Impedance (V_{CC} = 12.5 Vdc, P_{out} = 45 W, f = 470 MHz)	Z_{in}	–	1.4 + j4.0	–	Ohms
Series Equivalent Output Impedance (V_{CC} = 12.5 Vdc, P_{out} = 45 W, f = 470 MHz)	Z_{OL}*	–	1.2 + j2.8	–	Ohms

Notes:

1. P_{in} = 150% of Drive Requirement for 45 W output @ 12.5 V.

* ψ = Mismatch stress factor—the electrical criterion established to verify the device resistance to load mismatch failure. The mismatch stress test is accomplished in the standard test fixture (Figure 1) terminated in a 20:1 minimum load mismatch at all phase angles.

FIGURE 1-1

Typical D.C. and functional specifications from a RF power data sheet. The references in the "Notes" above to a test fixture and "Figure 1" pertain to the data sheet from which this figure was extracted.

A critical part of selecting a transistor is choosing one that has *breakdown voltages* compatible with the supply voltage available in an intended application. It is important that the design engineer select a transistor on the one hand that has breakdown voltages which will NOT be exceeded by the DC and RF voltages that appear across the various junctions of the transistor and on the other hand has breakdown voltages that permit the "gain at frequency" objectives to be met by the transistor. Mobile radios normally operate from a 12 volt source and portable radios use a lower voltage, typically 6 to 9 volts. Avionics applications are commonly 28 volt supplies, while base station and other ground applications such as medical electronics generally take advantage of the superior performance characteristics of high voltage devices and operate with 24 to 50 volt supplies. In making a transistor, breakdown voltages are largely determined by material resistivity and junction depths (see Figure 1-2). It is for these reasons that breakdown voltages are intimately entwined with functional performance characteristics. Most product portfolios in the RF power transistor industry have families of transistors designed for use at specified supply voltages such as 7.5 volts, 12.5 volts, 28 volts, and 50 volts.

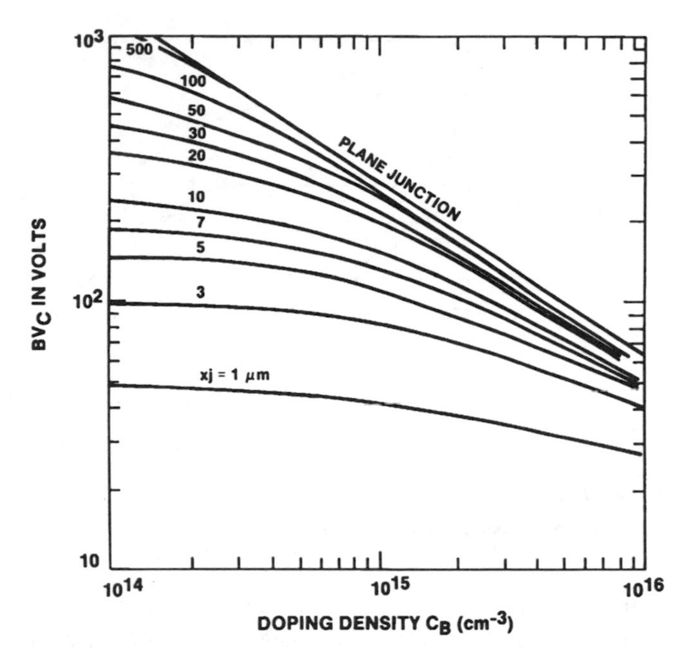

FIGURE 1-2
The effect of curvature and resistivity on breakdown voltage.

Leakage currents (defined as reverse biased junction currents that occur prior to avalanche breakdown) are likely to be more varied in their specification and also more informative. Many transistors do not have leakage currents specified because they can result in excessive (and frequently unnecessary) wafer/die yield losses. Leakage currents arise as a result of material defects, mask imperfections, and/or undesired impurities that enter during wafer processing. Some sources of leakage currents are potential reliability problems; most are not. Leakage currents can be material related such as stacking faults and dislocations or they can be "pipes" created by mask defects and/or processing inadequacies. These sources result in leakage currents that are constant with time and if initially acceptable for a particular application will remain so. They do not pose long term reliability problems.

On the other hand, leakage currents created by channels induced by mobile ionic contaminants in the oxide (primarily sodium) tend to change with time and can lead to increases in leakage current that render the device useless for a specific application. Distinguishing between sources of leakage current can be difficult, which is one reason devices for application in military environments require HTRB (high temperature reverse bias) and burn-in testing. However, even for commercial applications particularly where battery drain is critical or where bias considerations dictate limitations, it is essential that a leakage current limit be included in any complete device specification.

D.C. parameters such as h_{FE} and C_{ob} (output capacitance) need little comment. Typically, for RF devices, h_{FE} is relatively unimportant for unbiased power transistors because the functional parameter of gain at the desired frequency of operation is specified. Note, though, that D.C. beta is related to A.C. beta (see Figure 1-3). Functional gain will track D.C. beta particularly at lower RF fre-

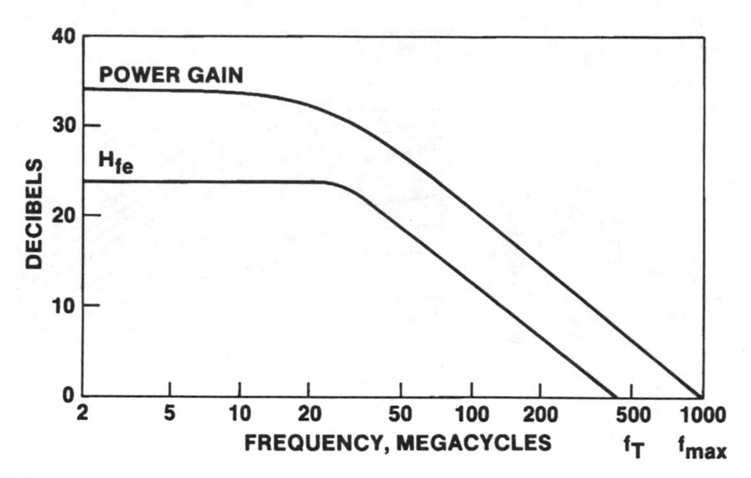

FIGURE 1-3
Relationship between transistor beta and operating frequency.

quencies. An h_{FE} specification is needed for transistors that require bias, which includes most small signal devices that are normally operated in a linear (Class A) mode (see Chapter 4). Generally RF device manufacturers do not like to have tight limits placed on h_{FE}. Primarily, the reasons that justify this position are:

a) Lack of correlation with RF performance
b) Difficulty in control in wafer processing
c) Other device manufacturing constraints dictated by functional performance specs which preclude tight limits for h_{FE}.

A good rule of thumb for h_{FE} is to set a maximum-to-minimum ratio of not less than 3 and not more than 4, with the minimum h_{FE} value determined by an acceptable margin in functional gain.

Output capacitance is an excellent measure of comparison of device size (base area) provided the majority of output capacitance is created by the base-collector junction and not parasitic capacitance arising from bond pads and other top metal of the die. Remember that junction capacitance will vary with voltage (see Figure 1-4) while parasitic capacitance will not vary. Also, in comparing devices, one should note the voltage at which a given capacitance is specified. No industry standard exists. The preferred voltage at Motorola is the transistor V_{cc} rating, i.e., 12.5 volts for 12.5 volt transistors and 28 volts for 28 volt transistors, etc.

MAXIMUM RATINGS AND THERMAL CHARACTERISTICS

Maximum ratings (shown for a typical RF power transistor in Figure 1-5) tend to be the most frequently misunderstood group of device specifications. Ratings for *maximum junction voltages* are straight forward and simply reflect the minimum values set forth in the D.C. specs for breakdown voltages. If the device in question meets the specified minimum breakdown voltages, then voltages less than the minimum will not cause junctions to reach reverse bias breakdown with the potentially destructive current levels that can result.

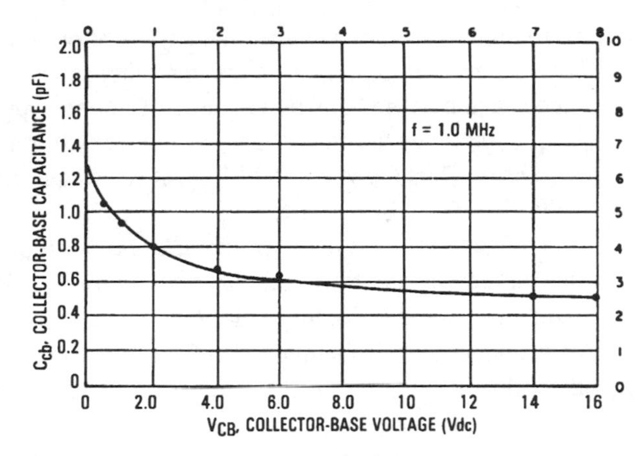

FIGURE 1-4
Relationship between junction capacitance vs. voltage for MRF.

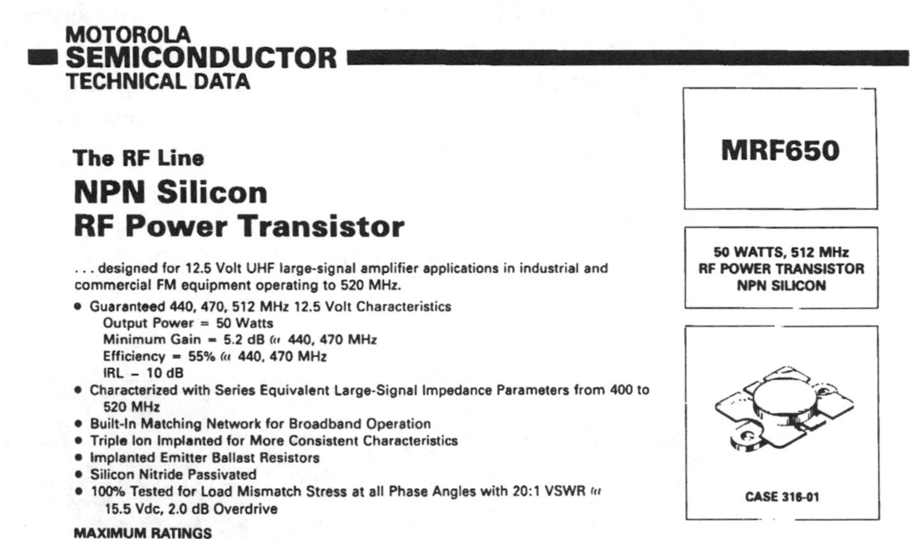

MOTOROLA
SEMICONDUCTOR
TECHNICAL DATA

MRF650

The RF Line
NPN Silicon
RF Power Transistor

50 WATTS, 512 MHz
RF POWER TRANSISTOR
NPN SILICON

. . . designed for 12.5 Volt UHF large-signal amplifier applications in industrial and commercial FM equipment operating to 520 MHz.

- Guaranteed 440, 470, 512 MHz 12.5 Volt Characteristics
  Output Power = 50 Watts
  Minimum Gain = 5.2 dB @ 440, 470 MHz
  Efficiency = 55% @ 440, 470 MHz
  IRL – 10 dB
- Characterized with Series Equivalent Large-Signal Impedance Parameters from 400 to 520 MHz
- Built-In Matching Network for Broadband Operation
- Triple Ion Implanted for More Consistent Characteristics
- Implanted Emitter Ballast Resistors
- Silicon Nitride Passivated
- 100% Tested for Load Mismatch Stress at all Phase Angles with 20:1 VSWR @ 15.5 Vdc, 2.0 dB Overdrive

CASE 316-01

MAXIMUM RATINGS

Rating	Symbol	Value	Unit
Collector-Emitter Voltage	V_{CEO}	16.5	Vdc
Collector-Emitter Voltage	V_{CES}	38	Vdc
Emitter-Base Voltage	V_{EBO}	4.0	Vdc
Collector-Current — Continuous	I_C	12	Adc
Total Device Dissipation @ T_C = 25°C Derate above 25°C	P_D	135 0.77	Watts W/°C
Storage Temperature Range	T_{stg}	–65 to +150	°C

THERMAL CHARACTERISTICS

Characteristic	Symbol	Max	Unit
Thermal Resistance, Junction to Case	$R_{\theta JC}$	1.3	°C/W

FIGURE 1-5
Maximum power ratings of a typical RF power transistor, the Motorola MRF650.

The value of BV_{CEO} is sometimes misunderstood. Its value can approach or even equal the supply voltage rating of the transistor. The question naturally arises as to how such a low voltage can be used in practical applications. First, BV_{CEO} is the breakdown voltage of the collector-base junction plus the forward drop across the base-emitter junction with the base open, and it is never encountered in amplifiers where the base is at or near the potential of the emitter. That is, most amplifiers have the base shorted or they use a low value of resistance such that the breakdown value of interest approaches BV_{CES}. Second, BV_{CEO} involves the current gain of the transistor and increases as frequency increases. Thus the value of BV_{CEO} at RF frequencies is always greater than the value at D.C.

The maximum rating for *power dissipation* (P_d) is closely associated with thermal resistance (θ_{JC}). Actually maximum P_d is in reality a fictitious number—a kind of figure of merit—because it is based on the assumption that case temperature is maintained at 25°C. However, providing everyone arrives at the value in a similar manner, the rating of maximum P_d is a useful tool with which to compare devices.

The rating begins with a determination of thermal resistance—die to case. Knowing θ_{JC} and assuming a maximum die temperature, one can easily determine maximum P_d (based on the previously stated case temperature of 25°C). Measuring θ_{JC} is normally done by monitoring case temperature (T_c) of the device while it operates at or near rated output power (P_o) in an RF circuit. The die temperature (T_j) is measured simultaneously using an infra-red microscope (see Figure 1-6) which has a spot size resolution as small as 1 mil in diameter. Normally, several readings are taken over the surface of the die and an average value is used to specify T_j.

It is true that temperatures over a die will vary typically 10–20°C. A poorly designed die (improper ballasting) could result in hot spot (worst case) temperatures that vary 40–50°C. Likewise, poor die bonds (see Figure 1-7) can result

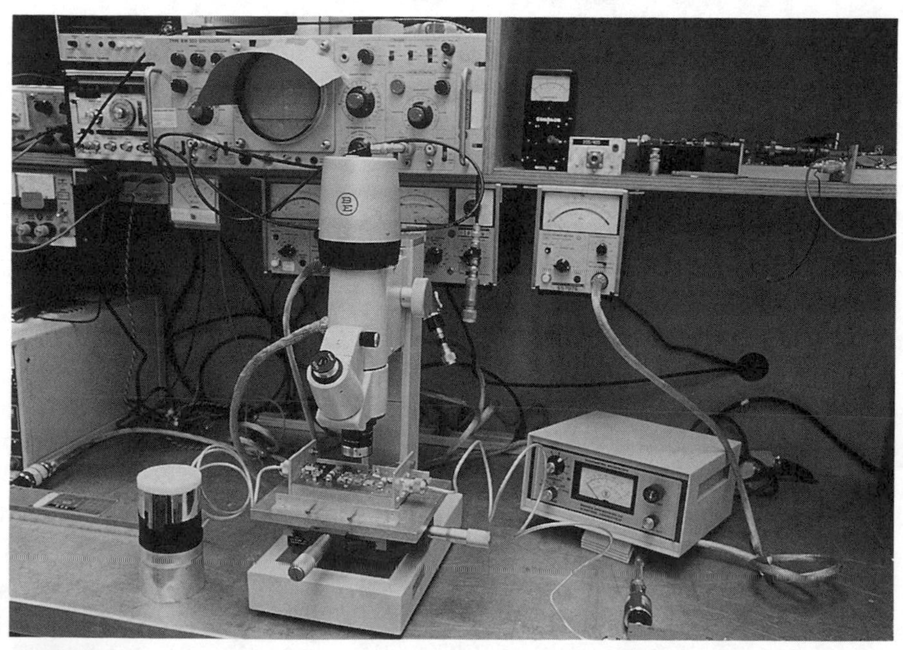

FIGURE 1-6
Measurement of die temperature using an infra-red microscope.

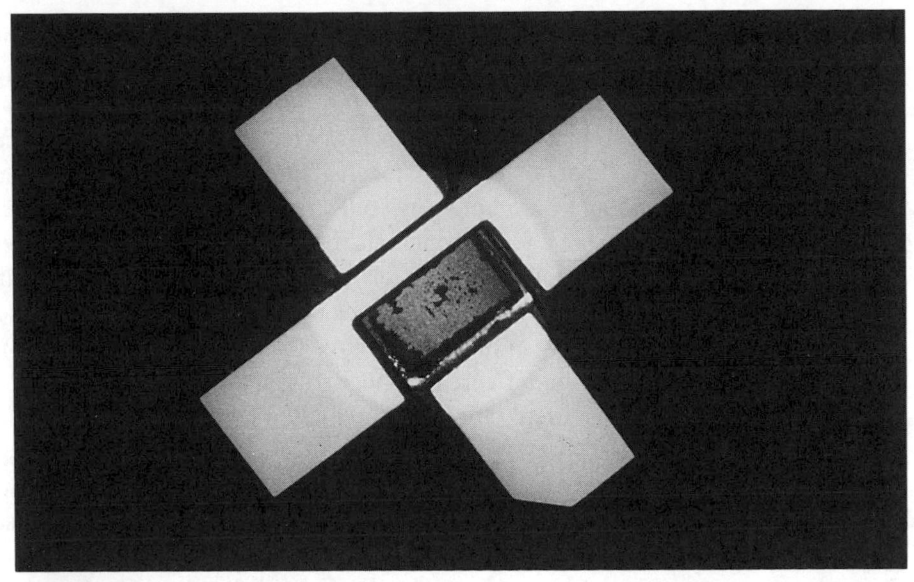

FIGURE 1-7
Voids appear as dark spots in X-ray photographs and will lead to "hot spots" in die temperatures.

in hot spots, but these are not normal characteristics of a properly designed and assembled transistor die.

By measuring T_c and T_j along with P_o and P_{in}—both D.C. and RF—one can calculate θ_{JC} from the formula $\theta_{JC} = (T_j - T_c)/(P_{in} - P_o)$. Typical values for an RF power transistor might be $T_j = 130°C$, $T_c = 50°C$; $V_{cc} = 12.5$ V, $I_c = 9.6$ A, P_{in} (RF) = 10 W, and P_o (RF) = 50 W. Thus $\theta_{JC} = (130 - 50)/(10 + \{12.5 \times 9.6\} - 30) = 80/80 = 1°C/W$.

Several reasons dictate a conservative value be placed on θ_{JC}. First, thermal resistance increases with temperature (and we realize $T_c = 25°C$ is NOT realistic). Second, T_j is not a worst case number. And, third, by using a conservative value of θ_{JC}, a realistic value is determined for maximum P_d. Generally, Motorola's practice is to publish θ_{JC} numbers approximately 25% higher than that determined by the measurements described in the preceding paragraphs, or for the case illustrated, a value of $\theta_{JC} = 1.25$ °C/W.

Now a few words are in order about die temperature. Reliability considerations dictate a safe value for an all Au (gold) system (die top metal and wire) to be 200°C. (See Chapter 5.) Once T_j max is determined, along with a value for θ_{JC}, maximum P_d is simply

$$P_d\,(\text{max}) = (T_j\,(\text{max}) - 25°C)/\theta_{JC}.$$

Specifying maximum P_d for $T_c = 25°C$ leads to the necessity to derate maximum P_d for any value of T_c above 25°C. The derating factor is simply the reciprocal of θ_{JC}!

Maximum collector current (I_c) is probably the most subjective maximum rating on the transistor data sheets. It has been, and is, determined in a number of ways each leading to different maximum values. Actually, the only valid maximum current limitations in an RF transistor have to do with the current handling ability of the wires or the die. However, power dissipation ratings may restrict current to values far below what should be the maximum rating. Unfortunately, many older transistors had their maximum current rating determined by dividing maximum P_d by collector voltage (or by BV_{CEO} for added safety) but this is not a fundamental maximum current limitation of the part. Many lower frequency parts have relatively gross top metal on the transistor die, i. e., wide metal runners and the "weak current link" in the part is the current handling capability of the emitter wires (for common emitter parts). The current handling ability of wire (various sizes and material) is well known; thus the maximum current rating may be limited by the number, size and material used for emitter wires.

Most modern high frequency transistors are die limited because of high current densities resulting from very small current carrying conductors, and these densities can lead to metal migration and premature failure. The determination of I_c max for these types of transistors results from use of Black's equation for metal migration[3] which determines a mean time between failures (MTBF) based on current density, temperature, and type of metal. At Motorola, MTBF is generally set at >7 years and maximum die temperature at 200°C. For plastic packaged transistors, maximum T_j is set at 150°C. The resulting current density along with a knowledge of the die geometry and top metal thickness and material allows the determination of I_c max for the device.

It is up to the transistor manufacturer to specify an I_c max based on which of the two limitations (die or wire) is paramount. It is recommended that the circuit design engineer consult the semiconductor manufacturer for additional information if I_c max is of any concern in his specific use of the transistor.

Storage temperature is another maximum rating that is frequently not given the attention it deserves. A range of –55°C to 200°C has become more or less an industry standard. And for the single metal, hermetic packaged type of device, the upper limit of 200°C creates no reliability problems. However, a lower high temperature limitation exists for plastic encapsulated or epoxy sealed devices. These should not be subjected to temperatures above 150°C to prevent deterioration of the plastic material.

POWER TRANSISTORS—FUNCTIONAL CHARACTERISTICS

The selection of a power transistor usually involves choosing one for a frequency of operation, a level of output power, a desired gain, a voltage of operation and preferred package configuration consistent with circuit construction techniques.

Functional characteristics of an RF power transistor are by necessity tied to a specific test circuit (an example is shown in Figure 1-8). Without specifying a circuit, the functional parameters of gain, reflected power, efficiency—even ruggedness—hold little meaning. Furthermore, most test circuits used by RF

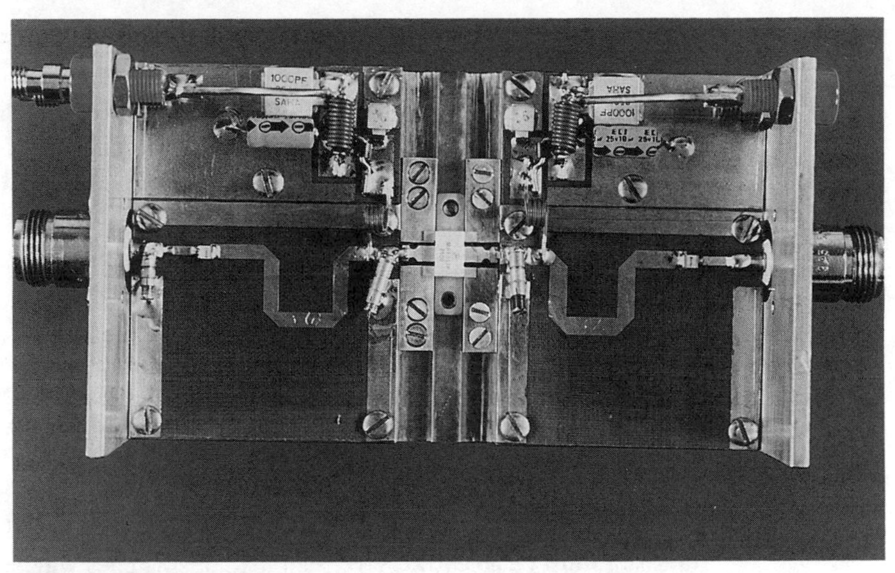

FIGURE 1-8
Test circuit for an RF power transistor.

transistor manufacturers today (even those used to characterize devices) are designed mechanically to allow for easy insertion and removal of the device under test (D.U.T.). This mechanical restriction sometimes limits achievable device performance which explains why performance by users frequently exceeds that indicated in data sheet curves. On the other hand, a circuit used to characterize a device is usually narrow band and tunable. This results in higher gain than attainable in a broadband circuit. Unless otherwise stated, it can be assumed that characterization data such as P_0 vs. frequency is generated on a point-by-point basis by tuning a narrow band circuit across a band of frequencies and, thus, represents what can be achieved at a specific frequency of interest provided the circuit presents optimum source and load impedances to the D.U.T.

Broadband, fixed tuned test circuits are the most desirable for testing functional performance of an RF transistor. Fixed tuned is particularly important in assuring everyone—the manufacturer and the user—of product consistency, i.e., that devices manufactured tomorrow will be identical to devices manufactured today.

Tunable, narrow band circuits have led to the necessity for device users and device manufacturers to resort to the use of "correlation units" to assure product consistency over a period of time. Fixed tuned circuits minimize (if not eliminate) the requirements for correlation. In so doing, they tend to compensate for the increased constraints they place on the device manufacturer. On the other hand, manufacturers like tunable test circuits because their use allows adjustments that can compensate for variations in die fabrication and/or device assembly. Unfortunately, gain is normally less in a broadband circuit than it is in a narrow band circuit, and this fact frequently forces transistor manufacturers to use narrow band circuits to make their product have sufficient attraction when compared with other similar devices made by competitors. This is called "specsmanship." One compromise for the transistor manufacturer is to use narrow band circuits with all tuning adjustments "locked" in place. For all of the above reasons, then, in comparing functional parameters of two or more devices, the data sheet reader should observe carefully the test circuit in which specific parameter limits are guaranteed.

For RF power transistors, the parameter of ruggedness takes on considerable importance. Ruggedness is the characteristic of a transistor to withstand extreme mismatch conditions in operation (which causes large amounts of output power to be "dumped back" into the transistor) without altering its performance capability or reliability. Many circuit environments, particularly portable and mobile radios, have limited control over the impedance presented to the power amplifier by an antenna (at least for some duration of time). In portables, the antenna may be placed against a metal surface; in mobiles, perhaps the antenna is broken off or inadvertently disconnected from the radio. Today's RF power transistor must be able to survive such load mismatches without any effect on subsequent operation. A truly realistic possibility for mobile radio transistors (although not a normal situation) is the condition whereby an RF power device "sees" a worst case load mismatch (an open circuit, any phase angle) along with maximum V_{cc} AND greater than normal input drive—all at the same time. Thus the ultimate test for ruggedness is to subject a transistor to a test wherein P_{in} (RF) is increased up to 50% above that value necessary to create rated P_o; V_{cc} is increased about 25% (12.5 V to 16 V for mobile transistors) AND then the load reflection coefficient is set at a magnitude of unity while its phase angle is varied through all possible values from 0 degrees to 360 degrees. Many 12 volt (land mobile) transistors are routinely given this test at Motorola Semiconductors by means of a test station similar to the one shown in Figure 1-9.

FIGURE 1-9
Test station for RF power transistors used by Motorola.

Ruggedness specifications come in many forms (or guises). Many older devices (and even some newer ones) simply have NO ruggedness spec. Others are said to be "capable of" withstanding load mismatches. Still others are guaranteed to withstand load mismatches of 2:1 VSWR to ∞:1 VSWR at rated output power. A few truly rugged transistors are guaranteed to withstand 30:1 VSWR at all phase angles (for all practical purposes 30:1 VSWR is the same as ∞:1 VSWR) with both overvoltage and overdrive. Once again, it is up to the user to match his circuit requirements against device specifications.

Then—as if the whole subject of ruggedness is not sufficiently confusing—the semiconductor manufacture slips in the ultimate "muddy the water" condition in stating what constitutes passing the ruggedness test. The words generally say that after the ruggedness test the D.U.T. "shall have no degradation in output power." A better phrase would be "no measurable change in output power." But even this is not the best. Unfortunately, the D.U.T. can be "damaged" by the ruggedness test and still have "no degradation in output power." Today's RF power transistors consist of up to 1000 or more low power transistors connected in parallel. Emitter resistors are placed in series with groups of these transistors in order to better control power sharing throughout the transistor die. It is well known by semiconductor manufacturers that a high percentage of an RF power transistor die (say up to 25–30%) can be destroyed with the transistor still able to deliver rated power at rated gain, at least for some period of time. If a ruggedness test destroys a high percentage of cells in a transistor, then it is likely that a second ruggedness test (by the manufacturer or by the user while in his circuit) would result in additional damage leading to premature device failure.

A more scientific measurement of "passing" or "failing" a ruggedness test is called ΔV_{re}, the change in emitter resistance before and after the ruggedness test. V_{re} is determined to a large extent by the net value of emitter resistance in the transistor die. Thus if cells are destroyed, emitter resistance will change with a resultant change in V_{re}. Changes as small as 1% are readily detectable, with 5% or less normally considered an acceptable limit. Today's more sophisticated device specifications for RF power transistors use this criteria to determine "success" or "failure" in ruggedness testing.

A circuit designer must know the input/output characteristics of the RF power transistor(s) he has selected in order to design a circuit that "matches" the transistor over the frequency band of operation. Data sheets provide this information in the form of large signal *impedance parameters,* Z_{in} and Z_{out} (commonly referred to as Z^*_{OL}). Normally, these are stated as a function of frequency and are plotted on a Smith Chart and/or given in tabular form. It should be noted that Z_{in} and Z_{out} apply only for a specified set of operating conditions, i.e., P_o, V_{cc}, and frequency. Both Z_{in} and Z_{out} of a device are determined in a similar way, i.e., place the D.U.T. in a tunable circuit and tune both input and output circuit elements to achieve maximum gain for the desired set of operating conditions. At maximum gain, D.U.T. impedances will be the conjugate of the input and output network impedances. Thus, terminate the input and output ports of the test circuit, remove the device and measure Z looking from the device—first, toward the input to obtain the conjugate of Z_{in}

and, second, toward the output to obtain Z_{OL} which is the output load required to achieve maximum P_o.

A network analyzer is used in the actual measurement process to determine the complex reflection coefficient of the circuit using, typically, the edge of the package as a plane of reference. A typical measurement setup is shown in Figure 1-10. Figure 1-11 shows the special fixture used to obtain the short circuit reference while Figure 1-12 illustrates the adapter which allows the circuit impedance to be measured from the edge of the package.

Once the circuit designer knows Z_{IN} and Z^*_{OL} of the transistor as a function of frequency, he can use computer aided design programs to design L and C matching networks for his particular application.

The entire impedance measuring process is somewhat laborious and time consuming since it must be repeated for each frequency of interest. Note that the frequency range permitted for characterization is that over which the circuit will tune. For other frequencies, additional test circuits must be designed and constructed, which explains why it is sometimes difficult to get a semiconductor manufacturer to supply impedance data for special conditions of operation such as different frequencies, different power levels or different operating voltages.

LOW POWER TRANSISTORS—FUNCTIONAL CHARACTERISTICS

Most semiconductor manufacturers characterize low power RF transistors for linear amplifier and/or low noise amplifier applications. Selecting a proper low power transistor involves choosing one having an adequate current rating, in the

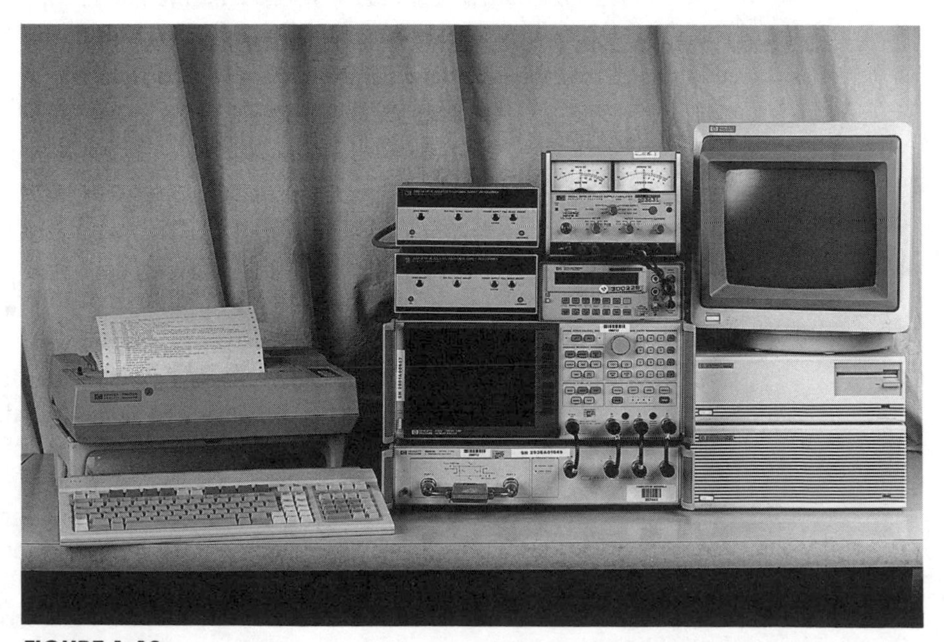

FIGURE 1-10
Impedance measurement setup.

FIGURE 1-11
Special fixture used to obtain short circuit reference.

FIGURE 1-12
Adapter that allows circuit impedance to be measured from the edge of the package.

"right" package, and with gain and noise figure capability that meets the requirements of the intended application.

One of the most useful means of specifying a linear device is by means of scattering parameters, commonly referred to as S-parameters which are in reality voltage reflection and transmission coefficients when the device is embedded into a 50 Ω system[4]; see Figure 1-13. $|S_{11}|$, the magnitude of the input reflection coefficient is directly related to input VSWR by the equation $VSWR = (1 + |S_{11}|) / (1 - |S_{11}|)$. Likewise, $|S_{22}|$, the magnitude of the output reflection coefficient is directly related to output VSWR. $|S_{21}|^2$, which is the square of the magnitude of the input-to-output transfer function, is also the power gain of the device. It is referred to on data sheets as "Insertion Gain." Note, however, that $|S_{21}|^2$ is the power gain of the device when the source and load impedances are 50 Ω. An improvement in gain can always be achieved by matching the device's input and output impedances (which are almost always different from 50 Ω) to 50 Ω by means of matching networks. The larger the linear device, the lower the impedances and the greater is the need to use matching networks to achieve useful gain.

Another gain specification shown on low power data sheets is called "Associated Gain." The symbol used for Associated Gain is G_{NF}. It is simply the gain of the device when matched for minimum noise figure. Yet another gain term is shown on some data sheets and it is called "Maximum Unilateral Gain." It's symbol is $G_{U\ max}$. As you might expect, $G_{U\ max}$ is the gain achievable by the transistor when the input and output are conjugately matched for maximum power transfer (and $S_{12} = 0$.). One can derive a value for $G_{U\ max}$ using scattering parameters:

$$G_{U\ max} = |S_{21}|^2 / \{(1 - |S_{11}|^2)(1 - |S_{22}|^2)\}.$$

Simply stated, this is the 50 Ω gain increased by a factor which represents matching the input and increased again by a factor which represents matching the output.

Many RF low power transistors are used as low noise amplifiers which has led to several transistor data sheet parameters related to noise figure. NF_{min} is defined as the minimum noise figure that can be achieved with the transistor. To achieve this NF requires source impedance matching which is usually different from that required to achieve maximum gain. The design of a low noise amplifier, then, is always a compromise between gain and NF. (For a more complete discussion of low power/low noise amplifier design, see Chapter 13.) A useful tool to aid in this compromise is a Smith Chart plot of constant gain and noise figure contours which can be drawn for specific operating conditions, typically

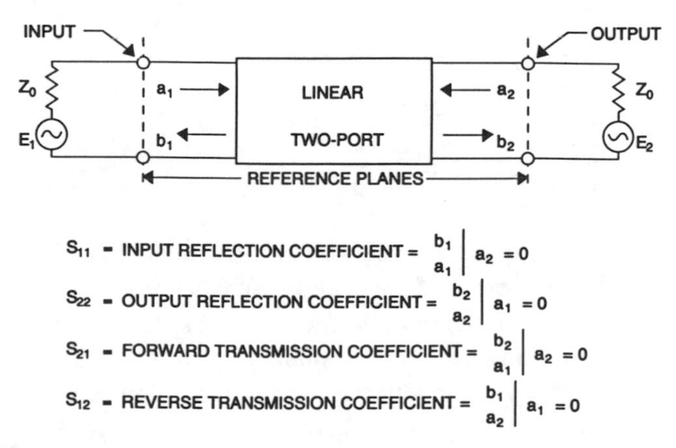

FIGURE 1-13
Two-port S-parameter definitions.

bias and frequency. A typical Smith Chart plot showing constant gain and NF contours is shown in Figure 1-14. These contours are circles which are either totally or partially complete within the confines of the Smith Chart. If the gain circles are contained entirely within the Smith Chart, then the device is unconditionally stable. If portions of the gain circles are outside the Smith Chart, then the device is considered to be "conditionally stable" and the device designer must concern himself with instabilities, particularly outside the normal frequency range of operation.

If the data sheet includes noise parameters[5], a value will be given for the optimum input reflection coefficient to achieve minimum noise figure. Its symbol is Γ_o, or sometimes $\Gamma_{opt.}$ But remember if you match this value of input reflection coefficient, you are likely to have far less gain than is achievable by the transistor. The input reflection coefficient for maximum gain is normally called Γ_{MS}, while the output reflection coefficient for maximum gain is normally called Γ_{ML}.

Another important noise parameter is noise resistance, which is given the symbol R_n and is expressed in ohms. Sometimes in tabular form, you may see this value normalized to 50 Ω, in which case it is designated r_n. The significance of r_n can be seen in the formula, $NF = NF_{min} + \{4r_n |\Gamma_s - \Gamma_o|^2\} / \{(1 - |\Gamma_s|^2) |1 + \Gamma_o|^2\}$, which determines noise figure NF of a transistor for any source reflection coefficient Γ_s if the three noise parameters—NF_{min}, r_n, and Γ_o (the source resistance for minimum noise figure)—are known. Typical noise parameters taken from the MRF942 data sheet are shown in Figure 1-15.

The locus of points for a given NF turns out to be a circle (the NF_{min} circle being a point); thus, by choosing different values of NF one can plot a series of noise circles on the Smith Chart. Incidentally, r_n can be measured by measuring noise figure for $\Gamma_s = 0$ and applying the equation stated above.

A parameter found on most RF low power data sheets is commonly called the current gain-bandwidth product. Its symbol is f_τ. Sometimes it is referred to

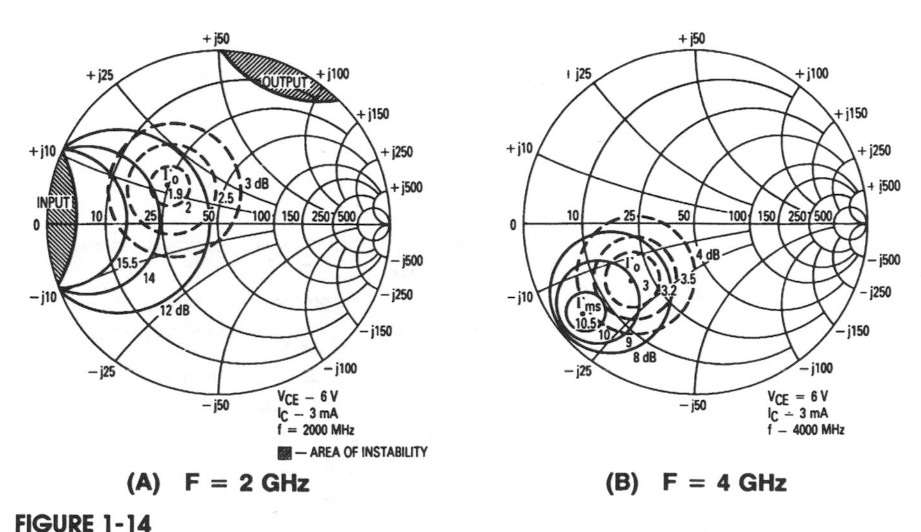

(A) F = 2 GHz (B) F = 4 GHz

FIGURE 1-14

Gain and noise contours. Solid circles represent gain and dotted circles represent noise figure.

MRF 942

V_{CE} (Vdc)	I_C (mA)	f (MHz)	NF_{min} (dB)	G_{NF} (dB)	Γ_o (MAG, ANG)	R_N (ohms)	$NF_{50\,\Omega}$ (dB)
6	3	1000	1.3	16	.36 ∠ 94	17.5	1.7
		2000	2.0	11	.37 ∠ −145	15.5	2.6
		4000	2.9	8.0	.50 ∠ 134	21.5	4.3
	15	1000	2.1	19	.25 ∠ 150	13	2.6
		2000	2.7	14	.26 ∠ 173	16.5	3.1
		4000	4.3	9.0	.48 ∠ −96	47	5.4

FIGURE 1-15

Typical noise parameters for the MRF942 transistor.

as the *cutoff frequency*, because it is generally thought to be the product of low frequency current gain and the frequency at which the current gain becomes unity. While this is not precisely true (see Figure 1-16), it is close enough for practical purposes[6]. And it is true that f_τ is an excellent figure-of-merit which becomes useful in comparing devices for gain and noise figure capability. High values of f_τ are normally required to achieve higher gain at higher frequencies, other factors being equal. To the device designer, high f_τ specs mean decreased spacings between emitter and base diffusions and it means shallower diffusions—things which are more difficult to achieve in making an RF transistor.

The complete RF low power transistor data sheet will include a plot of f_τ versus collector current. Such a curve (as shown in Figure 1-17) will increase with current, flatten, and then begin to decrease as I_c increases thereby revealing use-

WHERE $|h_{fe}|$ = MAGNITUDE OF SMALL-SIGNAL COMMON-EMITTER (CE) SHORT-CIRCUIT (SC) CURRENT GAIN, h_{fe}

h_{feo} = LOW-FREQUENCY VALUE OF h_{fe}

f_B = 3dB CUTOFF FREQUENCY FOR CE, SC CURRENT GAIN

f_τ = TRANSITION FREQUENCY = $|h_{fe}| \cdot f_{MEAS}$

WHERE $f_{MEAS.}$ = FREQUENCY OF MEASUREMENT

(NOTE: $2 \le |h_{fe}| \le \frac{h_{feo}}{2}$)

f_1 = FREQUENCY AT WHICH $|h_{fe}| = 1$

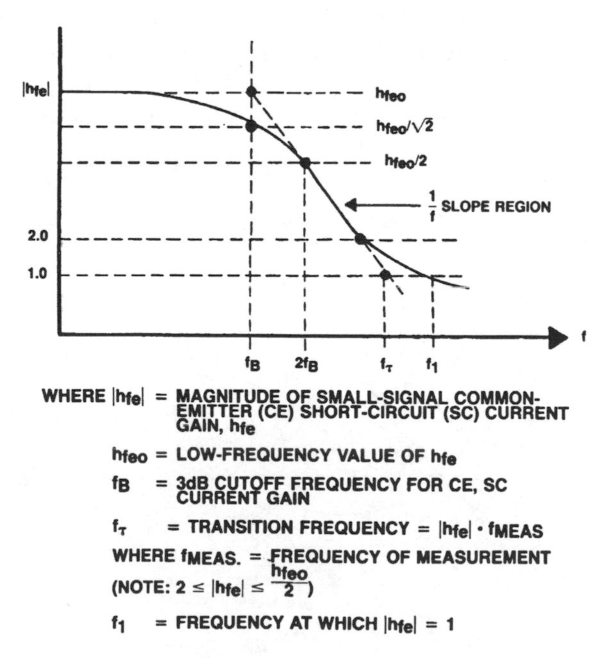

FIGURE 1-16

Small signal current gain vs. frequency.

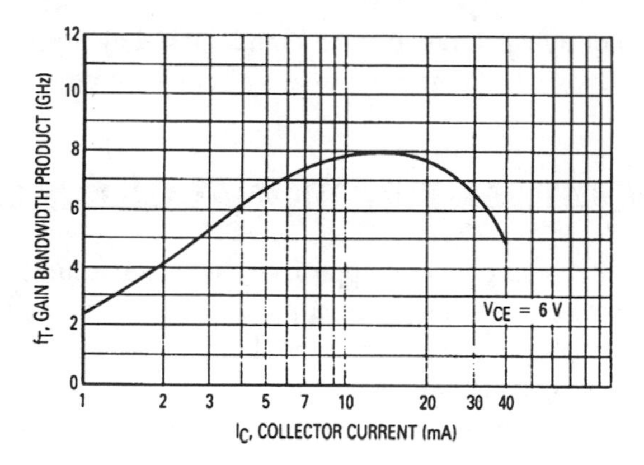

FIGURE 1-17
Gain-bandwidth product vs. collector current.

ful information about the optimum current with which to achieve maximum device gain.

Another group of characteristics associated with linear (or Class A) transistors has to do with the degree to which the device is linear. Most common are terms such as "P_o, 1 dB Gain Compression Point" and "Third Order Intercept Point" (or ITO as it is sometimes called). More will be said about non-linearities and distortion measurements in the section about linear amplifiers; however, suffice it to be said now that "Po, 1 dB Gain Compression Point" is simply the output power at which the input power has a gain associated with it that is 1 dB less than the low power gain. In other words, the device is beginning to go into "saturation," which is a condition where increases in input power fail to realize comparable increases in output power. The concept of gain compression is illustrated in Figure 1-18.

The importance of the "1 dB Gain Compression Point" is that this is generally accepted as the limit of non-linearity that is tolerable in a "linear" amplifier and leads one to the dynamic range of the low power amplifier. On the low end

FIGURE 1-18
Linear gain and the 1 dB compression point.

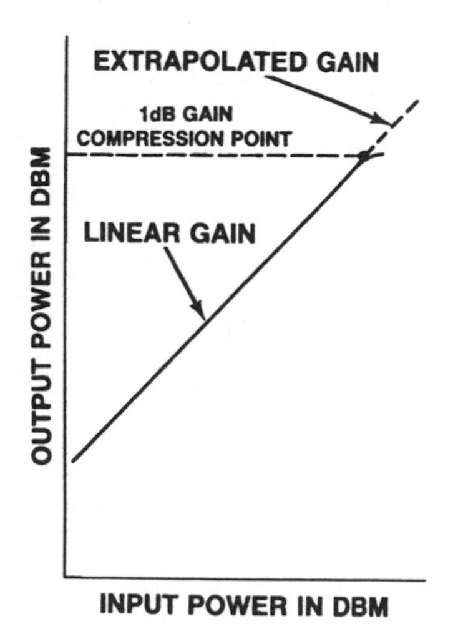

of dynamic range is the limit imposed by noise, and on the high end of dynamic range is the limit imposed by "gain compression."

LINEAR MODULES—FUNCTIONAL CHARACTERISTICS

Let's turn now to amplifiers and examine some specifications encountered that are unique to specific applications. Amplifiers intended for cable television applications are selected to have the desired gain and distortion characteristics compatible with the cable network requirements. They are linear amplifiers consisting of two or more stages of gain each using a push-pull cascode configuration. Remember that a cascode stage is one consisting of two transistors in which a common emitter stage drives a common base stage. A basic circuit configuration is shown in Figure 1-19. Most operate from a standard voltage of 24 volts, and are packaged in an industry standard configuration shown in Figure 1-20. Because they are used to "boost" the RF signals that have been attenuated by the losses in long lengths of coaxial cable (the losses of which increase with frequency), their gain characteristics as a function of frequency are very important. These are defined by the specifications of "slope" and "flatness" over the frequency band of interest. Slope is defined simply as the difference in gain at the high and low end of the frequency band of the amplifier. Flatness, on the other hand, is defined as the deviation (at any frequency in the band) from an ideal gain which is determined theoretically by a universal cable loss function. Motorola normally measures the peak-to-valley (high-to-low) variations in gain across the frequency band, but specifies the flatness as a "plus, minus" quantity because it is assumed that cable television system designers have the capability of adjusting overall gain level.

The frequency band requirements of a CATV amplifier are determined by the number of channels used in the CATV system. Each channel requires 6 MHz bandwidth (to handle conventional color TV signals). Currently available models in the industry have bandwidths extending from 40 to 550 MHz and will accommodate up to 77 channels, the center frequencies of which are determined by industry standard frequency allocations. (New state-of-the-art CATV amplifiers are currently being developed to operate at frequencies up to 1 GHz and 152 channels.)

FIGURE 1-19
Schematic diagram for basic CATV amplifier.

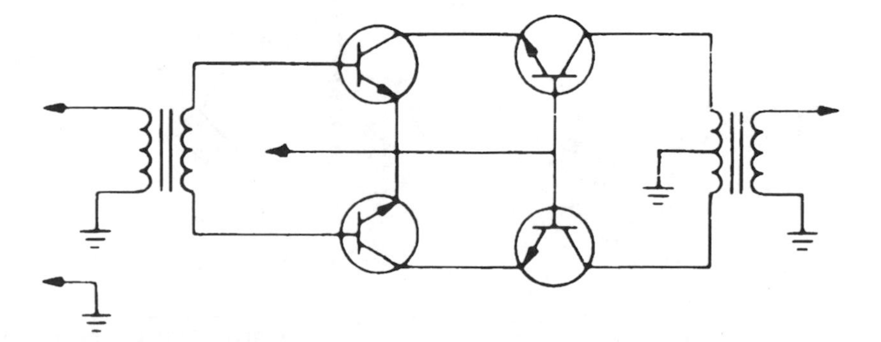

FIGURE 1-20
Standard CATV amplifer package (case # 714-04).

Because CATV amplifiers must amplify TV signals and they must handle many channels simultaneously, these amplifiers must be extremely linear. The more linear, the less distortion that is added to the signal and, thus, the better is the quality of the TV picture being viewed. Distortion is generally specified in three conventional ways: Second Order Intermodulation Distortion (IMD), Cross Modulation Distortion (XMD), and Composite Triple Beat (CTB). In order to better understand what these terms mean, a few words need to be said about distortion in general.

First, let's consider a perfectly linear amplifier. The output signal is exactly the same as the input except for a constant gain factor. Unfortunately, transistor amplifiers are, even under the best of circumstances, not perfectly linear. If one were to write a transfer function for a transistor amplifier, a typical input-output curve for which is shown in Figure 1-21, he would find the region near zero to be one best represented by "squared" terms, i.e., the output is proportional to the square of the input.[7] And the region near saturation, i.e., where the amplifier produces less incremental output for incremental increases in input is best represented by "cubed" terms, i.e., the output is proportional to the cube of the input. A mathematically rigorous analysis of the transfer function of an amplifier would include an infinite number of higher order terms. However, an excellent approximation is obtained by considering the first three terms, i.e., make the assumption we can write

$$F(x) = C_1x + C_2x^2 + C_3x^3,$$

where F is the output signal and x is the input signal. C_1, C_2, and C_3 are constants that represent the transfer function (gain) for the first, second and third order terms.

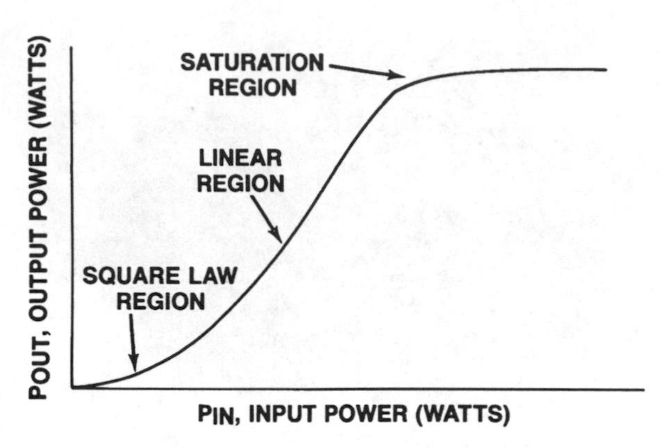

FIGURE 1-21
Transfer function for a typical transistor.

Now consider a relatively simple input signal consisting of three frequencies each having a different amplitude A, B, or C. (In the case of CATV amplifiers, there could be 50–60 channels each having a carrier frequency and associated modulation frequencies spread over a bandwidth approaching 6 MHz.) The input signal x then equals $A\cos\omega_1 t + B\cos\omega_2 t + C\cos\omega_3 t$. For simplicity, let's write this as Acosa + Bcosb +Ccosc, where $a = \omega_1 t$, $b = \omega_2 t$, and $c = \omega_3 t$. If we apply this input signal to the transfer function and calculate F(x), we will find many terms involving x, x^2, and x^3. The "x" terms represent the "perfect" linear amplification of the input signal. Terms involving x^2 when analyzed on a frequency basis result in signal components at two times the frequencies represented by a, b, and c. Also created by x^2 terms are signal components at sums and difference frequencies of all combinations of frequencies represented by a, b, and c. These are called second order intermodulation components. Likewise, the terms involving x^3 result in frequency components at three times the frequencies represented by a, b, and c. And there are also frequency components at sum and difference frequencies (these are called third order IMD). But in addition there are frequency components at a +,– b +,– c. These are called "triple beat" terms. And this is not all! A close examination reveals additional amplitude components at the original frequencies represented by a, b, and c. These terms can both "enhance" gain (expansion) or "reduce" gain (compression). The amplitude of these expansion and compression terms are such that we can divide the group of terms into two categories—"self-expansion/compression" and "cross-expansion/compression." Self-expansion/compression terms have amplitudes determined by the amplitude of a single frequency while cross-expansion/compression terms have amplitudes determined by the amplitudes of two frequencies. A summary of the terms that exist in this "simple" example is given in Table 1-1.

Before going into an explanation of the tests performed on linear amplifiers such as CATV amplifiers, it is appropriate to review a concept called "intercept point."[8] It can be shown mathematically that second order distortion products have amplitudes that are directly proportional to the square of the input signal level, while third order distortion products have amplitudes that are proportional

Table 1-1 Terms in Output for Three Frequency Signal at Input

First Order Components	**Comments**
$k1A \cos a + k1B \cos b + k1C \cos c$	Linear Amplification
Second Order Distortion Components	
$k_2A^2/2 + k_2B^2/2 + k_2C^2/2$	3 D.C. components
$k_2AB \cos(a+,-b) + k_2AC \cos(a+,-c) + k_2BC \cos(b+,-c)$	6 Sum & Difference Beats
$k_2A^2/2 \cos 2a + k_2B^2/2 \cos 2b + k_2C^2/2 \cos 2c$	3 Second Harmonic Components
Third Order Distortion Components	
$k_3A^3/4 \cos 3(a) + k_3B^3/4 \cos 3(b) + k_3C^3/4 \cos 3(c)$	3 Third Harmonic Components
$3k_3A^2B/4 \cos(2a+,-b) + 3k_3A^2C/4 \cos(2a+,-c) +$ $3k_3B^2A/4 \cos(2b+,-a) + 3k_3B^2C/4 \cos(2b+,-c) +$ $3k_3C^2A/4 \cos(2c+,-a) + 3k_3C^2B/4 \cos(2c+,-b)$	12 Intermodulation Beats
$3k_3ABC/2 \cos(a+,-b+,-c)$	4 Triple Beat Components
$3k_3A^3/4 \cos(a) + 3k_3B^3/4 \cos(b) + 3k_3C^3/4 \cos(c)$	3 Self Compression (k_3 is +) or Self Expansion (k_3 is –)
$3k_3AB^2/2 \cos(a) + 3k_3AC^2/2 \cos(a) +$ $3k_3BA^2/2 \cos(b) + 3k_3BC^2/2 \cos(b) +$ $3k_3CA^2/2 \cos(c) + 3k_3CB^2/2 \cos(c)$	6 Cross Compression (k_3 is +) or Cross Expansion (k_3 is -)

to the cube of the input signal level. Hence, it can be concluded that a plot of each response on a log-log scale (or dB/dB scale) will be a straight line with a slope corresponding to the order of the response. Fundamental responses will have a slope of 1, the second order responses will have a slope of 2, and the third order responses a slope of 3. Note that the difference between fundamental and second order is a slope of 1, and between fundamental and third order is a slope of 2. That is to say, for second order distortion, a 1 dB change in signal level results in a 1 dB change in second order distortion; however, a 1 dB change in signal level results in a 2 dB change in third order distortion. This is shown graphically in Figure 1-22. Using the curves of Figure 1-22, if the output level is 0 dBm, second order distortion is at –30 dBc and third order distortion is at –60 dBc. If we change the output level to –10 dBm, then second order distortion should improve to –40 dBc (–50dBm) but third order distortion will improve to –80 dBc (-90 dBm). Thus, we see that a 10 dB decrease in signal has improved second order distortion by 10 dB and third order distortion has improved by 20 dB.

Now for "intercept point." We define the intercept point as the point on the plot of fundamental response and second (or third) order response where the two straight lines intercept each other. It is also that value of signal (hypothetical) at which the level of distortion would equal the initial signal level. For example, if at our point of measurement, the second order distortion is –40 dBc and the signal level is –10 dBm, then the second order intercept point is 40 dB above –10 dBm or +30 dBm. Note in Figure 1-22 that +30 dBm is the value of output signal at which the fundamental, and second order response lines cross. The beauty of the concept of "intercept point" is that once you know the intercept point, you can determine the value of distortion for any signal level—

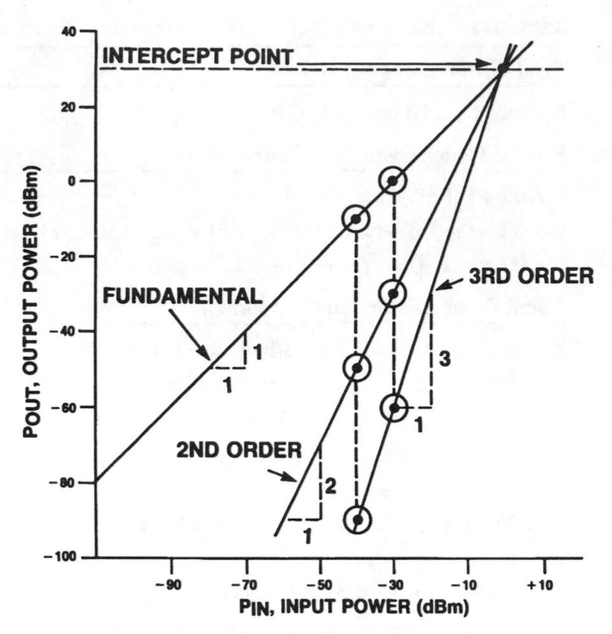

FIGURE 1-22

Fundamental, second order, and third order amplifier response curves.

provided you are in a region of operation governed by the mathematical relationships stated, which typically means IMDs greater than 60 dB below the carrier.

Likewise, to determine third order intercept point, one must measure third order distortion at a known signal level. Then, take half the value of the distortion (expressed in dBc) and add to the signal level. For example, if the signal level is +10 dBm and the third order distortion is –40 dBc, the third order intercept point is the same as the second order intercept point or 10 dBm + 20 dB = 30 dBm. Both second order and third order intercept points are illustrated in Figure 1-22 using the values assumed in the preceding examples. Note, also, that in general the intercept points for second and third order distortion will be different because the non-linearities that create second order distortion are usually different from those that create third order distortion. However, the concept of intercept point is still valid; the slopes of the responses are still 1, 2, and 3 respectively and all that needs to be done is to specify a second order intercept point different from the third order intercept point.

With this background information, let's turn to specific distortion specifications listed on many RF linear amplifier data sheets. If the amplifiers are for use in cable television distribution systems, as previously stated, it is common practice to specify Second Order Intermodulation Distortion, Cross Modulation Distortion, and Composite Triple Beat. We will examine these one at a time. First, consider Second Order Intermodulation Distortion (IMD). Remember these are unwanted signals created by the sums and differences of any two frequencies present in the amplifier. IMD is normally specified at a given signal output level and involves three channels—two for input frequencies and one to measure the resulting distortion frequency. The channel combinations are standardized in the industry but selected in a manner that typically gives a worst case condition for the second order distortion results. An actual measurement consists of creating output signals (unmodulated) in the first two channels listed and looking for the distortion products that appear in the 3rd channel. If one wishes to predict the second order IMD that would occur if the signals were stronger (or weaker), it is only necessary to remember the 1:1 relationship that led to a Second Order Intercept Point. In other words, if the specification guarantees an IMD of –68 dB Max. for a V_{out} = +46 dBmV per channel, then one would expect an IMD of –64 dB Max. for a V_{out} = +50 dBmV per channel, etc.

Cross Modulation Distortion (XMD) is a result of the cross-compression and cross-expansion terms generated by the third order non-linearity in the amplifier's input-output transfer function. In general, the XMD test is a measurement of the presence of modulation on an unmodulated carrier caused by the distortion contribution of a large number of modulated carriers. The actual measurement consists of modulating each carrier with 100% square wave modulation at 15.75 kHz. Then the modulation is removed from one channel and the presence of residual modulation is measured with an amplitude modulation (AM) detector such as the commercially available Matrix RX12 distortion analyzer. Power levels and frequency relationships present in the XMD test are shown in Figure 1-23.

Composite Triple Beat (CTB) is quite similar to XMD, except all channel frequencies are set to a specified output level without modulation. Then one channel frequency is removed and the presence of signal at that frequency is measured. The signals existing in the "off" channel are a result of triple beats (the mixing of three signals) among the host of carrier frequencies that are present in the amplifier. A graphical representation of the CTB test is shown in Figure 1-24.

European cable television systems usually invoke an additional specification for linear amplifiers which is called the DIN test. DIN is a German standard meaning "Deutsche Industrie Norm" (German Industrial Standard); the standard that applies for CATV amplifiers is #45004B. DIN45004B is a special case of a three channel triple beat measurement in which the signal levels are

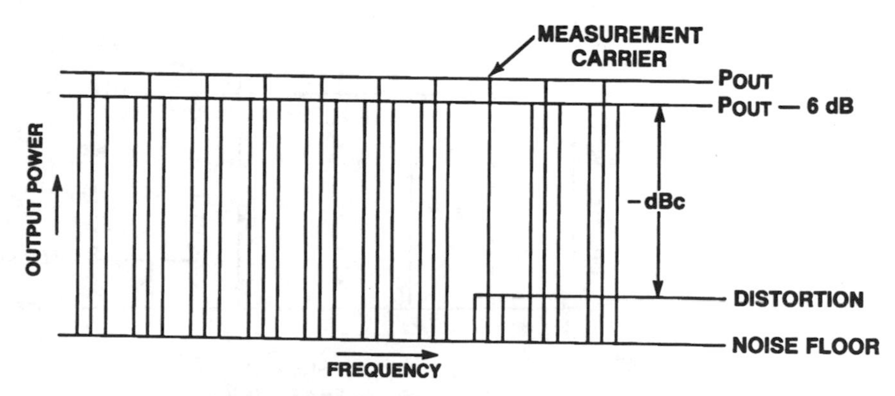

FIGURE 1-23

Frequency-power relationships for XMD.

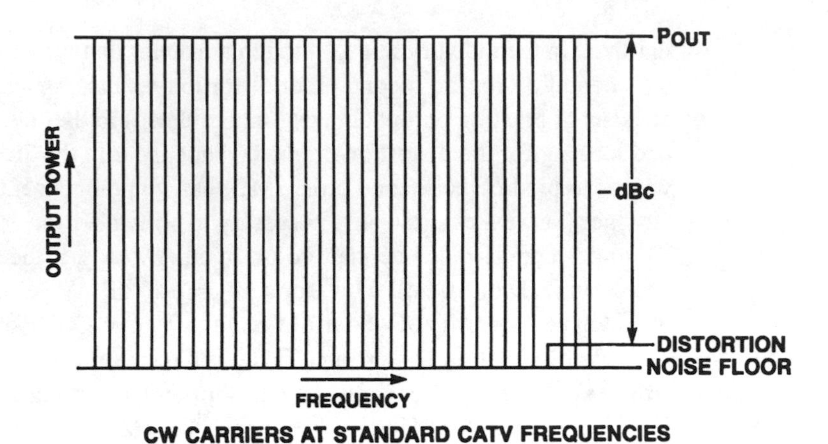

FIGURE 1-24
Frequency-power relationships for CTB.

adjusted to produce a –60 dBc distortion level. An additional difference from normal triple beat measurements is the fact that the levels are different for the three combining signals. If we call the four frequencies involved in the measurement F, F_1, F_2, and F_m, then F is set at the required output level that, along with F_1 and F_2 lead to a distortion level 60 dB below the level of F and F_1 and F_2 are adjusted to a level 6 dB below the level of F. Distortion is measured at the frequency F_m. Frequency relationships (used by Motorola) between F, F_1, F_2, and F_m are as follows: $F_1 = F - 18$ MHz; $F_2 = F - 12$ MHz and $F_m = F + F_2 - F_1$. Figure 1-25 illustrates the frequency and power level relationships that exist in the DIN test.

Linear amplifiers aimed at television transmitter applications will generally have another distortion test involving three frequencies. Basically it is another third order intermodulation test with power levels and frequencies that simulate a TV signal. Relative power levels and frequencies are shown in Figure 1-26.

Thermal resistance ratings of CATV modules (as well as power modules de-

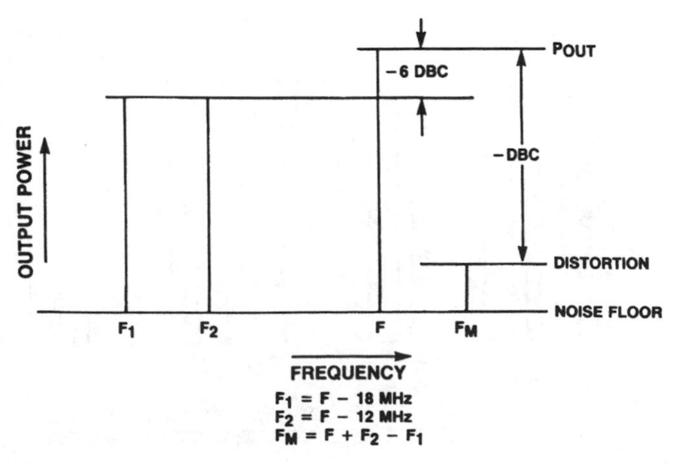

FIGURE 1-25
Frequency-power relationships for DIN45004B.

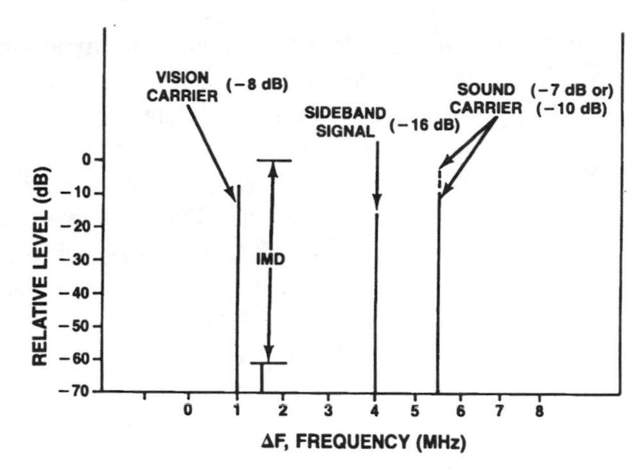

FIGURE 1-26
Third order IMD test for television.

scribed in the next section) are, perhaps, conspicuous by their absence. Because the amplifiers have several heat sources that are contained within the amplifier, it is necessary for the user to provide sufficient heat sinking to the case of the amplifier such that the operating case temperature never exceeds its maximum rating. Actual power dissipation can be determined by considering the operating voltage and the maximum current rating of the device. RF power output of most CATV modules is at most a few milliwatts, which means that most of the power consumed by the module is dissipated in the form of heat. Typically this power dissipation runs in the range of 5 watts for conventional modules such as the MHW5122A but can increase to 10 watts for a power doubler such as the MHW5185.

Because linear (and power) modules have inputs and outputs that are matched to standard system impedances (75 Ω for CATV amplifiers and 50 Ω for power amplifiers), test circuits and fixtures are generally less important than for discrete devices. Basically test fixtures for modules are simply means of making RF and D.C. power connections to the module being tested. It is important if you build your own test fixture that you carefully decouple the D.C. power lines and that you provide adequate heat sinking for the D.U.T. However, if the fixture is for linear modules involving low values of input and output VSWR, then it is extremely important, for accuracy, that the input and output networks (lines and connectors) be designed to exhibit return losses greater than 35 dB. Motorola modifies the RF connectors used in the fixture and, then, calibrates their fixtures to be sure that the fixture does not introduce errors in measuring module return loss.

POWER MODULES—FUNCTIONAL CHARACTERISTICS

Power modules are generally used to amplify the transmit signals in a two-way radio to the desired level for radiation by the antenna. They consist of several stages of amplification (usually common emitter, Class C except for some low level stages that are Class A) combined in a hybrid integrated assembly with nominally 50 Ω RF input and output impedances. Selection of a module involves choosing one having the proper operating voltage, frequency range, output power, overall gain, and mechanical form factor suitable for a particular application.

Power modules for mobile and portable radios also have unique specifications related to their applications. One of the most significant is that of stability. The stability of a module is affected not only by its design but also by many external factors such as load and source impedances, by the value of supply voltage, and by the amount of RF input signal. External factors influencing stability are highlighted in Figure 1-27. Combinations of these factors over a range of values for each factor must be considered to be certain the module will remain stable under typical conditions of operation. The greater the range of values for which stability is guaranteed, the more stable is the module. Of particular importance is the degree of load mismatch which can be tolerated as evidenced by the stated value of load VSWR (the larger the value, the better). Stability specifications are generally evaluated thoroughly during the preproduction phase and then guaranteed but not tested on a production basis.

Efficiency is becoming an increasingly important specification particularly in modules for portable radio applications. The correct way to specify efficiency is to divide the RF power out of the module by the total RF and D.C. powers that are put into the module. Efficiency is generally specified at rated output power because it will decrease when the module is operated at lower power levels. Be careful that the specification includes the current supplied for biasing and for stages other than the output stage. Overlooking these currents (and the D.C. power they use) results in an artificially high value for module efficiency.

Most power module data sheets include a curve of output power versus temperature. Some modules specify this "power slump" in terms of a minimum power output at a stated maximum temperature; others state the maximum permissible decrease in power (in dB) referenced to rated power output. It is important to note the temperature range and the other conditions applied to the specification before passing judgment on this specification.

Generally power modules, like linear modules, do not have thermal resistance specified from die to heatsink. For multiple stage modules, there would need to be a specific thermal resistance from heatsink to each die. Thermal design of the module will take care of internal temperature rises provided the user adheres to the maximum rating attached to the operating case temperature

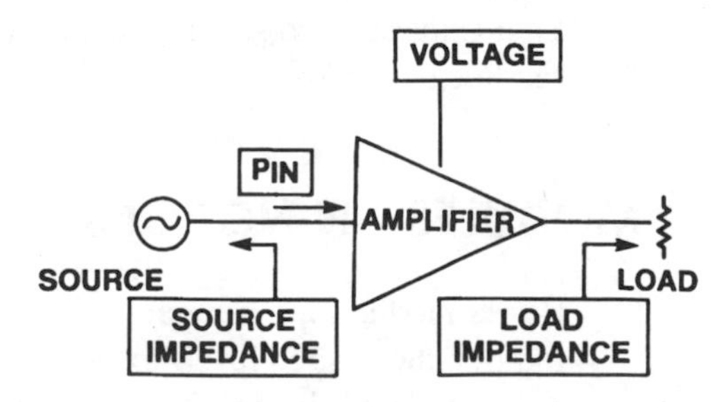

FIGURE 1-27
External factors affecting amplifier stability.

range. This is an extremely important specification, particularly at the high temperature end because of two factors. First, exceeding the maximum case temperature can result in die temperatures that exceed 200°C. This, in turn, will lead, at a minimum, to decreased operating life and, as a maximum, to catastrophic failure as a result of thermal runaway destroying the die. Second, hybrid modules have components that are normally attached to a circuit board and the circuit board attached to the flange with a low temperature solder which may become liquid at temperatures as low as 125°C. Again, the power to be dissipated can be determined by considering the RF output power and the minimum efficiency of the module. For example, for the MHW607, output power is 7 watts and input power is 1 mW; efficiency is 40% minimum. Thus the D.C. power input must be 7/0.4 =17.5 watts. It follows that power dissipation would be 17.5 – 7 = 10.5 watts worst case.

Storage temperature maximum values are also important as a result of the melting temperatures of solder used in assembly of the modules. Another factor is the epoxy seal used to attach the cover to the flange. It is a material similar to that used in attaching caps for discrete transistors and, as stated earlier, is known to deteriorate at temperatures greater than 150°C.

Modules designed for use in cellular radios require wide dynamic range control of output power. Most modules provide for gain control by adjusting the gain of one (or two) stages by means of changing the voltage applied to that stage(s). Usually the control is to vary the collector voltage applied to an intermediate stage. A maximum voltage is stated on the data sheet to limit the control voltage to a safe value. This form of gain control is quite sensitive to small changes in control voltage as is evidenced by viewing the output power versus control voltage curves provided for the user (an example is shown in Figure 1-28. An alternative control procedure which uses much less current is to vary the base-to-emitter voltage of the input stages (which are generally Class A) as illustrated in Figure 1-29. This is of particular significance in portables because of the power dissipated in the control network external to the module.

While not stated on most data sheets, it is always possible to control the

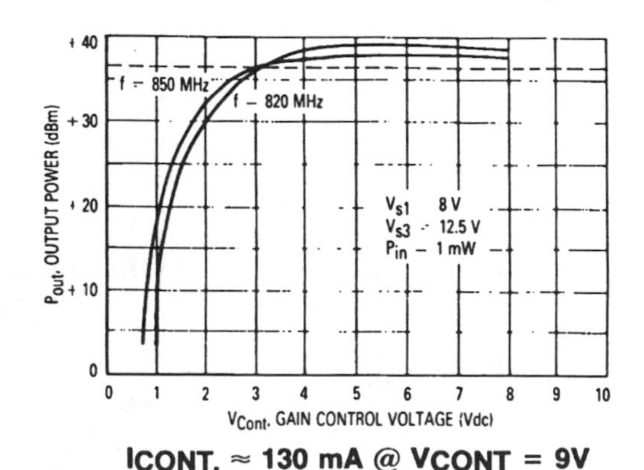

FIGURE 1-28
Output power vs. gain control voltage.

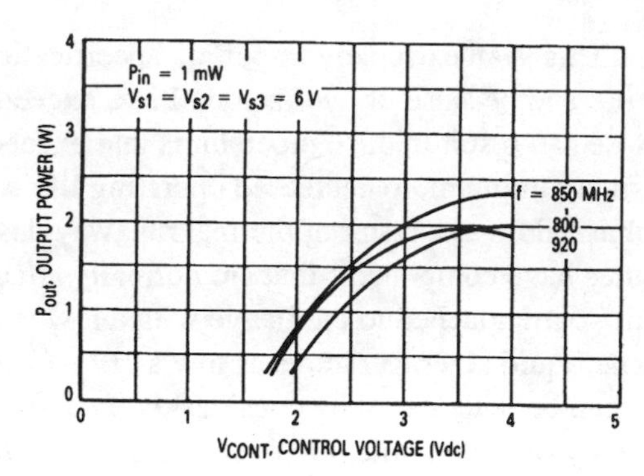

FIGURE 1-29
Output power vs. control voltage.

output power of the module by controlling the RF input signal. Normally this is done by means of a PIN diode attenuator. Controlling the RF input signal allows the module to operate at optimum gain conditions regardless of output power. Under these conditions, the module will produce less sideband noise, particularly for small values of output power, when compared to the situation that arises from gain control by gain reduction within the module.

Noise produced by a power module becomes significant in a duplexed radio in the frequency band of the received signal (see Figure 1-30). A specification becoming more prominent, therefore, in power modules is one that controls the maximum noise power in a specified frequency band a given distance from the transmit frequency. Caution must be taken in making measurements of noise power. Because the levels are generally very low (–85dBm), one must be

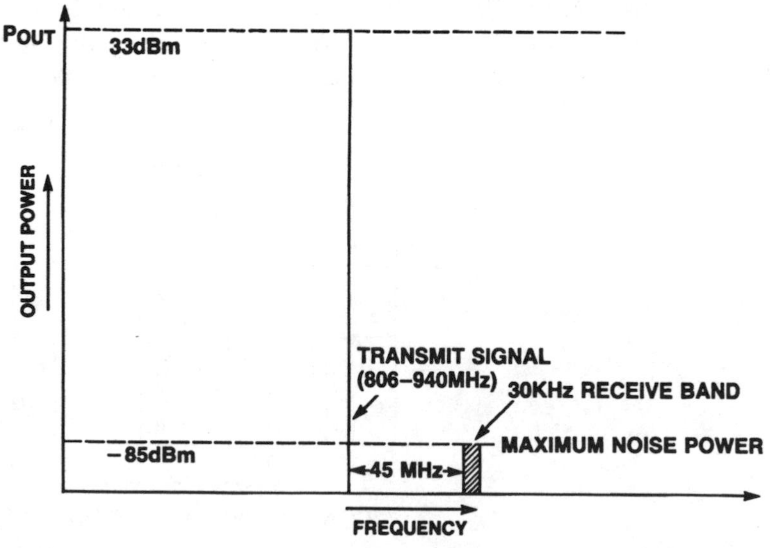

FIGURE 1-30
Noise power in receiving bandwidth.

assured of a frequency source driving the module that has extremely low noise. Any noise on the input signal is amplified by the module and cannot be discerned from noise generated within the module. Another precaution is to be sure that the noise floor of the spectrum analyzer used to measure the noise power is at least 10 dB below the level to be measured.[9]

DATA SHEETS OF THE FUTURE

World class data sheets in the next few years will tend to provide more and more information about characteristics of the RF device; information that will be directly applicable by the engineer in using the device. Semiconductor manufacturers such as Motorola will provide statistical data about parameters showing mean values and sigma deviations. For discrete devices, there will be additional data for computer aided circuit design such as SPICE constants. The use of typical values will become more widespread; and, the availability of statistical data and the major efforts to make more consistent products (six-sigma quality) will increase the usefulness of these values.

References

[1] *Motorola RF Data Book*, DL110, Revision 4, Motorola Semiconductor Sector, Phoenix, AZ, October, 1991.

[2] Alvin Phillips, *Transistor Engineering*, New York: Mc Graw-Hill Book Company, Inc.,1962.

[3] James R. Black, "RF Power Transistor Metallization Failure," *IEEE Transactions on Electron Devices*, September, 1970.

[4] "S-Parameters. . . Circuit Analysis and Design," Application Note #95, Hewlett-Packard, Palo Alto, CA,September, 1968

[5] Guillermo Gonzalez, *Microwave Transistor Amplifiers*, Englewood Cliffs, NJ: Prentice-Hall, Inc., 1984.

[6] "High-Frequency Transistor Primer," Part 1, Avantek, Santa Clara, CA, 1982.

[7] Ken Simons, *Technical Handbook for CATV Systems*, Jerrold Electronics Corporation, Hatboro, PA,1968.

[8] Franz C. McVay, "Don't Guess the Spurious Level," *Electronic Design*, February 1, 1967.

[9] Norm Dye & Mike Shields, "Considerations in Using the MHW801 and MHW851 Series RF Power Modules," Application Note AN-1106, Motorola Semiconductor Sector, Phoenix, AZ.

2

RF Transistor Fundamentals

WHAT'S DIFFERENT ABOUT RF TRANSISTORS?

Whether selecting a device to amplify RF signals to high power levels or low power levels where noise is a factor, today's engineers are totally committed to solid state devices. Only at extremely high power levels (in excess of typically 1 Kw) are vacuum tubes or other forms of amplification taken into consideration. The world of the bipolar transistor—and more recently the FET—has revolutionized the way RF engineers design circuits that amplify, oscillate, switch, or process RF signals in myriad ways.

Today's RF bipolar transistor and even RF FETs are manufactured using processes similar to those used to make low frequency transistors, regardless of whether they are low power or high power. So what's different about them? For one thing, they are made with epitaxial material to more precisely control material properties. More importantly, they are designed with "small" horizontal and vertical structures to permit them to function at RF frequencies. Finally, they are placed in special packages that are designed to minimize the effect of the package in high frequency applications.

The small horizontal structures of RF transistors consist of large amounts of emitter periphery "packed" into a given base area, with the end result being very high power densities in small areas. This results (for power devices) in special considerations being required to handle the power dissipation and maintain the die temperature below a safe maximum value (typically 200°C as discussed in Chapter 5).

Fundamental to RF transistor die and package design are the effects created by the electrical quantities called inductance and capacitance. These quantities have little effect in low frequency applications, but become increasingly paramount as frequency increases. Electrical engineers understand this phenomenon which manifests itself in the form of inductive reactance (which increases linearly with frequency) and capacitive reactance (which is inversely proportional to frequency) and learn to live with it in *all* RF applications.

Unfortunately, in the world of radio frequencies things are not always what they seem to be. Resistors take on the properties of inductance (and possibly some capacitance), capacitors take on the properties of resistance (and even inductance), and inductors become capacitors, etc. It is not at all uncommon for a capacitor (particularly one with leads) to reach a frequency above which it behaves like an inductor. Equivalent circuits are obviously frequency dependent. So if there really is "black magic" in RF, it's because of these kinds of component properties which tend to bewilder and confuse the user.

The first thing you will notice when you try to select an RF transistor is there are frequently different sets of devices for different frequency regions. Transistors having similar voltages of operation, similar gains, similar output powers, and even similar packages exist for different frequency ranges. Different transistors are made to operate from different voltage sources also. This is most notable in high power transistors where you can readily obtain transistors for use at 7.5 volts, 12.5 volts, 28 volts, and in some instances 50 volts. Generally it is easier in a silicon bipolar transistor to obtain gain at a given frequency with a high voltage transistor than it is with a low voltage transistor. However, high voltage transistors won't operate efficiently, nor will they deliver rated power out, if they are used at lower than rated voltages.

The second thing you will notice about RF transistors is their price. Discrete RF transistors generally cost significantly more money than discrete low frequency transistors whether they be low power or high power. We've already touched on one of the reasons—"tight" horizontal structures and "shallow" vertical structures that make the RF transistor die more difficult to manufacture. The second and probably more significant factor is the package cost. Again, the major cost in RF packages is associated with high power parts where the manufacturer must not only contend with minimizing package parasitics but must also provide adequate power dissipation capability for the high dissipated powers that are encountered.

Finally, what's different about RF transistors also has to do with their use. For low frequency, low power transistors, all the user has to do is to provide proper voltages to the elements to create an amplifier. For low frequency, high power transistors, the user only needs to provide voltages and properly mount the device to dissipate power. By contrast, the RF circuit designer must *match* the transistor at both input and output. And to make matters worse, the "match" is dependent on frequency! Single frequency matches are relatively simple; broadband matches can be quite complex and difficult to design but also difficult to implement.

TRANSISTOR CHARACTERISTICS IN SPECIFIC APPLICATIONS

Low Power

Several factors enter into the selection of a transistor for an RF application. The most obvious is the application itself. If the intent is to design a low noise amplifier, then the criteria for choice will be the frequency of operation and the value of noise figure. Probably the most practical consideration is to choose a transistor which the manufacturer has characterized with the necessary noise parameters which are the minimum noise figure at a frequency, the noise resistance R_n and the source resistance for minimum noise figure, Γ_o (see Chapter 1 and Chapter 13 for additional comments). Frequently the transistor manufacturer will plot gain and noise figure contours for a specified bias condition and frequency of operation. These are extremely helpful in making the necessary tradeoff between optimum gain and optimum noise figure when actually designing the low noise stage (also see Chapter 13).[1,2]

Choosing a transistor for other low power applications is generally simpler than for either low noise or high power because the choices are fewer. Most low power transistors have similar breakdown voltages although a few are designed for higher voltage use. Occasionally you will find a special low power transistor that is designed to operate at very low voltages and low current. But generally all you need to do is select a low power transistor which has sufficient current rating for your intended application and which has a high enough cutoff frequency, f_τ,[14] to provide the desired gain at the frequency of operation. If the application is switching, then the higher the cutoff frequency generally the faster will be the switching capability of the device.

A most important consideration when choosing a low power transistor is the package type. The same die is frequently offered in metal cans, plastic SOEs (stripline opposed emitter), surface mount, and hermetically sealed metal-ceramic packages. Generally the smaller the package, the lower are the package parasitics and the better will be the RF performance of the die, especially at higher frequencies. The choice is up to you. RF transistor packages will be discussed in more detail in Chapter 6.

High Power

High power (greater than 1 watt) RF transistors are offered in a wider variety of choices and, thus, present additional problems in device selection. The major distinctions are in voltage of operation, frequency of operation and output power. Other factors also enter into the selection process. Some of these are the linearity and bandwidth required (assuming the application is an amplifier), the efficiency, the ruggedness (which is the ability of the transistor to withstand unfavorable load environments—also see Chapter 1), the thermal requirements for reliability and, of course, the type of package.

The operating voltage is usually a pre-determined specification, but in some applications, such as fixed location transmitters, there may be a choice. In such cases, the designer must determine the advantages and disadvantages of low and high voltage designs. There is no considerable difference in the input impedance and matching in each case, but the output impedance is highly dependent on the voltage of operation and power output level. Thus, depending on the power level in question, one should select an operating voltage resulting in the lowest impedance transformation required to the load impedance (usually 50 Ω). In multistage designs, the drivers and predrivers are often operated at a lower supply voltage than the power amplifier stage partly due to their naturally higher output impedances. This results in a closer match to the input of the following stage.[3]

The choice with respect to frequency of operation is straight-forward. Manufacturers generally grade high power RF transistors by frequency as well as voltage. One must then choose a transistor which will have adequate gain at the desired frequency of use. It is always possible to use a "high" frequency transistor at a lower frequency of operation; however, in this case the user must concern himself with stability, ruggedness, and cost. RF transistors generally have a gain capability that decreases with increasing frequency. If they are used at frequencies below their normal range of operation, the gain will be higher and may create instabilities. High frequency transistors are built using shallower diffusions, lower collector resistivity, and less emitter ballasting—all the things necessary in device design to achieve greater amplification at higher frequencies. Unfortunately, these are also the opposites of device design to improve the ruggedness of a transistor. Gain and ruggedness at a given frequency are a tradeoff in device design. You give up one to get the other. Finally, high frequency transistors cost more than lower frequency transistors, all other factors being equal. So, choose a transistor that will give you desired gain at a frequency but not more (assuming you have a choice).[15]

The third major factor, output power, is also obvious. You must choose a transistor that will give you sufficient output power. In designing an amplifier lineup, you should always start at the output stage and work back from that point in selecting transistors. The gain available from the output transistor then sets the requirements for the driver stage, etc.[3, 4, 6]

BANDWIDTH CONSIDERATIONS IN SELECTING TRANSISTORS

Generally, circuit design determines bandwidth. However at higher frequencies, the Q of the input impedance of a power transistor increases thereby making it more difficult to achieve broad band circuit designs. As the power transistor gets larger and larger (higher and higher power) and is designed to operate at higher and higher frequencies, the result is a continual decrease in the input and output impedances of the device. Think of it this way: higher power transistors are simply increased numbers of low power transistors connected in parallel. Resistors in parallel result in a lower overall resistance; capacitors in parallel result in a higher overall capacitance. The net result is an input impedance for high power, high frequency transistors that is too low to be practical for circuit designers who have access only to the terminals of the transistor. A commonly used procedure for selecting an RF transistor is shown as a flow chart in Figure 2-1.

High power RF transistor manufacturers have alleviated the problem of low input impedance and high Q of high power, high frequency transistors by placing impedance matching networks inside the device package in close proximity to the die. The purpose of these matching networks is to not only raise the impedance of the transistor as seen at the edge of the package but also to transform the impedance values to reduce the reactive components and thereby decrease Q. By choosing an internally matched transistor for your circuit design, you will have less difficulty in achieving broad band circuits over the frequency range specified for device operation.

In general, bipolar transistors designed for VHF and rated for 40–50 watts or higher use internal matching techniques. At UHF the corresponding numbers are 10–20 watts and at 800 MHz about 5 watts. The internal matching networks

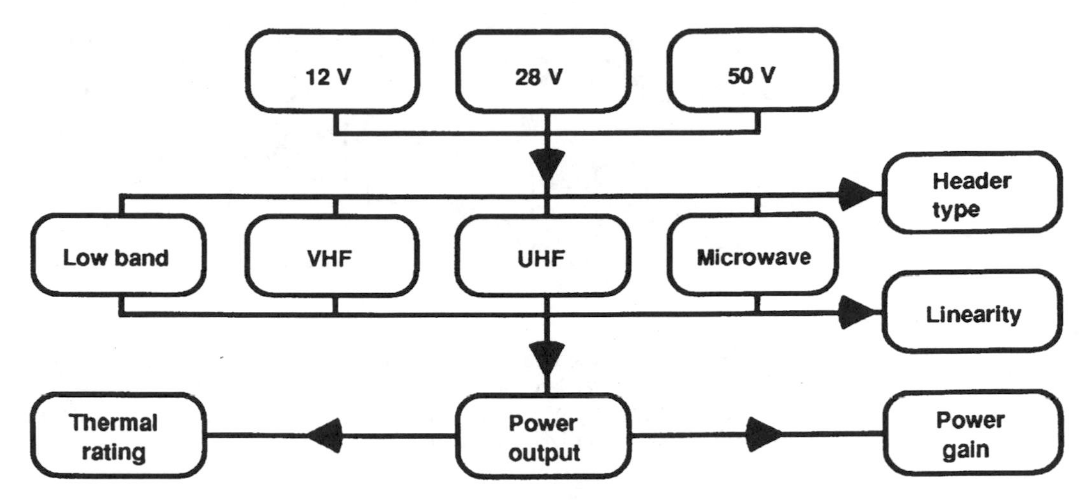

FIGURE 2-1
Flow chart of a possible procedure to be used in selecting an RF transistor for a specific application.

are low-pass filters usually optimized for the high end of the specified spectrum range, where the power gain and impedance levels are the lowest. Most RF power devices specified for operation below 1 GHz employ only internal input matching, but internal output matching is also adapted to higher power UHF transistors and most microwave devices. The input matching network normally consists of an LCL combination, where L is the distributed inductance of the die bonding wires and C is a metal oxide capacitor (MOS). The same guidelines are used for the output matching network designs.[7] Matching networks are illustrated in simplified form in Figure 2-2.

It is obvious that these internal matching networks place some bandwidth limitations on the device's operation, particularly at frequencies above the rated limits of operation. For example, a matched transistor designed for operation in the 225 to 400 MHz frequency range should perform well within this band. However above 400 MHz, the power gain will drop sharply and the base to emitter impedance will increase in its reactive component to a point where the given drive power cannot be transferred to the die itself. Somewhere at an even higher frequency the internal matching network will have a point of resonance, where the input impedance becomes extremely high and the device's power gain is minimal. Below the low end of the specified operating range, the internal matching network has a diminishing effect. However, at some intermediate frequency (100–200 MHz in this case) the matching network may result in an even lower input impedance than it would be without internal matching. This is due to the lesser effect of the series L's and the remaining shunt C. Going further down in frequency, the effect of the internal L's and C's will get to a point, where a normal input impedance is approached. From the above discussion, one can see that internally matched transistors may be difficult to use for bandwidths wider than those for which the transistor was originally designed. It is true there are certain design techniques for external circuitry to allow matched transistors to be used at lower frequencies and for extended bandwidths with somewhat

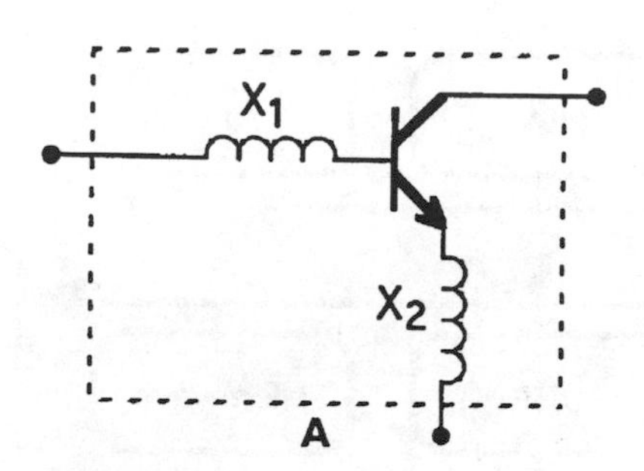

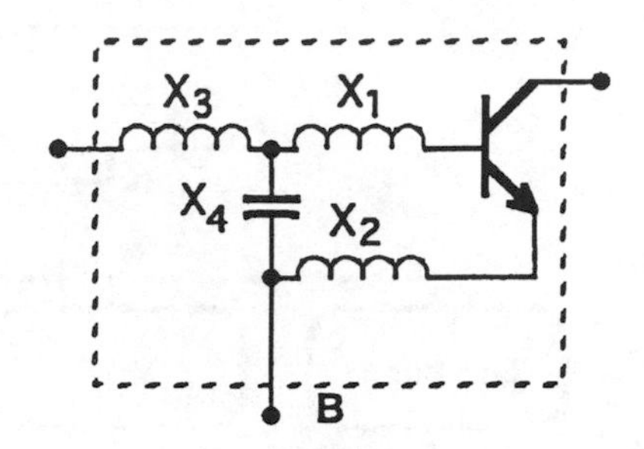

FIGURE 2-2

Electrical models of an unmatched transistor (A) and one employing internal input matching (B). X_1 and X_2 represent the standard base and emitter wire bonds. In (B), X_1 and X_3 represent wire bond loops whose heights must be closely controlled. X_4 is a MOS capacitor with typical values of 150–500 pF for UHF and up to 2000 pF for VHF.

compromised performance, but such matching circuitry is usually complex. Furthermore, one must know the device's impedance profile at these frequencies, which is not given in most data sheets.

MOSFETS VS. BIPOLARS IN SELECTING A TRANSISTOR

It appears that extremely wideband amplifier designs are possible only with MOSFETs (metal oxide semiconductor field effect transistor; much more will be said about RF power MOSFETs in subsequent chapters of this book). For RF power purposes, they have been available for approximately 15 years, although most of the technology breakthrough has occured within approximately the past five years from when this book was written. No internal impedance matching is employed with MOSFETs, except in rare cases at 800–900 MHz and higher frequencies. Such data sheet bandwidth specifications as 2–175 MHz, 100–500 MHz, and 30–90 MHz are misleading since all unmatched MOSFETs (as well as bipolar transistors) are operable down to D.C. if stability can be maintained. They can also be used at higher than the specified frequency limit, keeping in mind the normal 5 dB per octave power gain rolloff. Since the input impedance of a MOSFET is several times higher than that of a comparable bipolar transistor without internal input matching, multi-octave bandwidths can easily be realized with proper circuit design. Because a MOSFET is a high voltage device by its nature (high R_{DS}(on) compared to bipolars V_{CE}(sat)), the performance in low voltage operation may be challenged by its bipolar counterpart.[5, 8, 9, 10]

OTHER FACTORS IN RF POWER TRANSISTOR SELECTION

Efficiency and linearity are usually dictated by the type of modulation used in the signal to be amplified. The three most commonly used methods of modulation in communications equipment (see Chapter 4) are frequency modulation

(FM), amplitude modulation (AM), and single sideband (SSB). FM is the simplest and does not place any linearity requirements on the amplifying devices. In fact, the whole amplifier chain can be operated in Class C or any other class not requiring bias idle current for the devices. The low angle of conduction results in improved efficiency over Classes A and AB. The modulation is usually accomplished by varying the frequency of the oscillator generating the low level carrier to be amplified. The only requirement for an amplifying device chain of an FM transmitter is for the devices to have a proper power gain to provide a drive signal at a correct level for each stage. In some instances each amplifying stage, except possibly the power amplifier, is designed to have 1–2 dB of excess gain resulting in overdrive and saturation of the stages. This results in less critical power gain selection for the amplifying devices and reduces the possibility of any unwanted AM reaching the power amplifier to generate distortion. Typical line-up examples for designing an amplifier chain are shown in Figure 2-3. One must take the worst case data sheet values to determine the total power gain, etc.

Unlike 40 years ago, AM is today almost exclusively used for broadcast purposes. High power transmitters generally employ vacuum tubes, except for fairly recent designs which may be completely solid state. However, it is relatively common to find high power tube transmitters with solid state driver stages. Since SSB is a version of AM, suitable devices for it are discussed together with normal full-carrier AM. The only basic difference between AM and SSB is that in AM the carrier and both sidebands are transmitted. In SSB, the carrier and one sideband are suppressed in the transmitter, and the carrier is re-inserted in the receiver. Both AM and SSB require linearity of their amplifier stages. SSB requires the highest dynamic range of all modulated systems since

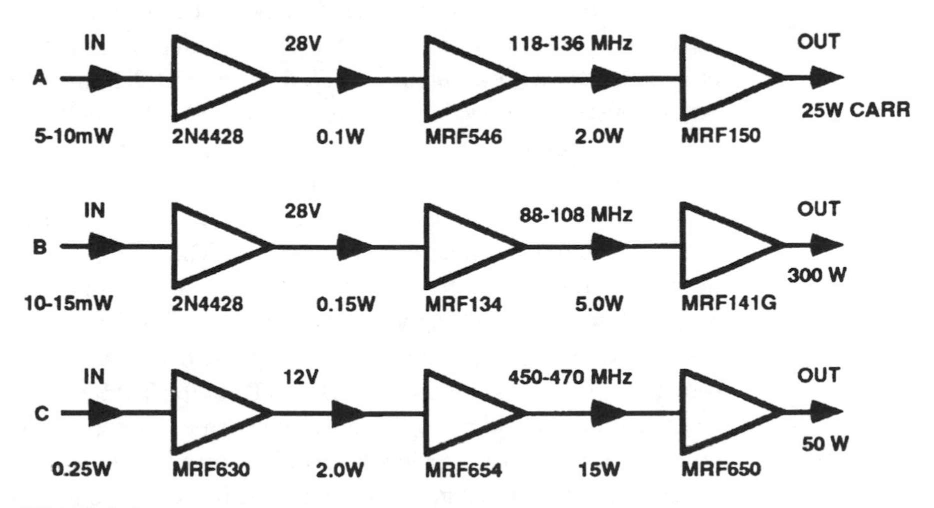

FIGURE 2-3

Typical RF transistor lineups. In A, the power amplifier and driver are high level amplitude modulated, requiring high breakdown voltages, such as customary for devices made for 50 volt operation. In B is a lineup for FM broadcast with all devices operating from a 28 volt supply. C represents a standard arrangement for UHF land mobile use.

the power level, theoretically, varies between zero and the peak power. Limitations to the dynamic range are not usually semiconductor device oriented, but are rather related to biasing or circuit design. There are two types of carrier amplitude modulation employed in solid state amplifier designs. In the so-called "low level" AM, the modulation is done in one or more of the low power driver stages, requiring linearity of the modulated stages as well as every stage thereafter. Thus, the amplifying devices must ideally be linear up to four times the carrier power with 100% modulation. The Class A drivers, which should be degraded to about 30% of their Class C ratings, must be further degraded by 400% for the carrier power level. For example, if a 100 watt Class C rated transistor is operated in Class A, its maximum power output can be 30–35 watts, resulting in a P_{out} level of only 7–9 watts at the carrier level. The same is true in SSB, but the devices are usually specified for the peak envelope power (PEP) level and the difference between PEP and average is only 1 to 2. In so-called "high level" AM, only the power amplifier stage and, sometimes, the driver are modulated and must be linear, whereas the predrivers can operate in a class of nonlinearity. The semiconductor industry specifies device linearity for SSB as distortion below one of the test tones (Military Standard (MIL STD) 1131, version A, test method 2204B) and in some cases below the peak power (Electronic Industry Association (EIA) method). There is a 6 dB difference in the specifications between the two test methods, the MIL STD being the more stringent. These distortion specifications do not directly apply to the carrier AM operation, but if there is a degradation in linearity, it would be only the result of thermal effects. The Federal Communications Commission (FCC) has specific regulations for AM and SSB transmitters depending on their frequency of operation and the power output levels. Since the device data sheet specifications for linear devices only refer to two tone testing, and give specs such as 30dB below the peak power, etc., it is up to the designer to determine at what power level the device can be operated to meet the FCC regulations. For example, if a device is specified for a power output at –30 dB intermodulation distortion, its power output must be degraded in order to achieve –35 dB or higher. These different methods of modulation along with their associated waveforms and spectrum coverages are described in Chapter 4.

In addition to amplitude modulated systems for voice transmission, linearity is even more important in video transmitters. There are a large number of solid state television transmitters and translators in use for bands 1 (54–88 MHz) and 3 (174–216 MHz), and an increasing number for UHF (up to 806 MHz in the USA). The aural portion of the signal employs frequency modulation as in FM broadcast voice transmission, and is usually amplified by a separate amplifier chain in order to achieve higher efficiency and to reduce the dissipation in the video amplifier chain. The low level video amplifier stages up to 100 watts or so, are mostly operated in Class A for good linearity and low phase distortion, which is critical in video transmission. There are a number of solid state devices available for such application from various manufacturers, giving the necessary distortion data measured under triple tone (beat) conditions. The

Class A stage(s) drive the final power amplifiers, which can be arranged in a number of ways for power combining to achieve multi-kilowatt power levels. These final power amplifiers are almost always operated in Class AB for increased efficiency and lower power consumption. It appears that there are not many solid state devices available specifically designed for Class AB video amplifier use. A designer must more or less determine the suitability of each device from its $P_{in} - P_{out}$ graphs or two tone test data for low frequency modulated applications. The synchronization pulses represent the peak power output, and the white picture level the lowest, or 5-10% of the peak. The required linearity for these is much easier to achieve than for the "black" level signals, which varies between the white level and the sync pulses in amplitude and has an average amplitude of 68–75% of the peak. All amplifying devices have a varying phase shift versus amplitude characteristic, but it is probably more predominant with solid state devices than with vacuum tubes. Thus, phase correction circuitry is required and it can be most easily done in the low level drivers.

References

[1] Guillermo Gonzalez, *Microwave Transistor Amplifiers*, Englewood Cliffs, NJ: Prentice-Hall, Inc., 1984.

[2] "S-Parameter Design," Application Note #154, Hewlett-Packard Co., Palo Alto, CA, 1972.

[3] Krauss, Bostian, Raab, *Solid State Radio Engineering*, New York: John Wiley & Sons, Inc., 1980.

[4] Irving M. Gottlieb, *Solid State High-Frequency Power*, Reston, VA: Reston Publishing Co., 1978.

[5] H. O. Granberg, "Power MOSFETs versus Bipolar Transistors," Application Note AN-860, Motorola Semiconductor Products Sector, Phoenix, AZ.

[6] Jack Browne, "RF Devices Gain High Power Levels," *Microwaves & RF*, November, 1987.

[7] "Controlled-Q RF Technology—What it Means, How its Done," Engineering Bulletin EB-19, Motorola Semiconductor Products Sector, Phoenix, AZ.

[8] Gary Appel, Jim Gong, "Power FETs for RF Amplifiers," *RF Design*, September/October, 1982.

[9] H. O. Granberg, "RF Power MOSFETs," Article Reprint AR-165S, Motorola Semiconductor Products Sector, Phoenix, AZ.

[10] Roy Hejhall, "VHF MOS Power Applications," Application Note AN-878, Motorola Semiconductor Products Sector, Phoenix, AZ.

[11] *MOSPOWER Applications Handbook*, Siliconix Inc., Santa Clara, CA,1984.

[12] All RF MOSFET Data Sheets, Motorola Semiconductor Sector, Phoenix, AZ.

[13] Various Applications Notes, *The Acrian Handbook*, Acrian Power Solutions, 490 Race Street, San Jose, CA,1987, pp. 622-674.
[14] "High-Frequency Transistor Primer," Avantek, Inc. Santa Clara, CA, 1982.
[15] *RF Device Data*, DL110, Rev. 4, Motorola Semiconductor Products Sector, Phoenix, AZ, 1990.

3

FETs and BJTs: Comparison of Parameters and Circuitry

TYPES OF TRANSISTORS

Because solid state RF power transistors now consist of two basic types of devices, namely bipolar junction and field effect, it is appropriate to discuss and compare their parameters and performance. In certain applications the bipolar junction transistor (BJT) will without a doubt yield superior performance, whereas in other areas a field effect transistor (FET) will do a better job. There are only two types of BJTs commercially available today. These are based on silicon technology and are either NPN or PNP polarities. PNP transistors (despite their inferior performance over NPN types) are primarily used in land mobile communications equipment requiring a positive ground system. All UHF and higher frequency devices are of the NPN polarity due to their higher mobility of electrons as majority carriers, which translates to higher f_τ and improved high frequency power gain.

There are far more types of FETs commercially available for RF power use. These include a late newcomer SIT (static induction transistor) which is a version of a depletion mode junction FET and the MESFET (metal gate Schottky FET). The latter is usually made of gallium arsenide and is also a depletion mode type. Another depletion mode device is the standard junction FET, which is only practical in low power use for pre-drivers and mixers, etc. The vertical channel silicon MOSFET is the most common RF power FET. It comes in a number of varieties of die structures, each having slightly different characteristics in R_{DS}(on) and the various capacitances. The vertical channel MOSFET has been on the market since around 1975, and has seen numerous improvements regarding its perfomance and manufaturability. There is also a lateral channel power MOSFET in existence. It consists of a series of small signal FETs connected in parallel on a single chip. Due to its lateral channel structure, it consumes more die area for a given power rating than the vertical channel device and, therefore, is less cost effective. However, the lateral FET features extremely low feedback capacitance (C_{RSS}) which results in increased stability and higher gain at high frequencies. Both these silicon MOSFETs are enhancement mode devices, meaning their gates require positive voltages with respect to the sources in order for the drain-source channel to conduct. Conversely a depletion mode FET conducts when the gate and source are at an equal potential, and requires a negative gate voltage for turn off (depletion).[3]

COMPARING THE PARAMETERS

One main difference between a BJT and a MOSFET in RF amplifier use is the need for base/gate bias voltage. A BJT only requires base bias when linear operation is desired and there is very little difference in its power gain between a biased (Class A, AB, or B) and an unbiased condition (Class C). In an unbiased enhancement mode FET, the gate input voltage swing must overcome the gate threshold voltage to turn the FET "on" with its positive peaks. Some FETs have their gate threshold voltages specified as high as 6 volts. If the D.C. gate voltage is brought closer to its threshold level, a smaller voltage swing is necessary to overcome it. Since in each case the gate-source RF impedance is about the same, the actual power gain can vary as much as 5–6 dB depending on the initial threshold voltage and the frequency of operation. For linearity, a FET also needs to be biased to some idle current in Class A or AB operation. The bias source may be a simple resistor divider since no D.C. current is drawn, whereas a BJT requires a constant voltage source of 0.65–0.70 volts with a current capability of I_C(peak)/h_{FE}. A summary of specific characteristics of each device type is presented in Table 3-1. Note that the table focuses only on silicon MOSFETs in the FET category and some of the characteristics may not apply to JFETs and other depletion mode FETs. Similar electrical sizes for each are assumed for the impedance comparison.[1, 2, 3, 4]

Most RF power design engineers accustomed to circuit design with BJTs are slowly beginning to look at the FET designs and learn about the differences in parameters and behavior between the two types of semiconductors. Circuit design with each type is very similar. The same RF design practices, such as grounding, filtering, bypassing, and creating a good circuit board layout apply in each case. Precautions must be taken with each type device, when designed into a particular application. The FETs are sensitive to gate rupture. Rupture can be caused by excessive D.C. potential or an instantaneous transient between the gate and the source. This can be compared to exceeding the voltage rating of a capacitor, which usually results in a short or leakage. A power FET can be "restored" in some instances by applying a voltage lower than the rupture level between the gate and the source. It must be at a sufficient current, but not higher than 1–1.5 A to clear the gate short. A higher current would fuse one of the bonding wires to the area of the short on the die. A number of cells will always be destroyed, but with larger devices, such as 30 W and higher no difference in performance may be noticed.[3] Long term reliability after such an operation may be jeopardized, and is not recommended in cases where very high reliability is required.

A weak spot with the BJTs is a possibility for thermal runaway. Devices with diffused silicon emitter ballast resistors are less susceptible to thermal runaway than devices having nichrome resistors. The diffused silicon resistors have a

TABLE 3-1 Bipolar transistor and RF power MOSFET characteristics when used as RF amplifiers.

Characteristic	Bipolar	MOSFET
Z_{in}, R_s/X_s (2.0 MHz):	3.80 – j2.0 Ohms	19.0 -j3.0 Ohms
Z_{in}, R_s/X_s (150 MHz):	0.40 + j1.50 Ohms	0.60 -j0.65 Ohms
Z_{OI} (Load Impedance):	Nearly equal for each transistor, depending upon supply voltage and power output.	
Biasing:	Not required, except for linear operation. High current (I_C/h_{FE}) constant voltage source necessary.	Required for linear operation. Low current source, such as a resistor divider is sufficient. Gate voltage can be varied to provide an AGC function.
Linearity:	Low order distortion depends on electrical size of the die, geometry and h_{FE}. High order IMD is a function of type and value of emitter ballast resistors.	Low order distortion worse than with bibolars for a given die size and geometry. High order IMD better due to lack of ballast resistors and associated nonlinear feedback.
Stability:	Instability mode known as half f_o troublesome because of varactor effect in base-emitter junction. Lower ratio of feedback capacitance versus input impedance.	Superior stability because of lack of diode junctions and higher ratio of feedback capacitance versus input impedance.
Ruggedness:	Usually fails under high current conditions (overdissipation). Thermal runaway and secondary breakdown possible. h_{FE} increases with temperature.	Overdissipation failure less likely, except under high voltage conditions. g_{FS} decreases with temperature. Other failure modes: Gate punch through.
Advantages:	Wafer processing simpler, making devices less expensive. Low collector-emitter saturation voltage makes low voltage operation feasible.	Input impedance more constant under varying drive levels. Better stability, better high order IMD, easier to broadband. Devices and die can be paralleled with certain precautions. High voltage devices easy to implement.
Disadvantages:	Low input impedance with high reactive component. Internal matching required to increse input impedance. Input impedance varies with drive level. Devices or die can not be easily paralleled.	Larger die required for comparable power level. Nonrecoverable gate puncture. High drain-source saturation, which makes low voltage, high power devices less practical.

slight positive temperature coefficient, while the nichrome resistors have near zero coefficient. However, the diffused resistors are nonlinear with current, and devices using them are less suitable for applications requiring good linearity. The main reason for the thermal runaway with the BJTs is the increasing h_{FE} with temperature, while the g_{FS} of a MOSFET goes down, trying to turn the device off. In contrast, the gate threshold voltage decreases by about 1mV/°C, which makes the temperature profile of a gate biased device dependent on the initial value of g_{FS} and the voltage of operation.

The figures of merit of a BJT and FET are defined as emitter periphery/base area and gate periphery/channel length respectively. In practical terms these relate to the ratio of feedback capacitance to the input impedance since finer geometries result in lower feedback capacitances. This only applies to common emitter and common source configurations. Thus it would appear that higher figure of merit devices are more stable than those with low figure of merit. This would actually be true, except that in the first case the power gain is also higher, which can cause instabilities as a result of stray feedback or at a high frequency, where the feedback capacitance produces positive feedback due to phase delays. There is still another instability mechanism with the BJT, which is a result of a varactor effect in its diode junctions, mainly the collector-base. It is commonly known as "half f_o," which is usually a steady spurious signal half the frequency of the excitation. Due to the lack of junctions in a FET, this phenomenon is unknown in MOSFET power circuits.

Regarding impedance matching, the largest difference can be noticed in the base-emitter and gate-source impedances. At D.C. the MOSFET has an infinite gate-source impedance, whereas the BJT exhibits the impedance of a forward biased diode. At higher frequencies, depending on the device's electrical size, the gate-source capacitance (C_{ISS}) (enhanced by the Miller effect), together with the wire bond inductances, etc., will form a complex impedance which may be lower than that of BJT's. The output capacitance (C_{OB}/C_{OSS}) is almost equal for each type device of equivalent electrical size. The output capacitance has a large effect on the efficiency of an amplifier, as it must be charged to around twice the supply voltage and discharged again during each cycle of the operating frequency, and the power used in the charging process is dissipated in the amplifying device. At a single frequency, a part but not all of the capacitance can be tuned out, since its value varies with the output voltage swing. The power loss due to the output capacitance, for example, for a single ended BJT amplifier can be defined as: $P_s = (2C_{ob})(V_{CC})^2(f)$ where P_s = power loss, f = frequency and the efficiency equals $P_{out}/(P_{out}+P_s)$. We can see that the power loss is in direct relation to the capacitance and to the square of the supply voltage. Thus, a higher operating voltage does not always result in higher efficiency as commonly thought. Efficiency will be discussed in more detail in Chapter 4.

Equivalent parameters and their designations for bipolar transistors and MOSFETs are compared in Table 3-2. The table gives a designer accustomed only to bipolar circuitry an idea of comparable FET data sheet figures and vice versa. Note that all parameters are not applicable to both types of devices.

Here are some points to remember:[5]

1) Do not subject a BJT to an I_{EBO} condition at currents higher than about 1% of the current at which h_{FE} is specified. Higher currents may permanently degrade the device's h_{FE}.
2) Do not measure beakdown voltages at a current lower than the I_{CES}/I_{DSS} specification of the device. Irrational readings may be obtained as a result.
3) Do not attempt to measure the V_{GS} of a MOSFET. Permanent gate damage will occur.
4) Never measure the breakdown voltage of a MOSFET with the gate open. Permanent gate damage may occur.

TABLE 3-2 "Equivalent" parameters of bipolar and MOSFET transistors.

Bipolar	MOSFET	
BV_{CEO}	BV_{DSO}	Breakdown avalanche voltage, measured with the base open. Not specified or measurable with MOSFETs. In case of any drain-gate leakage, the gate can charge to voltages exceeding the V_{GS} rating.
BV_{CES}	BV_{DSS}	Breakdown avalanche voltage, measured with the base and emitter or gate and source shorted. Normal method of measuring MOSFET breakdown voltage.
BV_{CBO}	BV_{DGO}	Breakdown avalanche voltage, measured with the emitter open. Not specified or measurable with MOSFETs. Gate-source rupture voltage could be exceeded.
BV_{EBO}	V_{GS}	Reverse breakdown voltage of the base-emitter junction. Not specified or measurable with MOSFETs unless done carefully at low current levels. Gate rupture can be compared to exceeding a capacitor's maximum voltage rating.
V_B (forward)	V_{GS}(th)	Not specified or necessary in most cases for BJTs. For a MOSFET, this parameter determines the turn-on gate voltage, and must be known for biasing the device.
I_{CES}	I_{DSS}	Collector-emitter or drain-source leakage current with base and emitter or gate and source shorted. BJT and FET parameters are equivalent, and normally the only effects of leakage are wasted D.C. power, increased dissipation and long term reliability.
I_{EBO}	I_{GS}	Base-emitter reverse leakage current and gate-source leakage current. Not normally given in BJT data sheets, but important for MOSFET biasing. Both affect their associated device's long-term reliability.
V_{CE}(SAT)	V_{DS}(SAT)	Device saturation voltage at D.C.. Not usually given in BJT data sheets, but important in certain applications. With power MOSFETs this parameter is of great importance. The MOSFET numbers are higher than those for BJTs and are dependent on several factors in processing the die.
h_{FE}	g_{FS}	These are parameters for low frequency current and voltage gain, respectively. In a MOSFET, the g_{FS} is an indication of the device's electrical size. To a certain extent, it depends on device type and die geometry.
f_T	(f_T)	Unity current or voltage gain frequency. Not given in many BJT or MOSFET data sheets. The value can be two to five times greater for the MOSFET for equivalent geometry and electrical size.
G_{PE}	G_{PS}	Power gain in common-emitter or common-source configuration. This figure is roughly the same for both types of devices. It is normally regarded as current gain for the BJT and voltage gain for the MOSFET.
C_{ib}	C_{iss}	Base-emitter or gate-source capacitance. Rarely given for BJTs. In RF power FETs, the C_{iss} has a greater effect on the gate-source impedance.
C_{ob}	C_{oss}	Collector-emitter or drain-source capacitance. Both are usually specified, and are approximately equal in value for a given device rating and voltage. Both are combinations of MOS and diode capacitance.
C_{rb}	C_{rss}	Collector-base or drain-gate capacitance. Rarely specified for BJTs. Normally referred to as the feedback capacitance for MOSFETs.

CIRCUIT CONFIGURATIONS

Common emitter and common base circuit configurations are most widely used in RF amplifiers and are discussed here in detail. However, a common collector circuit can be used as an RF amplifier in addition to its normal application as a wideband emitter follower. Since the circuit does not have voltage gain, the power amplification must take place through impedance transformation. In a common collector configuration, the input impedance is high compared to common emitter and common base, but the output impedance is extremely low due to the less than unity voltage gain, resulting in a higher ratio of impedance transformation to 50 Ω. Amplifiers with the emitter follower principle have been succesfully designed for power levels up to 100–150 W and up to frequencies of 50–60 MHz, but beyond this frequency range the power gain falls off rapidly probably due to the high impedance ratio matching networks required and their associated losses. In addition the transistor itself is likely to have added losses in the form of high RF currents at the emitter. Although it has been proven that an emitter follower works as a tuned RF power amplifier, it is unlikely that it will see considerable commercial use in this application.

Until a few years ago, there was a clear separation between common emitter and common base circuit applications. Common emitter was used for low frequency to UHF amplifiers and common base was for frequencies above UHF up to microwaves. Today, devices intended for common base operation at low frequencies and common emitter operation at microwaves are a reality. Which way should a designer go? Which configuration is the best for your application?

The common emitter and common base circuits have very different gain characteristics. In Figure 3-1 the β curve closely follows the power gain curve and the same is true of the α curve. In each the 6 dB per octave slope (closer to 5 dB in practice) is a result of the decreasing α and β with increasing frequency. Frequency dependence of α and β is primarily a result of parasitic capacitances, inductances, and resistances in the die itself. This is also the case with MOSFETs, in which the g_{FS} represents β. Again their frequency dependence resembles that of BJTs. The term f_τ, which is the frequency at which $\beta = 1$, is not really applicable to FETs, although commonly used. In a MOSFET the unity gain frequency is determined by the ratio of Z_{in}/C_{RSS}, as discussed earlier. In theory this is much the same as with BJTs, except that the effects of parasitic resistances are smaller. A definiton for f_τ for FETs can be approximated as:

$$g_{FS}\ \{2\pi[C_{ISS} + (g_{FS} \times C_{RSS})]\}$$

However, for an equal power rating the MOSFET requires almost twice the die area of a BJT, which somewhat equalizes the gain-bandwidth perfomance of the

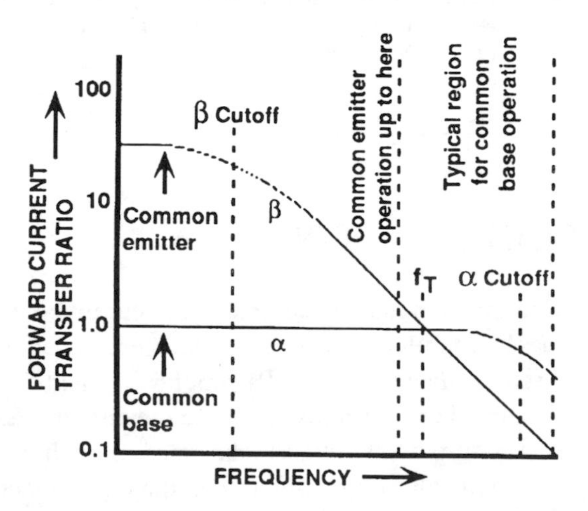

FIGURE 3-1

Forward current transfer ratio versus frequency for common emitter and common base circuit configurations.

two. The plot of α (Figure 3-1) represents the common gate gain curve of a FET and could be called g_{FG}, although practice has shown that it may not be feasible to operate power MOSFETs in common gate configuration as explained later in this chapter.[1, 3, 6]

Only a few designers have access to a semiconductor curve tracer, which is one of the basic tools for measuring transistor D.C. parameters. Some of these parameters can tell an experienced designer much about the RF performance of a transistor. For example in a MOSFET, g_{FS} is a good indication of its electrical size. The quantity g_{FS} is also related to power gain, but to compare devices for power gain C_{ISS} and C_{RSS} must also be known. Similarly the electrical size of a BJT can be determined by current up to the point where β is linear. Figure 3-2 shows methods to measure leakage currents, breakdown voltages and h_{FE}/g_{FS}

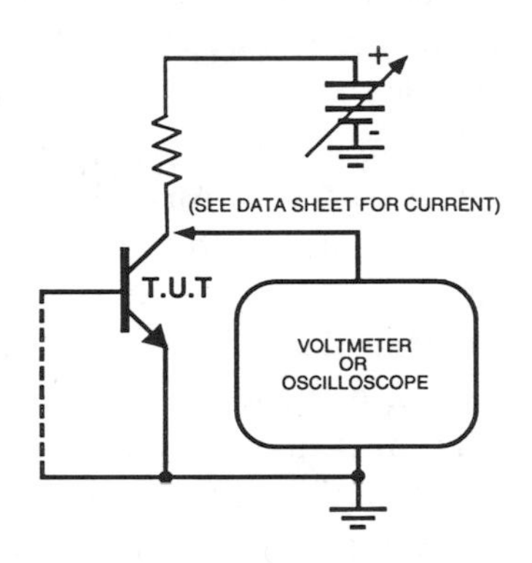

(A) TRANSISTOR BREAKDOWN VOLTAGE MEASUREMENT

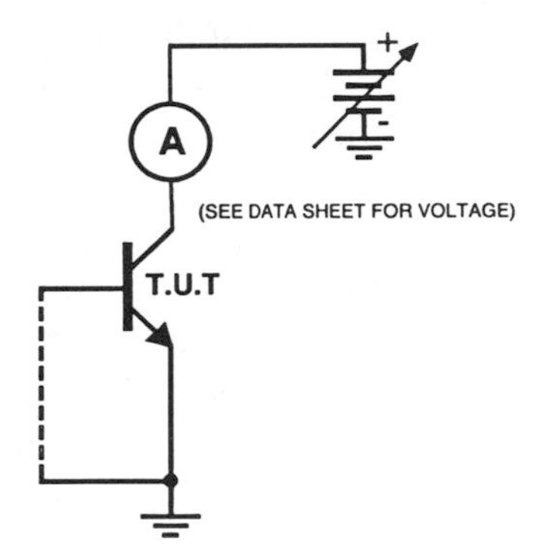

(B) TRANSISTOR LEAKAGE MEASUREMENT

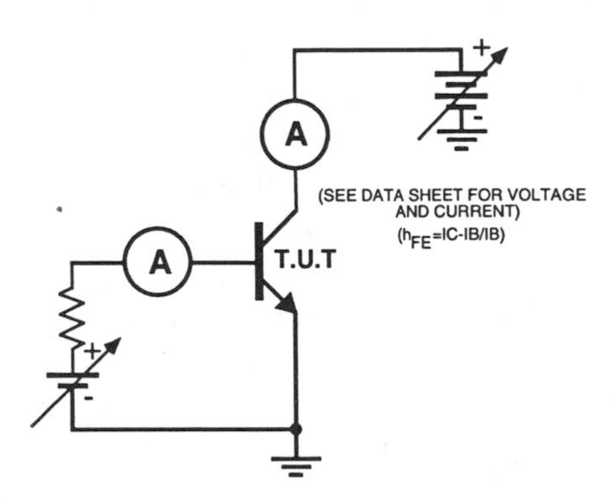

(C) TRANSISTOR h_{FE} MEASUREMENT

FIGURE 3-2

Measurement techniques for the most common transistor D.C. parameters without the use of a curve tracer. A and B can be readily adapted for measurements on MOSFETs whereas C must be modified by replacing the ammeter at the base by a voltmeter from gate-to-ground, then $g_{FS} = \Delta I_D/\Delta V_{GS}$.

using only power supplies and meters. Leakage current measurements should not be performed with an ohmmeter, as seems to be a frequently attempted practice, since the internal impedance and the voltage across the input terminals of the ohmmeter are generally not known.

COMMON EMITTER AND COMMON SOURCE

The common emitter (CE) and common source (CS) are the most widely used circuit configurations. They exhibit good stability, good linearity and high power gain up to UHF. CE and CS are the only circuit configurations where the input and output are out of phase. This enhances the stabilty, except for the half f_0 mode and at frequencies where the feedback capacitance delays are close to 180°. If the common emitter or source inductance is increased, the power gain will go down due to the negative feedback generated by the reactance. Thus it is very important to keep common element inductance as low as physically possible for proper operation of the device. The gain is inversely proportional to the frequency and increases approximately 5 dB per octave until the β cutoff is reached, at which point the gain may be as high as 30-40 dB. An example of a common emitter circuit is shown in Figure 3-3. It can be directly adapted to MOSFETs, but in that case since $I_B = 0$, $I_D = I_E$.

Lumped constant matching elements in narrow band circuits are practical up to VHF, but at frequencies over 300–400 MHz, microstrip techniques, or a combination of microstrip and transformer impedance matching techniques are normally used. If broadband performance is desired, a push-pull configuration makes the impedance matching easier to implement to a 50 Ω interface due to the initially higher device impedance levels. In multistage systems the interstage impedance matching is usually done at lower than 50 Ω levels, and in some instances very little impedance transformation is required. This may result in better broad band performance than deploying 50 Ω interfaces between each stage, but the latter has the advantage that each stage can be individually tested in a standard 50 Ω set-up. Up to VHF and low UHF, the input impedance of a MOSFET is high compared to that of a BJT but at higher frequencies they will reach similar values and the matching procedures become almost identical.

In practice virtually all multi-octave amplifier designs independent of the frequency spectrum and device type are of a push-pull circuit configuration. Another advantage with push-pull is that the power levels of two devices are

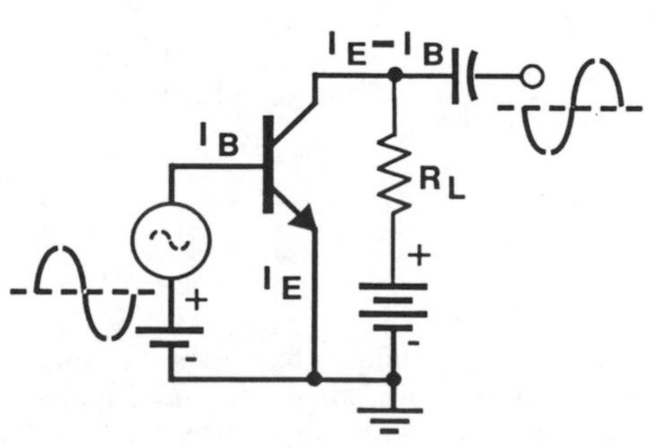

FIGURE 3-3

Common emitter circuit configuration. Note that it is the only one with phase reversal between the input and the output.

automatically combined for higher power output levels which allows the use of electrically smaller individual devices for a given power output. RF power transistors housed in push-pull headers have been available since the mid-1970s, but only since the development of high frequency FETs has the concept of push-pull packages become popular. (RF transistor package types will be discussed in detail in Chapter 6.) Both FETs and BJTs are now available in push-pull headers, most of them in the so called “Gemini” type. The term Gemini (twins) refers to two individual and independent transistors mounted on a common flange next to each other. The Gemini package is manufactured in several physical sizes, the largest being able to dissipate up to 500–600 watts. An obvious advantage with any push-pull transistor, whether in a single push-pull header or in a Gemini package, is the close electrical proximity of the two dice. This greatly enhances the device performance of a push-pull circuit, where the important factor is a low emitter-to-emitter (source-to-source) inductance and not the emitter-to-ground inductance. In all Gemini housed devices, the emitter (or source) is connected to the mounting flange, which is considered to be the electrical D.C. ground.

There are no significant differences in the efficiencies of amplifiers using either FETs or bipolars, although it is believed that the higher saturation voltage of FETs would make them less efficient. This may, however, be true only at low operating voltages (12 V and lower). At higher frequencies the device output capacitance has a much larger effect on efficiency; however, part of it can be “tuned out” in narrow band circuits as mentioned earlier.

COMMON BASE AND COMMON GATE

Common base circuits with BJTs are widely used at UHF and microwave frequencies due to the higher α cutoff than the β cutoff characteristics. This means that higher power gains are possible at these frequencies in common base than in common emitter configuration. The power gain of a common base amplifier increases if any base to ground inductance is added because it generates positive feedback. If more inductance is added, the gain will increase to a point of instability and finally lead to a condition of steady oscillation usually at a frequency where the matching networks resonate. All common base transistors have some positive feedback, generated by the inductances of the base bonding wires and the internal part of the base lead. However, this inductance is generally low enough not to generate sufficient positive feedback to create instability. As in the common emitter circuit configuration, a common base transistor’s gain is inversely proportional to the frequency of operation. It also has the same slope of approximately 5 dB per octave, but only up to the α cutoff. Below α cutoff gain flattens out to 12–15 dB and remains at that level down to D.C.

There is no feed through input power in a common base amplifier circuit, so the power output is actual and not $P_{in}+P_{out}$ as in a CE amplifier circuit. This probably improves the device ruggedness (ability to withstand load mismatches) in the form of reduced dissipation. A typical common base circuit is shown in Figure 3-4. Since the total current flows through the emitter, the input matching network or an emitter D.C. return choke must be able to carry I_B+I_C.

FIGURE 3-4

The common base circuit configuration has the lowest input impedance and no phase reversal between the input and the output.

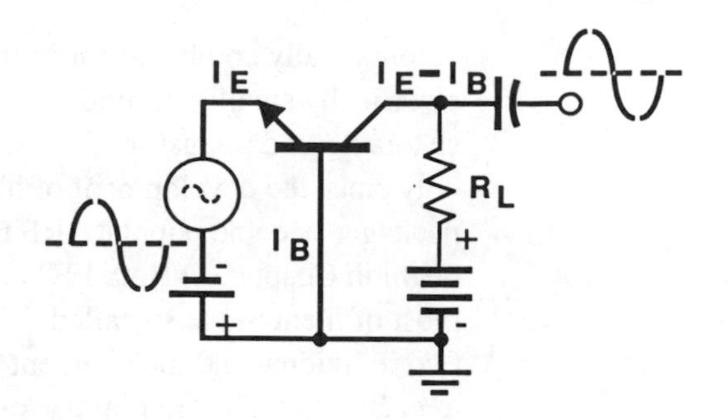

In a common base circuit the normal output capacitance (C_{ob}) and the feedback capacitance (C_{rb}) are reversed. Fortunately their values are about equal, except at low bias voltages where C_{rb} can be several times higher than C_{ob}. Under normal drive conditions there should not be much difference in the output capacitance or impedance between common emitter and common base circuits. However, the highly nonlinear C_{rb} reportedly creates increased tendencies for the well known half f_o phenomenon. With MOSFETs when operated as a common gate amplifier the situation is totally different. Their feedback capacitance (C_{rss}) has a value many times lower than the output capacitance (C_{oss}). When these are reversed, it makes the actual feedack capacitance high in respect to the input and output capacitances, creating an unstable condition. Even if the common gate inductance can be minimized, stability may not be achievable. The input impedance is lower than in a common source circuit because of the high value of feedback capacitance enhanced by the Miller effect.

Stable single frequency or narrow band circuits with fractional octave bandwidths are possible using the common base configuration, but wideband circuits are difficult to design if internal matching is required. Neutralization is employed in some instances to improve the stability, but it is not easy to implement except in push-pull designs. In high power circuits, biasing to a linear mode is somewhat difficult as an opposite polarity supply is required at the emitter. In addition there is a rectification effect which tends to reduce the bias voltage with RF drive. In small signal circuits, where the class of operation is mostly Class A, this can be accomplished with an amount of by-passed base-to-ground resistance to generate a self bias. Push-pull common base circuits are not commonly seen at higher power levels, which usually operate at high UHF or higher frequencies. One reason may be that the 180° phase shift is difficult to achieve and hold except for very narrow bandwidths. However, push-pull common base circuits are widely employed at power levels up to 0.5–1 watts in applications such as cable TV amplifiers, where an unbypassed common base resistance can be used for self biasing to a linear mode of operation. In each configuration—common emitter and common base—the push-pull design offers the same advantages, the most important of which is the noncritical base or emitter common mode inductance. The power gain and stability of the push-pull circuit depends to a large extent on base-to-base inductance. The MOSFET would always have to be biased to a level close to or greater than the gate threshold voltage in order to overcome $V_g(th)$ with RF input drive (excluding Class D and other switchmode systems). The bias source must be able to carry the full drain current, which at a gate threshold voltage of 4–5 volts would amount to a considerable level of dissipation. With BJTs the voltage is only 0.6–0.7 V and, thus, much more tolerable. The common gate MOSFET circuit could be useful in relatively low power applications, in circuits where neutralization can be easily realized and its high AGC range (power gain/ gate voltage) can be an advantage.

Some of the disadvantages of the common base amplifier circuit are: requirement for two D.C. power supplies for Classes A, AB and B, poor linearity due to regeneration, low input impedance, no possibility to implement negative feedback (except in push-pull), and high susceptibility for half f_o instability.

COMMON COLLECTOR AND COMMON DRAIN

A common collector (emitter follower) circuit (shown in Figure 3-5) is widely used where high input and low output impedance levels are desired. As in a common base configuration, there is no phase reversal between the input and the output. The emitter follower has a voltage gain of less than unity, and amplification is obtained from the current gain through impedance transformation. The output impedance is directly related to the input impedance divided by the current gain (h_{FE}). Conversely the input impedance equals the output load multiplied by $h_{FE.}$ This makes the emitter follower less suitable for RF power amplifiers than the two other circuit configurations since variations in the load impedance are directly reflected back to the input. For this reason it is most widely used as a wideband buffer amplifier to drive low impedance or capacitive loads. Especially in a complementary configuration, which provides active "pull-up" and "pull-down" in the output, the circuit offers one of the best drivers for capacitive loads. Some of the applications include CRT video drivers and MOSFET gate drivers in Class D/E amplifier systems.

A common drain or source follower circuit configuration represents the emitter follower in bipolar circuits.[1] As in the emitter follower, the input impedance is high and output impedance low. The input capacitance (drain-to-gate) is low compared to common source and common gate circuits, and considerably lower for the FET than for a bipolar of comparable electrical size. This low

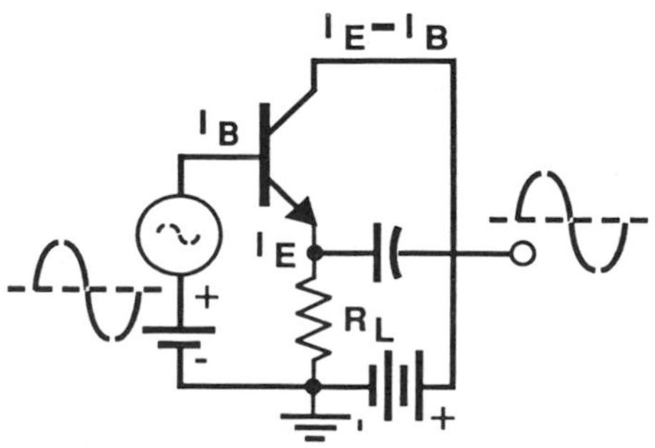

FIGURE 3-5
The common collector circuit configuration has the highest input impedance and lowest output impedance. No phase reversal exists between the input and the output.

input capacitance in the FET is because of the absence of the forward biased collector-base diode junction.

A source follower also has a voltage gain of less than unity, and since it is not a current amplifier, one cannot talk about current gain either. However, the amplification takes place through impedance transformation as in a bipolar circuit. Because of the extremely high input impedance, which varies more with frequency than does the input impedance in common source and common gate circuits, heavy resistive loading at the gate is necessary for any type broadband application. Negative feedback is not necessary, nor is it easy to implement due to equal phases of the input and output. A common source circuit exhibits exceptional stability for these reasons, but excessive stray inductances in the circuit lay-out can lead to low frequency oscillations. Unlike the emitter follower, variations in the load impedance in a source follower are not reflected to the input. This makes the source follower suitable for RF power amplifier applications at least up to VHF. Push-pull broadband circuits for a frequency range of 2–50 MHz have been designed for 200 to 300 watt power levels. Their inherent characteristics are good linearity, stability and gain flatness without the need for leveling networks. High power linear amplifiers are probably the most suitable application for this mode of operation. The AGC range is comparable to that in common source, but a higher voltage swing is required. In high voltage operation it must be noted that the gate rupture voltage can be easily exceeded since during the negative half cycle of the input signal the gate voltage can approach the level of V_{DS}.

4

Other Factors Affecting Amplifier Design

CLASSES OF OPERATION

The performance of an amplifier depends on how it is biased. We have already seen in Chapter 1 that low power transistors are characterized Class A and that many high power amplifiers are characterized Class C. It is important that the user of RF transistors understand the classes of operation and what significance the class plays in determining amplifier characteristics and in the choice of transistors for a specific application.

Basic classes of operation for an amplifier[1, 2, 3, 4, 10] are shown in Figure 4-1. Each class is limited to a specified portion of the input signal during which current flows in the amplifying device. Class A requires that current flow for all 360° (all the time) of the input signal (assumed to be in the form of a sine wave). Likewise, Class C requires current to flow for less than half the time, or less than 180°. The definitions of classes apply regardless whether the amplifier is a vacuum tube or a transistor, or whether it is a bipolar transistor or a FET. The significance of the class of operation has to do with the amplitude linearity of the amplification process. It is important to note that only Class A amplifiers are linear in that the output signal (in the ideal case) is a faithful reproduction of the input signal.

One biases a transistor, for example, in the center of its linear region for Class A operation. Generally the required bias current is close to a value equal to ½ the maximum current.[18] Once biased in this condition, and provided the input signal is kept small enough to prevent the transistor from being driven out of the linear region, the output signal will be a faithful reproduction of the input signal with appropriate amplification. In a Class C amplifier, current flows in the output circuit only during the peak swings of the input signal. The result is a highly amplitude distorted output signal consisting of bursts of current for short durations of the input waveform.

In reality, as we have already seen in Chapter 1, even practical Class A amplifiers are to some degree non-linear. The degree of non-linearity is generally specified and controlled for all so-called linear amplifiers. Class AB amplifiers are frequently used in situations requiring high power "linear" amplification of the RF signal, but in most cases special circuitry is required to improve the linearity of the overall amplification process. Such circuits take the form of

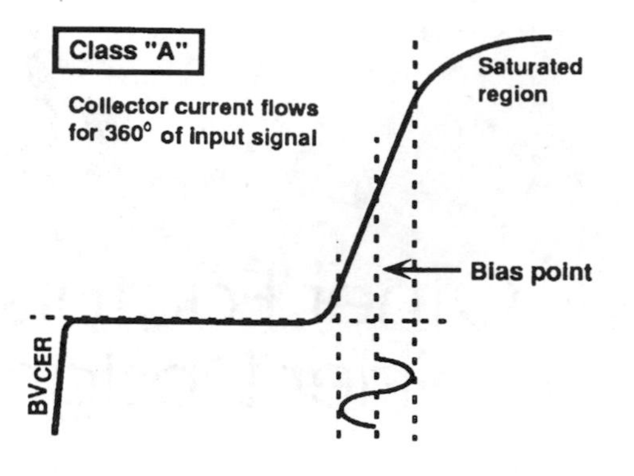

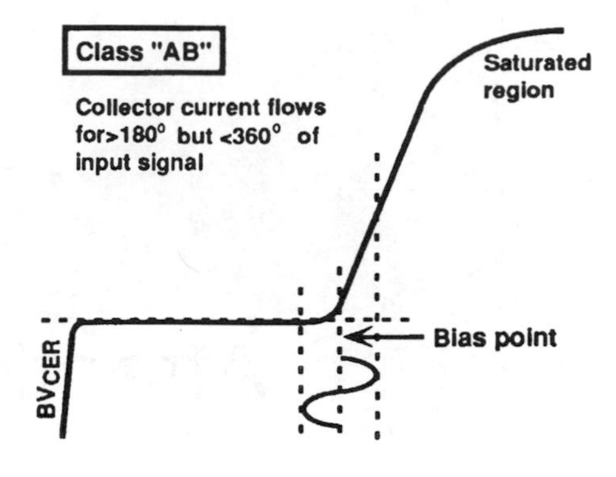

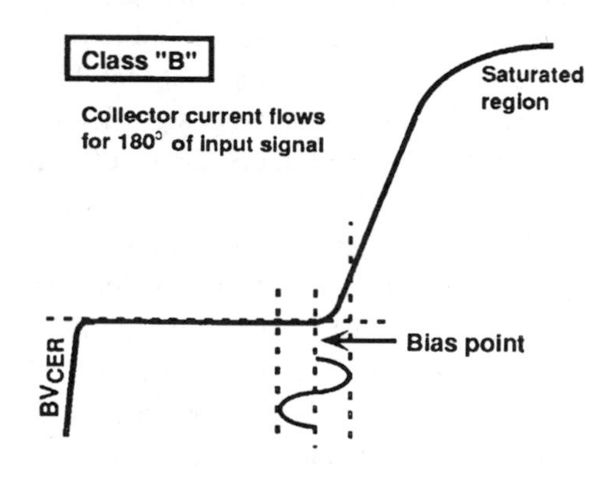

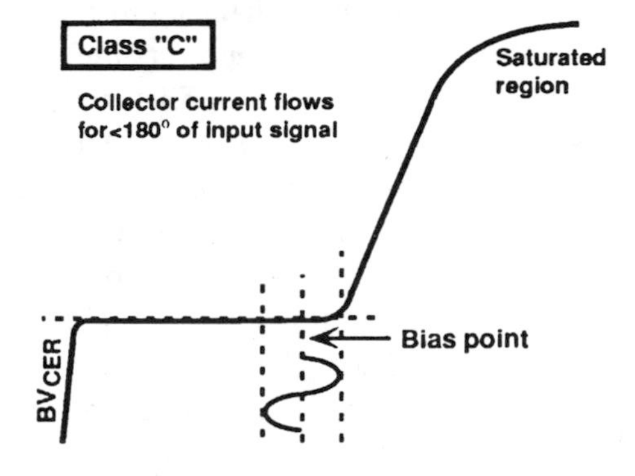

FIGURE 4-1
Classes of amplifier operation.

feedback networks and, sometimes, more complex configurations referred to as "feed-forward" circuits in which a distorting high-power amplifier is combined with a low power correction amplifier along with phasing networks to produce a high power, highly linear system.

The reason why all amplifiers are not Class A linear has to do with circuit efficiency. The theoretical maximum possible efficiency of a truly Class A amplifier is only 50%. However, the efficiency of a Class C amplifier can vary from approximately 80% to nearly 100% depending on output.[1] Because efficiency is important in most amplifier applications, particularly those involving high power, circuit designers tend to use the class of operation that gives best efficiency and still meet the requirements of preserving the information contained within the RF signal. A summary of maximum theoretical efficiencies in various classes of operation is shown in Table 4-1.

As shown in Table 4-1, there are other classes of operation, which are not as widely used as classes A, B, AB and C. These are Classes D and E, usually

TABLE 4-1 Maximum theoretical efficiencies for basic classes of amplifier operation.

Class	Configuration	Efficiency, η (%)	Comments
A	All	50	
B	All	78.5	
AB	All	50–78.5	(Depending on angle of conduction)
C	Nonsaturating	85–90	(Depending on angle of conduction)
D	All	100	(Assumes infinite switching speed)
E	——	100	(Assumes no overlap for the output RF currents and voltages)

categorized as high efficiency modes of operation.[1,6] Both are so-called switch-mode systems and use MOSFETs almost exclusively due to the absence of charge storage effects inherent in BJTs. This makes faster switching speeds and lower phase delays possible. Even the low frequency MOSFETs are capable of switching in a few nanoseconds providing that their input capacitances (C_{iss}) can be charged and discharged at a sufficiently fast rate. Theoretical efficiencies are 100%, but in practice they are limited to 90–95% due to non-ideal switching times, device output capacitances, etc. There are two basic types of Class D amplifiers: 1) a current switching amplifier, which must be driven with a square wave signal; and, 2) a voltage switching amplifier, which can be driven with either a square wave or sine wave input, of which the latter is the more common one. With sine wave drive, the gate voltage swing must be large enough to ensure a complete saturation and cut-off of the FET. The input and output waveforms are approximately identical in each case except that the current and voltage waveforms are reversed. The current switching, Class D amplifier is preferred for more demanding applications since its duty cycle is easily defined and is not affected by the amplitude of input drive.

Class E is basically a variation of Class D with an LC network added to its output. It compensates for part of the FET's output capacitance and helps to reduce overlap between the switching currents and voltages, thus improving the efficiency. The improvement can be on the order of 5-10%. However, it is a relatively narrow band system due to the LC network, whereas plain Class D can operate at bandwidths of several octaves. Output power of Class D/E amplifiers is limited by the switching speeds of MOSFETs and by the capacitive loads presented to driver stages. A graph of estimated practical power levels versus carrier frequencies of Class D/E amplifiers is presented in Figure 4-2.

FORMS OF MODULATION

A primary use of radio frequency signals is to transfer information from one point to another. This is the basic function of all communication systems and the difference created by using radio frequencies as "carriers" of the information is to permit the communication system to be wireless. Radio frequencies propagate in space and permit information to be transferred from a transmitter to a receiver when information is "added" to the RF signal by a process called *modulation*.

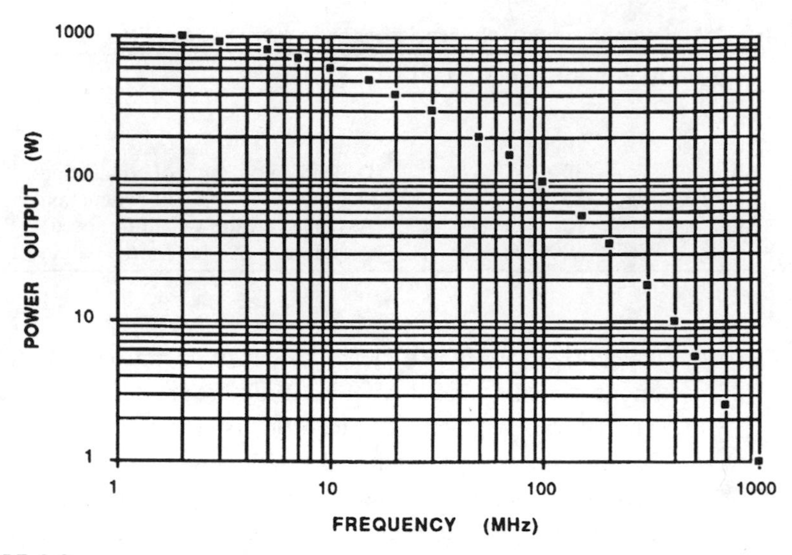

FIGURE 4-2

Estimated maximum power levels with push-pull or single ended Class D/E amplifiers based on present technology.

There are two basic kinds of modulation used to put information on an RF "carrier." The first, and most fundamental, is called *amplitude modulation*, in which the amplitude of the RF signal is made to change as a function of time by a modulating signal composed of a band of frequencies, much lower in value than the frequency of the "carrier." Such modulation is illustrated in Figure 4-3.[1, 2, 4, 7] Both the "time domain" signal and its counterpart in the "frequency domain" are illustrated in Figure 4-3. It should be noted that the frequency bandwidth or spectrum required for amplitude modulation is precisely twice the highest permissible modulating frequency. In this type AM, a full power carrier is modulated up to a maximum of 100% (in practice up to 60-80% depending on the application). With 100% modulation, the RF peak power would reach a value four times the unmodulated carrier power, or

$$\left(\frac{V_{pp}}{V_{carr}}\right)^2 (P_{carr})$$

(assuming a sine wave modulating signal).

There are two ways the amplitude modulation can be accomplished. In the so-called *low level* amplitude modulation, all the driver stages are modulated, which means that these stages must have linear transfer characteristics in addition to those of the final stage. In *high level* AM, the final stage is collector/drain modulated via a series element such as a transformer, thereby eliminating the necessity for linearity in the output stage as well as the driver stages. Practical applications require the immediate driver stage to also be modulated in order to reach modulation percentages higher than 70–80%. This requirement generally is created by the high saturation voltages and diode voltage drops that exist in the final stage of the high level modulating system.

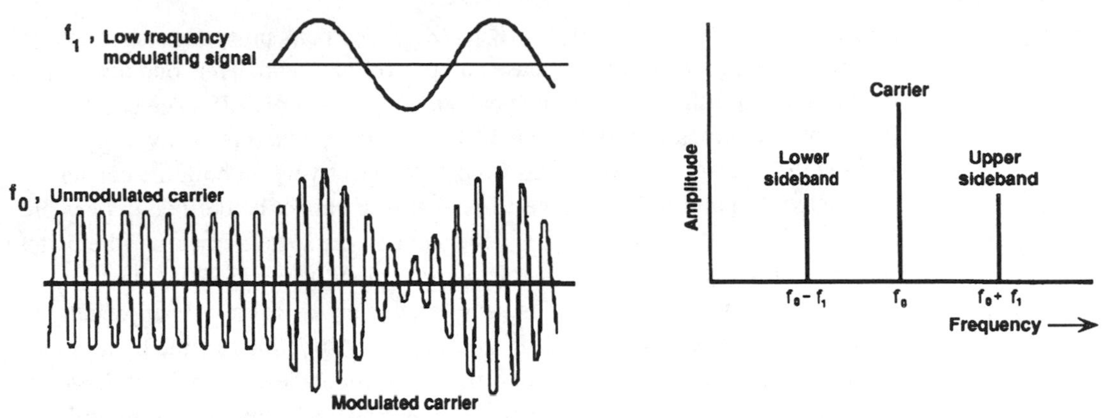

The separation of sidebands from the carrier is determined by the modulating frequency.

FIGURE 4-3
Amplitude modulation waveforms and sideband positions.

Another type of AM uses a suppressed carrier and is commonly called *double sideband AM* (shown in Figure 4-4). Since both sidebands contain identical information except relative to the phase, one sideband is usually removed by filtering or phase cancellation in order to conserve the spectrum. Without modulation there is ideally no RF output, which would be otherwise proportional to the modulation amplitude. The envelope seen in Figure 4-4 can be either a *double sideband* (DSB) single tone signal or a *single sideband* (SSB) two tone signal,[1,2,12] which appear identical. As in the carrier AM, the testing, distortion measurements, etc., are done with a sine wave reference, which is also used to set the specifications. The suppressed carrier signal is generated in low level stages (balanced modulators) or in higher level stages using the phasing method. In either case all amplifying stages after the sideband generator must be highly linear. Class A is recommended for the low level stages and Class AB for the ones above 1-2 W. A Class AB push-pull amplifier in general gives results comparable to a single ended Class A circuit. Referring to Figure

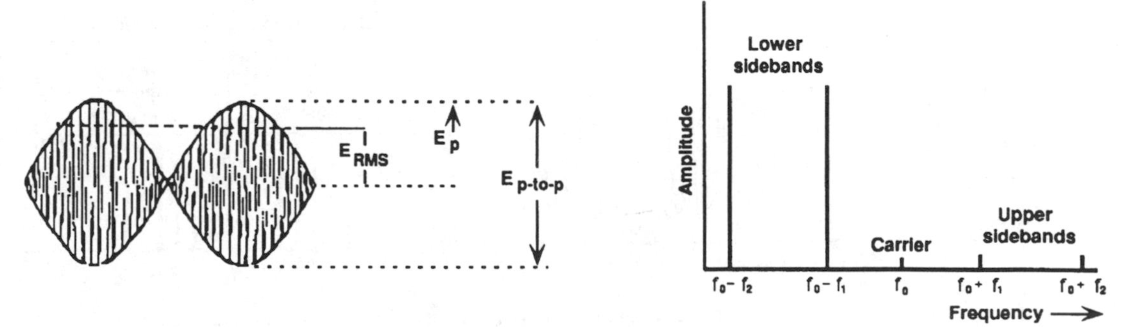

The separation of sidebands from the carrier is determined by the modulating frequency.

FIGURE 4-4
Single sideband (SSB) modulation envelope and sideband positions.

4-4, the $E_{RMS} = E_p/\sqrt{2}$ and $PEP = E^2_{RMS}/Z_{out}$. The peak power occurs when the two test frequencies are in phase and the voltages add. Thus one tone (peak) contains one fourth (–6dB) of the peak envelope power (PEP). The combined power of both tones would then be 3dB below PEP, which is the average power, making the average to PEP relationship 1:2 respectively. In both the carrier modulated AM and SSB, average-to-peak power cannot be defined under voice modulated conditions because it is dependent on the system linearity, the pattern of the modulating voice and possible compression factors (SSB).

A second kind of modulation is called *frequency modulation*[1,7] in which the frequency of the RF signal is made to change as a function of time by a modulating signal likewise composed of a band of frequencies, again much lower in value than the fundamental frequency of the "carrier." This form of modulation is illustrated in Figure 4-5. The frequencies created by frequency modulation (as shown in the "frequency domain" portion of Figure 4-5) extend in theory to zero frequency below and to infinity frequency above the unmodulated carrier frequency. In theory, the amplitude of frequencies extending below and above the carrier beyond a certain point approaches zero and the bandwidth required for frequency modulation can be limited to the minimum to maximum frequency deviation created by the modulation signal. A term called the *modulation index*, defined as the maximum deviation in frequency divided by the maximum modulation frequency, will determine the amplitude of each frequency in the so-called frequency spectrum. Frequency modulation is easy to implement since it is always done at a very low signal level, usually at the stage generating the carrier. The only concerns are the linearity of the modulating signal versus the frequency deviation and the maximum deviation as a result of the highest modulating frequency as well as the amplitude of the modulating signal.

When information is contained in the form of amplitude modulation of the RF signal, linear amplification is required in order to preserve the information during the amplification process. If the information is contained in the form of frequency modulation, then non-linear amplification is possible without conse-

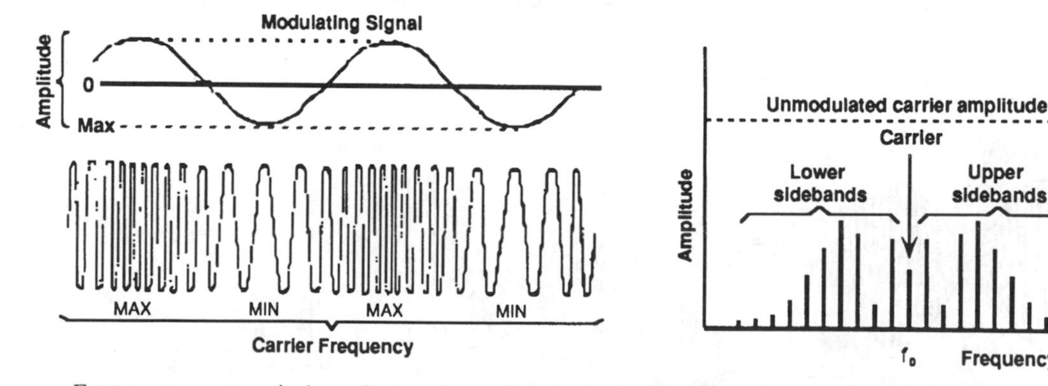

Frequency spectrum is dependent on the modulation index (frequency deviation/modulating frequency)

FIGURE 4-5

Frequency modulation carrier wave frequency variations and resultant sidebands.

quent loss of information in the amplification process. Because Class C amplifiers are more efficient than Class A amplifiers, they are the amplifying "class of choice" for most RF signals involving frequency modulation. Likewise, Class A or AB amplifiers (in spite of their lower efficiency) are necessary in the amplification of RF signals that contain amplitude modulated information. RF power transistors are characterized either Class C or Class A/AB depending on their intended application. Transistors designed for use in analog two-way radio systems are almost always characterized Class C because the form of modulation used in these systems is FM. Those intended for use in television systems are characterized Class A or AB because the form of modulation used in these systems is AM.

A logical question would be "can Class A characterized transistors be used as Class C amplifiers?" And the obvious answer is yes. Likewise, one might wonder if Class C characterized transistors can be used in Class A amplifiers. Again the answer is yes, provided certain conditions are met. These involve a "derating" of the Class C transistor to a lower power level. The amount of derating depends on the class of operation. If a Class C characterized transistor is used in a truly linear Class A amplifier, it should be derated by a factor of 4. That is to say, if the transistor is capable of delivering 60 watts Class C, it should not be used Class A at a power level greater than 15 watts. If the use is Class AB, a safe derating factor would be 3.

There are two factors that make it necessary to derate RF power transistors characterized Class C when it is desired to use them in a more linear mode. First, linear classes of operation require bias. It is not uncommon for a Class AB high power transistor to be biased at several amperes of current. This bias results in a large amount of power dissipated in the device. Second, the efficiency of more linear forms of amplification decreases as the linearity increases. This means for the same amount of output power, the power dissipated in the transistor will increase. Dissipated power raises the die temperature of a device (for a given heat sink temperature) and as will be discussed in Chapter 5, the die temperature for silicon devices should not be allowed to exceed 200°C.

BIASING TO LINEAR OPERATION

Linearity, Testing and Applications

All solid state devices and vacuum tubes intended for linear operation must have a certain amount of "forward bias" (D.C. idle current) in order to place their operating points in the linear region of the transfer curve (see Figure 4-6). Perfect linearity means that the power output follows the power input in a linear fashion, e.g., a P_{in} of 1W produces a P_{out} of 10 W, 2 W results in a P_{out} = 20 W, etc. Or simply this means the power gain must be constant from almost zero to the maximum P_{out} level. This can also be expressed as gain compression in dB or as third order intercept point, which is widely used in low power and CATV applications. In large signal applications for voice communications, the linearity is usually measured as intermodulation distortion (IMD) using two test frequencies (tones) spaced 1kHz apart as a standard. In testing amplifying devices for linear use in television, two or three test frequencies can be employed (depending on the specifications) and their spacings are in the MHz range. Three test frequencies (triple beat) are common with low power device specifications and are standard in CATV device testing, where distortion levels are very low. This also allows a wider spectrum to be analyzed, which better simulates multichannel systems.[11] (These concepts of linearity are also discussed in Chapter 1.)

The distortion level expressed as IMD is easier to relate to actual numbers and is the quantity usually desired. It is the method by which linearity is initially measured and can then be converted to third order intercept if necessary. The test frequencies are viewed on a spectrum analyzer screen and the distortion products (third, fifth, seventh order, etc.) appear on each side of the test tones. Their amplitudes can be read directly and are expressed either in dB below one of the tones (MIL STD) or below the peak power (EIA standard). There are numerous ways to generate the test tones, some of which are given in[2, 8]. Conversion to third order intercept can be done as:

$$IP^3 = P_{out} + \frac{IMD}{2},$$

where IP^3 = third order intercept point, P_{out} = power output (one tone, dBm), IMD = third order intermodulation distortion below one tone (dB). Reversing the equation: $IMD = 2(IP^3 - P_{out})$. Example: If an amplifier has an IP^3 of +20 dBm and the P_{out} = +5 dBm/tone, the third order IMD = 2[20–(+5)] = 30 dB below one of the +5 dBm tones. Either the power input or the power output can be used for the power reference. In circuits having an insertion loss such as mixers, the P_{in} is generally used as a reference and in circuits with power gain, the P_{out} is preferred due to a smaller factor of possible error.[2, 9]

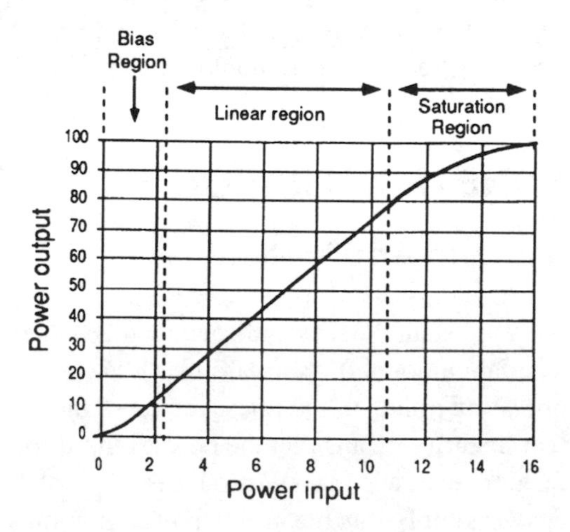

FIGURE 4-6

A typical input-output transfer curve of a solid state amplifier. Nonlinearity can be seen in the bias region. The purpose of the forward bias is to move the operating point to the linear portion of the curve.

Bipolar devices require a constant voltage source, whereas MOSFETs can be biased with simple resistor divider networks. Both will get more complex, however, if temperature stability is required. In addition enhancement mode MOSFETs always require some amount of gate bias voltage to overcome the gate threshold (see Chapter 3). Exceptions are MOSFETs operated in Class D or in other switchmode classes. In addition to applications requiring amplifier linearity (discussed earlier), examples include all amplitude modulated systems for communications and broadcast, nuclear magnetic resonance, magnetic resonance imaging, digital cellular telephone, and signal sources for instrumentation.

One of the requirements[17] for transistor linearity is the flatness of f_τ versus I_c (see Figure 1-17). When the collector current varies, it results in a variation of f_τ and, consequently, in a variation in power gain. The low I_c area is not very critical and produces only cross-over distortion, which in most cases can be reduced by increasing the bias idle current. If the "knee" from zero current to maximum f_τ is sharp, a smaller amount of bias or idle current is required. MOSFETs will produce a similar f_τ versus I_D curve, except that their low current "knee" is not as sharp as that of a BJT, which explains their requirement for higher bias idle currents.

The input signal can drive the transistor to peak current levels that are significantly higher than the bias current. Thus the slope of the f_τ curve from the bias current level to the maximum current caused by the input signal determines the transistor's linearity performance at high current. A certain amount of reduction with increasing current can be tolerated (see Figure 1-17) without noticeable nonlinearities, but excessive "sloping down" would cause early saturation of the amplifier and result in "flat topping" of the output modulation peaks. Note also that the measurements of f_τ versus I_c are usually done under pulse conditions, which excludes thermal effects. Thus the f_τ versus I_c curve shows less "sloping down" than will be experienced in actual use of the transistor.

Bipolar Linear Amplifiers

Since the base current of a bipolar transistor is equal to I_c (peak)/h_{FE}, the base bias supply must be able to supply this current without considerable excursions in the base-emitter voltage between the no-signal and the maximum signal conditions. This requires a constant voltage source, as variations of a few millivolts represent a large portion of the nominal 0.63–0.67 volt typical value. Depending on the specification of a specific application, various degrees of requirements are set for the base bias voltage source. In some instances a large value capacitor can be connected across the bias voltage supply to further reduce its A.C. impedance. However, this makes the impedance dependent on the frequency of modulation, and is a good and practical solution only in applications where the modulating frequency is in the medium to high audio frequency range. One of the simplest biasing circuits for bipolar transistors[5, 13, 14] is shown in Figure 4-7. It uses a clamping diode to provide a low impedance voltage source. The diode forward current must be greater than the peak base current of the transistor. This current is adjusted with R2 and the resistance of

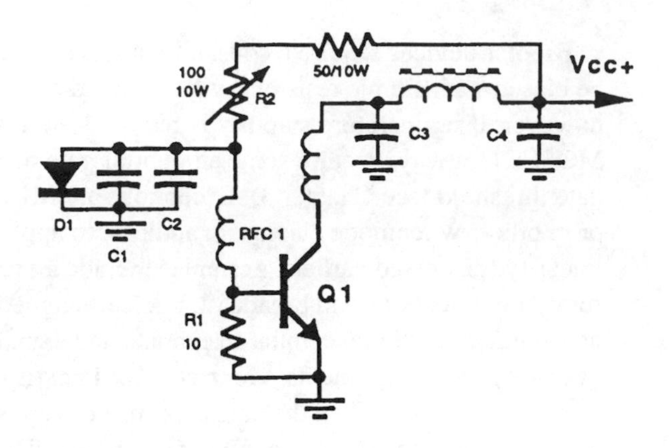

FIGURE 4-7

A simple biasing circuit using a clamping diode. It is inefficient since a minimum of I_B (peak) must go through D1. Q1 is the transistor to be biased.

RFC1 and R1 is used to reduce the actual base voltage to a slightly lower value than the forward voltage of D1. The diode can be mechanically connected to the heat sink or the transistor housing to perform a temperature compensating function to Q1. This technique works adequately, although for perfect temperature tracking, Q1 and D1 should have similar D.C. parameters. One disadvantage with the circuit shown in Figure 4-7 is its inefficiency especially in biasing high power devices, since $(V_{cc} - V_b)xI_b(max)$ will always be dissipated in the dropping resistors.

The reduced efficiency of the circuit shown in Figure 4-7 can be overcome by amplifying the clamping diode current with an emitter follower[14] as shown in Figure 4-8. Two series diodes (D1and D2) are required since one has to compensate for the $V_{BE}(f)$ drop in Q1. In this case low current signal diodes can be used and their forward current is equal to I(bias)/h_{FE}(Q1). For best results, Q1 should have a linear h_{FE} up to the peak bias current required and in higher power systems it must be cooled by some means. Ideally Q1 and one of the series diodes should remain at ambient temperature, whereas the other diode

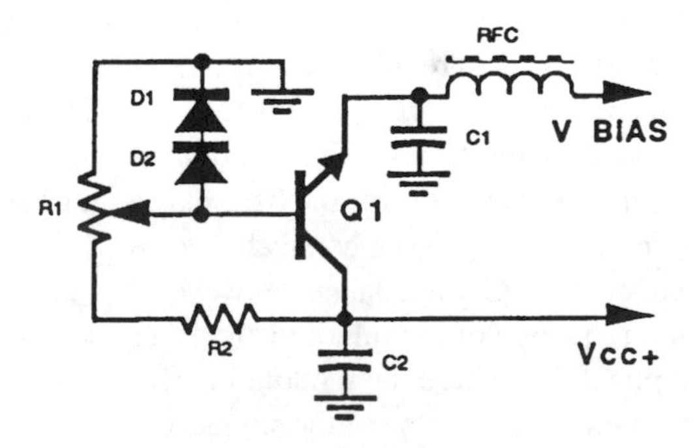

FIGURE 4-8

As in Figure 4-7, a clamping diode scheme is also used, but a low current flows through the diodes and is then amplified by an emitter follower.

(D1 or D2) can be used for temperature compensation of the RF device. An effective and fast responding system is obtained if the diode (having long leads) is located near the RF transistor. The leads can be suitably formed allowing the body of the diode to be pressed against the ceramic lid of the RF transistor and fastened in place with thermally conductive epoxy. R1 is used to set the bias idle current and R2 limits its range of adjustment. The value of R2 depends on the supply voltage employed. The function of C1 and the RF choke (RFC) is simply to prevent the RF signal from getting into Q1.

Another fairly simple bipolar bias source[15] is shown in Figure 4-9. Its output voltage equals the base-emitter junction drop of Q1 plus the drop across R3. R1 must be selected to provide sufficient base drive current for Q2, set by its h_{FE}. Normally this current is in the range of a few milliamperes, and Q1 can be any small signal transistor in a package configuration that can be easily mounted to the heat sink or RF transistor housing for temperature compensation. The only requirement is that its $V_{BE}(f)$ at that current must be lower than that of the RF transistor at its bias current level. The maximum current capability depends on Q2 and R2. The power dissipation of Q2 can be up to a few watts and in most cases should be heat sunk, but must be electrically isolated from ground. The value of R2 can be calculated as: $V_{CE} - V_{CE}(sat))/I_b$. C1 through C3 are a precaution to suppress high frequency oscillations, but may not be necessary depending on the transistors used and the physical circuit lay-out. Output source impedances for this circuit, when used in conjunction with a 300 W amplifier, have been calculated as low as 200–300 mΩ.

More sophisticated bias sources can include an integrated circuit voltage regulator.[5] In most instances a pass transistor is required for current boost and to lower the source impedance. There are high current regulators available today,

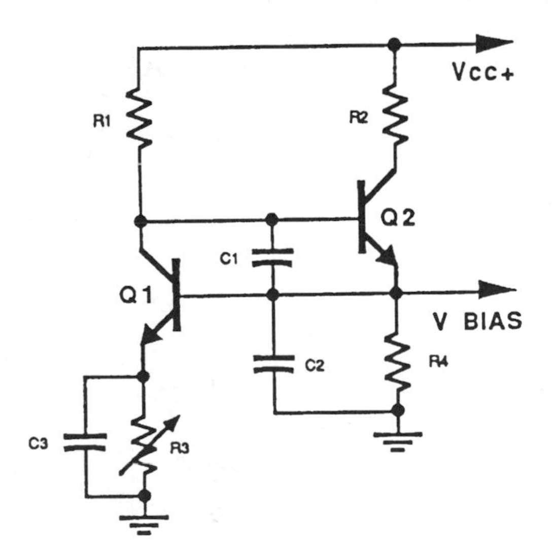

FIGURE 4-9
This bias circuit features the lowest source impedance of the less complex bias circuits. Therefore, it is recommended for high power device biasing and for other demanding applications.

such as the LM317, LM337, etc., but their suitability for applications such as this is not known. The circuit in Figure 4-10 uses a 723 regulator, which is available from several manufacturers with a variety of prefixes. It has been used for bipolar bias sources since the early 1970s and, more recently, for MOSFET biasing as well. The 723 is specified for a minimum V_{out} of 2 volts, but with certain circuit modifications can be lowered to less than 0.5 V.

The main advantages of the bias source shown in Figure 4-10 are: 1) it provides the lowest source impedance at a relatively low cost, 2) the bias voltage remains independent of variations in the power supply voltage, and 3) temperature compensation is easy to implement. In Figure 4-10, D1 performs this function and should be in thermal contact with the heat source. The same technique discussed with the circuit shown in Figure 4-8 can also be adapted here. Depending on the current requirement and the pass transistor used, Q1 may have to be cooled. It has a positive temperature coefficient to the bias voltage, but the temperature coefficient is negligible compared to the negative coefficient of D1. This permits Q1 to be attached to the main heat sink. R1 and D2 are only necessary if the RF amplifier is operated at a supply voltage higher than 40 V, which is the maximum rating for the regulator.

Biasing of MOSFETs

Since MOSFETs have gate threshold voltages up to 5–6 volts, they require some gate bias voltage in most applications. They can be operated in Class C (zero gate bias), but at a cost of low power gain. In such case the input voltage swing must have an amplitude sufficient to overcome the gate voltage from zero to over the threshold level. The drain efficiency is usually higher than in other classes of operation. Especially if overdriven, the class of operation can

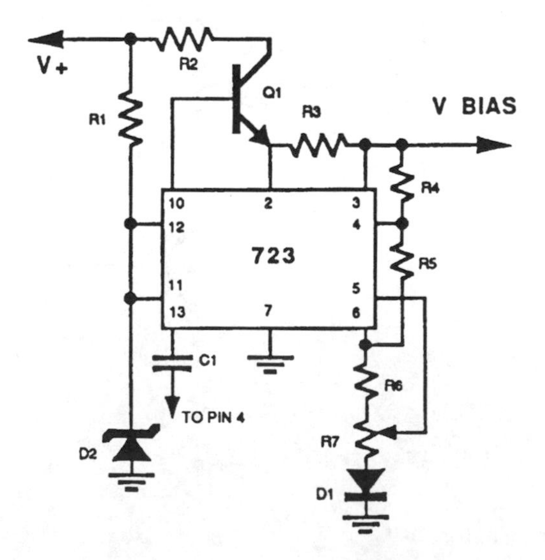

FIGURE 4-10

An integrated circuit bias source. This circuit also provides regulation against supply voltage variations. The source impedance mainly depends on the h_{FE} of Q1.

approach Class D. Zero bias is often used in amplifiers intended for signals that do not need linear amplification (such as FM signals and some forms of CW signals) and efficiencies in excess of 80% are not uncommon. In Class B, the gate bias voltage is set just below the threshold, resulting in zero drain idle current flow. The power gain is higher than in Class C but the drain efficiency is 10–15% lower. Class B is also suitable only for non-linear amplification. Between classes of operation, one must decide whether the system has power gain to spare and how important is efficiency. At higher frequencies, such as UHF, a good compromise may be Class B or even Class AB. In Class AB the gate bias voltage is somewhat higher than the device threshold, resulting in drain idle current flow. The idle current required to place the device in the linear mode of operation is usually given in a data sheet. In this respect MOSFETs are much more sensitive to the level of idle current than are bipolar transistors. They also require somewhat higher current levels compared to bipolars of similar electrical size.

The temperature compensation of MOSFETs can be most readily accomplished with networks consisting of thermistors and resistors. The ratio of the two must be adjusted according to the thermistor characteristics and the g_{fs} of the FET. The changes in the gate threshold voltage are inversely proportional to temperature and amount to approximately 1 mV/°C. These changes have a larger effect on the I_{DQ} of a FET with high g_{fs} than one with low g_{fs}. Unfortunately the situation is complicated by the fact that g_{fs} is also reduced at elevated temperatures, making the drain idle current dependent on two variables. In spite of the dependence, this method of temperature compensation can be designed to operate satisfactorily and is repeatable for production. The thermistor is thermally connected into a convenient location in the heat source in a manner similar to that described for the compensating diodes with bipolar units discussed earlier. An example of a simple MOSFET biasing circuit[16] as described here is shown in Figure 4-11.

Most MOSFET device data sheets give V_{GS}(th) versus I_D data, but the values are only typical, and in some cases g_{fs} can vary as much as 100% from unit to unit. Thus, in production the devices should have g_{fs} values that are within 20% of each other. Otherwise each amplifier must be individually checked for

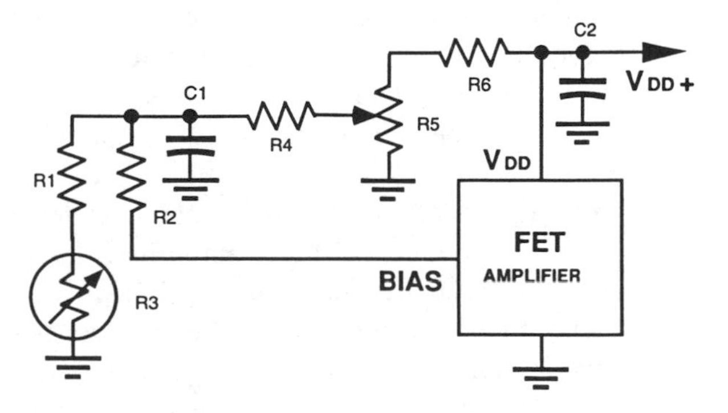

FIGURE 4-11

A simple MOSFET bias circuit using a thermistor-resistor network for temperature compensation.

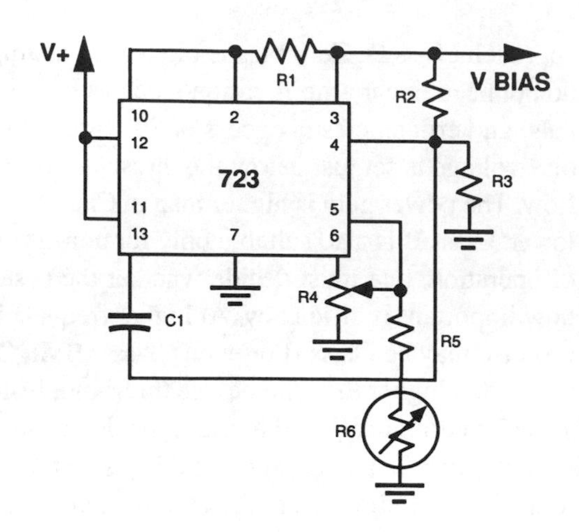

FIGURE 4-12

A more sophisticated MOSFET bias system with an integrated circuit voltage regulator. It also employs a thermistor for temperature tracking and provides supply voltage regulation like the circuit in Figure 4-10.

temperature tracking. Some manufacturers such as Motorola supply RF power FETs with specified ranges of g_{fs} matching.

The circuit in Figure 4-12 shows a typical MOSFET bias voltage source using the 723 IC regulator,[5] which was earlier presented for bipolar transistor biasing. Since a MOSFET draws no gate bias current, except in the form of leakage, the pass transistor (Q1) has been omitted and D1 replaced by R5-R6 combination. The values of other passive components have also been modified to produce a maximum output of 8 volts. The temperature slope is adjusted by the ratio of the series resistor (R5) and the thermistor (R6). In addition to maintaining a constant bias voltage, this circuit also features bias voltage regulation against changes in the power supply voltage.

Figure 4-13 shows a closed loop system for MOSFET biasing. It provides an automatic and precise temperature compensation to any MOSFET regardless of its electrical size and g_{fs}. No temperature sensing elements need be connected to the heat sink or to the device housing. In fact FETs with different gate threshold voltages can be changed in the amplifier without affecting the level of the idle current. This means that the gate threshold voltage can vary within wide limits over short or long periods of time for a variety of reasons. In addition to temperature, other factors affecting V_{gs}(th) might be moisture, atmospheric pressure, etc.

The principle of operation of the circuit shown in Figure 4-13 is as follows: the idle current of the MOSFET amplifier is initially set to Class A, AB, or anywhere in between these bias limits by R7, which also provides a stable voltage reference to the negative input of the operational amplifier U1. Current flows through R1 with a consequent voltage generated across it. This voltage is fed to the positive input of U1, which results in the output of U1 following it in polarity but not in amplitude. Due to the voltage gain in U1, which operates in a D.C. open loop mode, its output voltage excursions are much higher than those

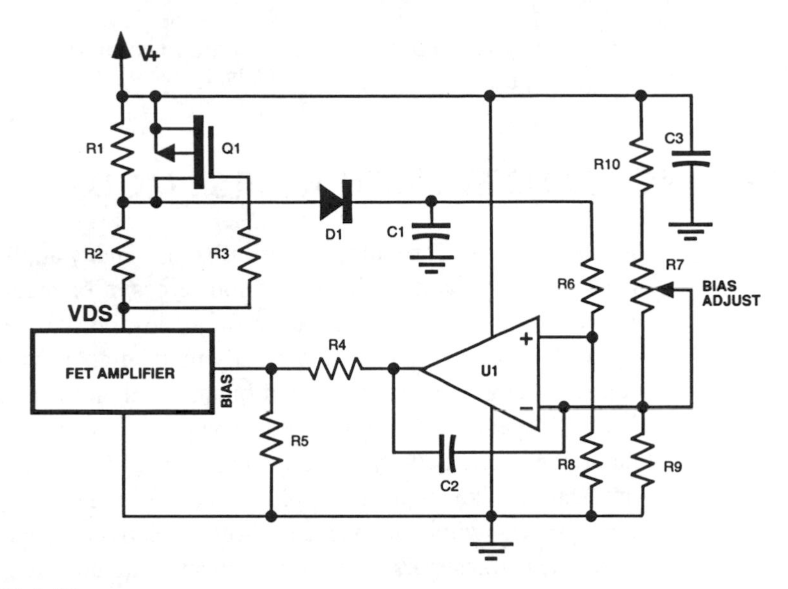

FIGURE 4-13

An automatic bias tracking system for MOSFET power amplifiers. It provides automatic temperature compensation without sensors as well as versatility for substituting a variety of electrical sizes of FETs operating at any supply voltage.

generated across R1. Thus, if the current through R1 tends to increase for any reason, part of the output voltage of U1 fed to the amplifier gate bias input will adjust to a lower level, holding the current through R1 at its original value. A similar self adjustment will take place in the opposite direction as well.

The values for the resistive voltage divider R4-R5 have been selected for a suitable range, sufficient to control the amplifier FET gate with the full voltage swing at the output of U1. When the amplifier is RF driven, the current through R1 increases and the bias voltage to the amplifier tends to decrease along with the voltage to the positive input of U1. At the same time, however, Q1 will start conducting, which lowers the effective value of R1 since Q1 is in parallel with it. The turn-on gate voltage for Q1 is obtained from the voltage drop across R2. Typical values for R1 and R2 are 5–10 Ω and 0.1–0.2 Ω respectively. The values of R1 and R2 must be selected according to the characteristics of Q1 along with the exact application and the current levels in question. The higher the current drawn by the amplifier, the harder will Q1 be turned on. For example if R1 is 5 Ω and Q1 is fully turned on with its r_{DS}(on) of 0.2 Ω, the effective value of R1 will vary between 5 Ω and less than 0.2 Ω, depending on the current drawn. Thus, with the current variable resistor (Q1-R1) it is possible to keep the output of U1 and the resulting amplifier bias voltage relatively stable under varying current conditions. The circuit is ideal for Class A amplifiers, where the drain current remains constant regardless of the RF drive. Q1, R2 and R3 can be omitted for Class A amplifiers and the value of R1 can be made as low as 0.05–0.1 Ω.

The circuits presented in this section are of a basic nature. They may require refinement or modification according to specific applications. While all circuit examples have been tested and most are in common use, component values are not given, but are available in [5, 13, 14, 15, and 16].

OPERATING TRANSISTORS IN A PULSE MODE

RF energy in the form of pulses is utilized in many applications including medical electronics, laser excitation, various types of radar, etc. All of these vary in specifications regarding the carrier frequency, pulse repetition rate and duty cycle. The carrier frequency is usually much higher than the pulse repetition rate, resulting in the generation of bursts of RF at the carrier frequency whose lengths depend on the pulse width. The pulse repetition rates are typically in the audio range and duty cycles range between 0.05 and 10%. For low duty cycle applications like radar, special devices have been developed to operate at higher peak powers, while the average power is relatively low leading to low dissipation. These transistors (UHF to microwave) are almost exclusively of the bipolar type.

BJTs in general have higher peak power capabilities than MOSFETs, but their peak power performance can be further improved by adjusting the emitter ballast resistor values to lower than normally required for CW. The epitaxial layer that controls the transistor's saturated power, is also made thinner than normal since the problem of ruggedness is partly eliminated due to the low average power. With increasing pulse widths, the dissipation will increase and at pulse widths of 1 millisecond and wider the device can be considered to operate like a CW signal. This is due to the fact that the temperature time constant of a medium size RF power die is around 1 ms, beyond which more heat will be transferred into the bulk silicon and through it to the transistor housing.

A graph of thermal resistance versus pulse width is shown in Chapter 5 (Figure 5-5). If the pulses are short but the repetition rate approaches 1 kHz (1 ms period), the same effect is created. Transistors made exclusively for pulse operation can produce peak power levels of 5–6 times the CW rating for a die of a similar size. With standard transistors designed for CW, the multiplying factor is more on the order of 3 to 4. MOSFETs can be used for pulsed power operation, but they have some disadvantages as well as advantages over the BJTs. The disadvantages include "pulse drooping," which means that the trailing end of the pulse has a lower amplitude than the leading end. It is caused by the decreasing g_{FS} of a MOSFET with temperature. Corrective circuitry can be used to compensate for this, but this adds to the circuit complexity. The advantages include smaller phase delays and faster rise and fall times. Thus, it depends on the application and it is up to the designer to decide which device is the most suitable.

There are several ways to generate RF pulses, but the one illustrated in Figure 4-14 is probably the most common. Figure 4-14 also shows a measuring setup to measure input, reflected and output powers plus a way to derive the demodulated pulse to permit its examination by means of an oscilloscope. The video output on the power meters shown in Figure 4-14 is used to monitor the pulses for droop and instabilities.

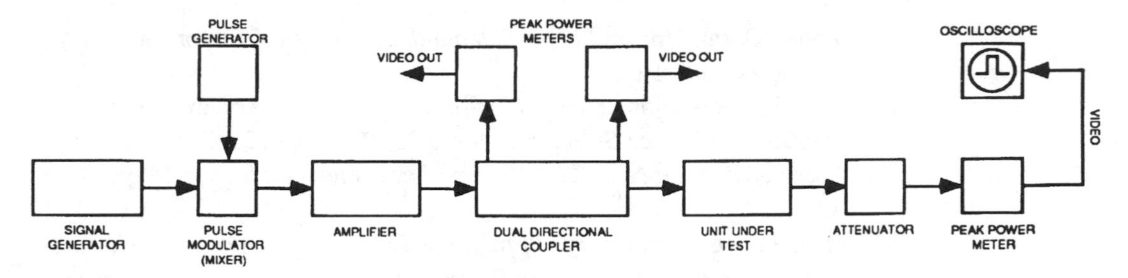

FIGURE 4-14
An example of a more sophisticated set-up to generate pulses, measure power gain, and measure peak power. Pulse shape can also be visually analyzed.

Additional considerations in the design of a pulsed amplifier are 1) energy storage near the device, and 2) minimizing the inductance in the emitter leads. These two items affect the rise time of the pulse and prevent droop resulting from voltage decay during the duration of the pulse. Some tradeoff in these areas will be required because as the emitter inductance to ground is reduced, wideband matching is made more difficult. Also a minimum amount of inductance is required in the collector circuit to achieve adequate decoupling. It is possible, however, to achieve pulses with rise times on the order of tens of nanoseconds with devices that deliver up to several hundred watts of power over bandwidths of at least 20 to 30%.

References

[1] Kraus, Bostian, Raab, *Solid State Radio Engineering,* New York: John Wiley & Sons, Inc., 1980.

[2] William E. Sabin and Edgar O. Schoenike, *Single-Sideband Systems and Circuits,* New York: McGraw-Hill Book Company, Inc., 1987.

[3] William I. Orr, *Radio Handbook,* 19th Edition, Indianapolis: Editors and Engineers (Howard W. Sams & Co.), 1972.

[4] Edward C. Jordan, *Reference Data for Engineers: Radio, Electronics, Computers and Communications,* Seventh Edition, Indianapolis: Howard W. Sams & Co., 1988.

[5] *RF Device Data,* DL110, Rev 4, Volume II, Motorola, Inc. Semiconductor Products Sector, Phoenix, AZ.

[6] H. O. Granberg, "Applying Power MOSFETs in Class C/E RF Power Amplifier Design," *RF Design,* June 1985.

[7] "Understanding the Principles of Amplitude and Frequency Modulation," *MSN,* July, 1989.

[8] Engineering Bulletin EB-38, Motorola Semiconductor Sector, Phoenix, AZ.

[9] Simons K, "Technical Handbook for CATV Systems," Third Edition, Jerrold Electronics Corporation, Hatboro, PA.

[10] "Power Circuits—DC to Microwave," Technical Series SP-51, RCA Solid State, Somerville, NJ, 1969.

[11] Morris Engelson, "Measuring IMD by Properly Using the Spectrum Analyzer," Tektronix, Inc., P.O. Box 500, Beaverton, OR 97077.

[12] Stoner, Goral, *Marine Single-Sideband,* Indianapolis: Editors and Engineers, Ltd., 1972.

[13] Various Applications Notes, *The Acrian Handbook,* Acrian Power Solutions, 490 Race Street, San Jose, CA,1987, pp. 622-674.

[14] Application Note AN-779, Motorola Semiconductor Sector, Phoenix, AZ..

[15] M. J. Koppen, *Electronic Applications Laboratory Report,* ECO7308, Philips Components, Discrete Semiconductor Group, 1974.

[16] H. O. Granberg, "Wideband RF Power Amplifier," *RF Design,* February, 1988.

[17] J. Mulder, *Electronic Applications Laboratory Report,* ECO 7114, Philips Components, Discrete Semiconductor Group, 1971.

[18] Nathan O. Sokal, "Classes of RF Power Amplifiers A Through S, How They Operate, and When to Use Each," *Proceedings of RF Expo East,* 1991.

5

Reliability Considerations

DIE TEMPERATURE AND ITS EFFECT ON RELIABILITY

Solid state RF power designs in the past were done with little concern for thermal properties. This may have been acceptable with relatively low power transistors (~ 50 watts) which were the only types available twenty years ago. With today's RF devices, capable of ten times those power levels, concerns have arisen about die temperature and its effect on reliability. As a result, thermal aspects must be studied in detail. A low thermal resistance ($\theta_{JC} = \Delta$ temperature from junction to case) is essential for a high power transistor in order to keep the junction temperature as low as possible. (See Chapter 1 for additional discussion of "thermal characteristics.")

Remember that both conventional silicon bipolar transistors and silicon FETs normally have "bottom-side" collector or drain connections to the die. This dictates that the transistor die must be isolated electrically and requirements to isolate the collector or drain from the heat sink normally dictates the use of an insulating material with good thermal conductivity as the interface between the transistor die and the mounting flange or stud. This material must have its thermal expansion coefficient close to that of silicon and must withstand temperatures in excess of 1000°C (to withstand brazing temperatures in attaching leads and a heat sink). It must also provide electrical isolation with minimum capacitance which could affect electrical performance. Beryllium oxide (BeO) has been most extensively used for this purpose, since it meets the requirements stated above and is fairly inexpensive compared to, for example, industrial diamond, which has the best overall characteristics exclusive of cost.

From the viewpoint of reliability, junction temperature of a solid state device should not exceed 200°C. Several factors have led to establishing this value as a safe upper limit for die temperature.[7] First, it is well below the temperature at which silicon becomes intrinsic. Second, above 200°C diffusion currents in silicon die begin to increase rapidly. Third, above 200°C metal migration increases rapidly. Fourth, it appears to be a realistic temperature consistent with thermal resistances and power dissipation needed in high power RF devices. However, there is nothing really magic about the value of 200°C. Silicon die will perform above this temperature; but it will perform even better below this temperature. Lower temperatures will only tend to enhance reliability.

The one predictable failure mechanism in RF transistors is metal migration, which is a result of high current densities at high temperatures. This phenomenon was identified in the late 1960s and was reported to the industry in a classic paper written by Dr. Jim Black of Motorola SPS and published in the Journal of the IEEE.[9] Dr. Black's experiments were confined to aluminum metal but subsequent work in the industry established appropriate constants to use with other materials such as gold. Dr. Black developed an equation that relates mean time before failure (MTBF) to temperature, current density and type of material. Figure 5-1 shows a plot of MTBF as a function of temperature for three kinds of top metal used in RF semiconductors. This equation is used today by semiconductor manufacturers to establish a safe maximum current level (for a predetermined value of MTBF—generally seven or more years) for a semiconductor device. It is apparent from Figure 5-1 that metal migration increases rapidly with temperature and is at least two orders of magnitude better (at a given temperature) for Au in comparison with Al.[5]

More will be said about other reliability factors later in this chapter. For now, suffice it to say that because die temperature is a factor in reliability, it is important to select a transistor with an adequate thermal resistance rating. The circuit designer must then provide sufficient heat sinking to the device to allow the case temperature to be maintained below a maximum value such that the die temperature never exceeds 200°C.

A die temperature of 200°C will result in an MTBF of several years for most devices, if it is assumed that the temperature distribution along the die is even, which seldom is the case. With a medium power size die (100 watts), the temperature difference between the coolest areas and the hot spots can be as much as 40–50°C at the 200°C average level. Thus, it is easy to exceed the critical

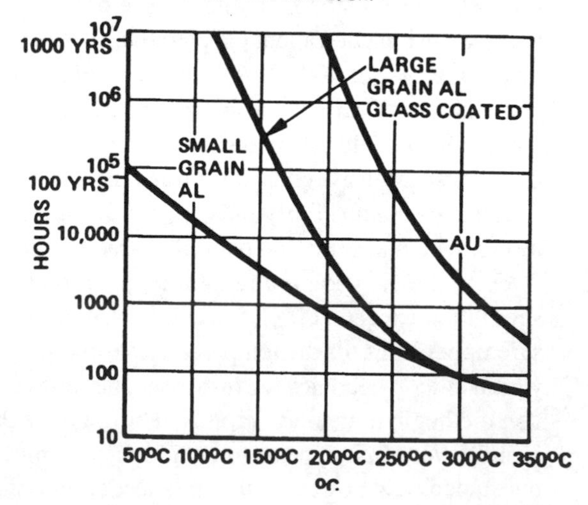

FIGURE 5-1
Failure rate (MTBF) vs. Temperature.

junction temperature in case of slight overdrive or load mismatch conditions, resulting in an unexpected device failure. It is considered good engineering practice to limit the maximum junction temperature to 150°C. Most military designs limit this number to 125°C. Junction temperature can be calculated from the formula:

$$T_J\,(\text{Ave.}) = P_d\,\theta_{JC} + T_C$$

where: T_J = junction temperature, θ_{JC} = thermal resistance (from data sheet), P_d = power dissipation, T_C = device case temperature[5]. Note that power dissipation enters into the equation as discussed in some detail in Chapter 1. This leads to a consideration of device efficiency. The efficiency of an amplifier is largely dependent on the output matching, including quality of the components in the matching network. Also, the saturation voltage and especially the output capacitance have an effect on efficiency. The saturation voltage is more or less determined by the emitter ballast resistor values for bipolar transistors and by the resistivity of the starting material for bipolars and FETs and is probably the most fixed parameter. There are larger variations in the output capacitance depending on the die geometries and types of emitter ballast resistors for the bipolar transistors. An "overkill" in the selection of the electrical size of the device would automatically result in a higher output capacitance, whereas a smaller device with its lower output capacitance would result in a higher efficiency. It is a common belief that higher voltage operation always results in a higher efficiency than a low voltage operation at a given power level. This is not the case, except after a certain point, where the increased current levels begin to dominate in the form of high IR losses, especially in high power 12 volt systems. The efficiency is directly related to the device's output capacitance, and related to the square of the supply voltage.

θ_{JC} can be measured using an infrared microscope (discussed in Chapter 1). A modern version of this measurement method uses the same basic principle, but displays a color picture of the die on a screen, each color shade representing a specific temperature. Measurements of devices intended for use at HF or VHF can be done at D.C., where the device is simply biased up to a known dissipation around its normal operating conditions. At UHF and higher frequencies the feed structure and the metal pattern of the die usually have sufficient inductance to make the power distribution between the cells or emitter/source fingers uneven, which would necessitate the measurements to be made under actual RF operating conditions. The IR measurements are done only when the device is characterized. The thermal resistance, or a relative number that is representative of it, can also be measured with a ΔV_{BE} test. This is done by measuring the forward voltage drop of the base-emitter junction and then forward biasing the junction with a high current pulse of a specified length and current. By noting the V_{BE} prior to applying the power pulse and after, one can calculate ΔV_{BE}, then divide by the temperature coefficient (approximately 2 mV°C) to obtain a corresponding ΔT_J. If one then divides the temperature rise by the power of the applied D.C. pulse, the thermal resistance is obtained.

A similar technique can be used with FETs by measuring the ΔV drop across the intrinsic diode. The junction temperature measured in this manner is more or less an average value and is generally lower than that corresponding to the hottest spot on the die. The failure of the ΔV_{BE} method to measure hotspot temperatures is its main limitation and, thus, is not considered a very accurate method to determine θ_{JC}. It is mostly used in production testing to detect units having bad die bonds.

We must remember that the thermal resistance of all materials increases with increasing temperature. In a transistor, the die itself and the BeO insulator are the weakest links. The θ_{JC} numbers given in data sheets are for 25°C case temperature, but in actual use of a device we are talking about case temperatures around 75°C typically, where the θ_{JC} would be some 25% higher. Figure 5-2 shows θ_{JC} versus temperature of a combined thermal resistance of Si and BeO. Although each side of the BeO disc is at a different temperature potential, and the numbers are dependent on the BeO thickness and die area, the graph provides accurate enough data for reference purposes. For a given package the θ_{JC} is in direct relation to the die area or more exactly the active area. In other words, the larger the die, the lower the θ_{JC}. Transistors designed for higher power levels have larger dice in general than low power transistors, but UHF and microwave devices with their denser geometries have smaller active areas than lower frequency units for a given power level. As an example, an 80 watt VHF transistor has typically 20–22 K $mils^2$ of die area. From Figure 5-3 we can find the die θ_{JC} as 0.3°C/W (plot B). When a 60 mil (1.5 mm) thick BeO insulator is added to the thermal chain (plot A), the total θ_{JC} becomes 0.7–0.8°C/W.

Of equal importance to the transistor junction to case thermal resistance is the thermal resistance between the transistor case and the heatsink. Each of the interfaces and layers of material in the heat flow path must be carefully investigated to ensure a proper thermal design. Figure 5-4 is a thermal flow chart of a transistor–heat sink combination. The thermal resistance numbers for case-to-heat sink (θ_{CS}) vary for different package configurations as shown below:

Large Gemini: 0.07–0.1°C/W.
μ500/μ600: 0.08–0.1°C/W.
0.50″ flange: 0.1°C/W.
Standard push–pull: 0.15°C/W.
0.380″ flange: 0.2°C/W.
Small Gemini: 0.2–0.3°C/W.
0.50″ stud: 0.35°C/W.
0.380″ stud: 0.5°C/W.
0.25″ stud: 0.9°C/W.

The heat sink is responsible for getting rid of the heat to the environment by convection and radiation. Because of all the many heat transfer modes occurring in a finned heat sink, the accurate way to obtain the exact thermal resistance of the heat sink would be to measure it. However, most heat sink manufacturers today provide information about their extrusions concerning the thermal resistance (θ_{SA}) per unit length. This information can be used to

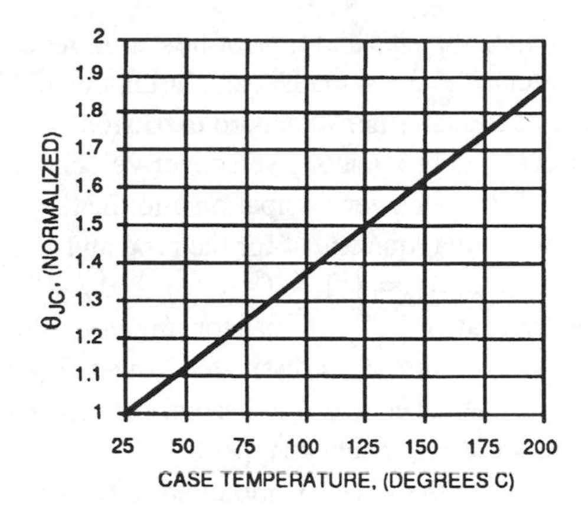

FIGURE 5-2

Approximate θ_{JC} vs. case temperature of the combined thermal resistance of Si and BeO in an RF power transistor.

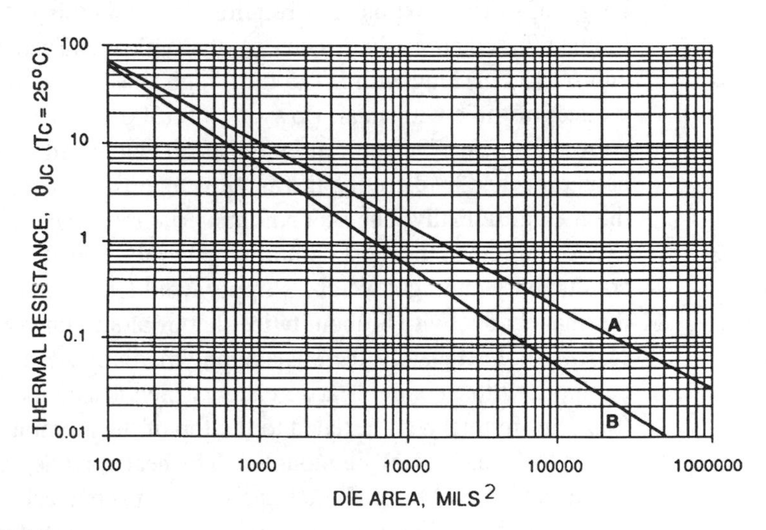

FIGURE 5-3

A chart showing the θ_{JC} for various sizes of dice (plot B) and the combined θ_{JC} for the die and a 60 mil (1.5 mm) thick BeO insulator (plot A).

FIGURE 5-4

The thermal flow path of a transistor mounted to a heat sink. The lighter shading of the thermal resistance symbols represents the diminishing heat radiation farther away from the mounting area.

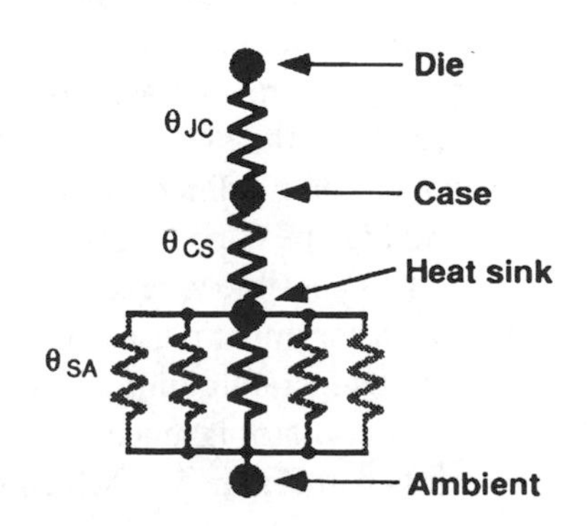

calculate the size and type of heat sink required without having to start "from scratch." "From scratch" calculations would benefit only a designer planning to use an industry non–standard extrusion or a custom cast design. Some manufacturers even provide θ_{SA} versus air velocity charts.

When the power dissipation and the thermal resistances of all interfaces are known, the requirement for the type and size of the heat sink required can be figured as: $\theta_{SA} = \{(T_j - T_A)/P_d\} - (\theta_{JC} + \theta_{CS})$, where θ_{SA} = thermal resistance of the heat sink ,T_J = transistor junction temperature (upper limit), T_A = ambient temperature, P_d = dissipated power, θ_{JC} = transistor thermal resistance (from device data sheet), θ_{CS} = thermal resistance case–to–heat sink (as given above for a variety of package types).

Assume the device is mounted to the heat sink with a thermal compound interface (such as Dow Corning 340 or equivalent, which is essential in mounting of all power semiconductors) and the θ_{CS} is 0.1°C/W and T_J = 150°C, T_A = 50°C, θ_{JC} = 0.6°C/W and the power dissipation (P_d) = 100 W. Then, the θ_{SA} is {(150 – 50)/ 100} – (0.60 + 0.10) = 0.30°C/W. This means that a sufficient length of suitable extrusion is required to obtain this θ_{SA} value, which does not include forced air cooling. The θ_{SA} of a typical extrusion 4.5″(11.5 cm) wide and 6″(15 cm) long with 1″(25 mm) high fins can be lowered by approximately a factor of three with an air flow of 10 ft (300 cm)/second. The forced air cooling is most efficient if turbulence can be created within the fins. This can be approached by directing the air flow against the cooling fins instead of along them longitudinally. The heat sink material must have good thermal conductivity. Aluminum is the most common material for heat sinks because of its good conductivity and light weight. Copper would, of course, be better because its thermal conductivity is about twice that of aluminum, but it is heavier and more expensive. Fortunately, there is a happy medium. An aluminum heat sink can be equipped with a copper heat spreader, which is a copper plate of around 0.25″ (6.3 mm) in thickness fastened to the top of the aluminum heat sink, and against which the transistor(s) are mounted. The heat spreader should extend at least one inch (25 mm) beyond the transistor package in each direction. Although an additional thermal interface is created, the area is relatively large and there will still be a considerable improvement in the total θ_{SA} of the structure.[4]

In pulse operation, (previously discussed in Chapter 4), several relaxations can be applied to an RF power amplifier design. These include reduced cooling requirements, smaller power supplies, reduction in passive component (especially capacitor) sizes and selection of a lower power transistor for a given peak power. There are transistors especially developed for pulse operation, most of which are the bipolar type. Their emitter ballast resistors and epitaxial layers have been modified to make it possible for the device to handle high currents and high peak powers, but they would be very fragile in CW operation.

It is possible to reach peak powers of 3–4 times the CW rating of a standard transistor under pulsed conditions depending on the pulse width and duty cycle. The temperature time constant of an RF power die is about 1 mS, but that of the BeO insulator is much longer. From this we can create a series of curves as shown in Figure 5–5, which shows the θ_{JC} at various pulse widths versus duty

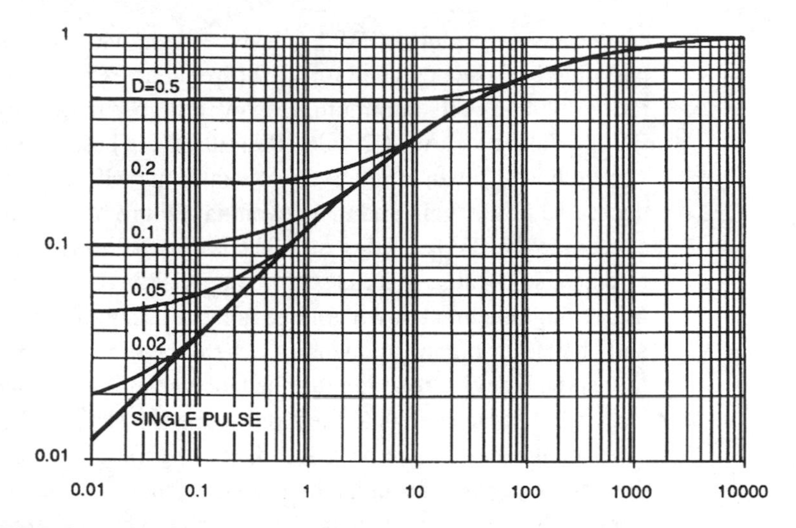

FIGURE 5-5
The θ_{JC} of a typical RF power transistor in pulse operation relative to the pulse width and duty cycle.

cycle of a typical high power transistor. This data is provided for most low frequency MOSFETs intended for switching applications.

In pulse operation of a transistor, a heat sink is not required to dissipate power on a continuous basis. If the pulse width is longer than a few seconds, it may be necessary to consider the thermal time constant of the heat sink as well. The steady state θ_{SA} is reached in 5–30 minutes, again depending on the pulse width and duty cycle. Since the temperature time constant of the heat sink is very slow, the calculations shown earlier for CW are still applicable.

OTHER RELIABILITY CONSIDERATIONS

In addition to excessive die temperature, there are other failure mechanisms in transistors. Some of these failure mechanisms are inherent in the transistor construction and, hopefully, are taken into account by the semiconductor manufacturer. For example, the type of wire used in bonding to the die can result in formation of intermetallic compounds (if the top metal on the die and the wire are dissimilar) or, if the wrong impurity is used in aluminum wire, it can lead to early breakage of the wire at the heel of the bond (metal fatigue resulting from expansion and contraction of the wire with temperature). Other failure mechanisms are associated with the use of the transistor. Some of these are exceeding the reverse base–emitter breakdown (BV_{EBO}), exceeding the collector-emitter breakdown (BV_{CER}/BV_{DSS}) and exceeding the maximum allowable dissipation rating (P_D). We must remember that the rating of the latter parameter is highly temperature dependent and one must include the derating factor when making power dissipation calculations (see Chapter 1).

One of the most common failures in bipolar transistors occurs when the base-emitter reverse voltage (BV_{EBO}) is exceeded.[3] Most silicon bipolar transistors, particularly those designed for use at frequencies below 500 MHz, have BV_{EBO} ratings of 4–6 volts minimum. However, higher frequency transistors may have BV_{EBO} values as low as 2 volts. Permitting the base-emitter total voltage to exceed BV_{EBO} even at relatively low currents will result in degradation of h_{FE}, decreased power gain and in an eventual failure of the unit. The failure mode of exceeding the BV_{EBO} rating of a BJT is one that most circuit designers find hard to diagnose because there is seldom any visible damage to the transistor die. Exceeding this breakdown voltage is most likely to occur in Class C operation and at VHF or higher frequencies, where the transistor power gain is relatively low and the device base input impedance level is moderately high, resulting in large RF voltage swings at the input to the transistor.[8] The possibility for excessive voltage swings can be reduced, but not completely avoided, by returning the base to ground through a low Q inductance. This technique is used in most circuits shown on device data sheets and is to be preferred over the use of only a resistor as is the case in some amplifier designs.

Three different base return configurations are shown in Figure 5-6. Figure 5-6A represents a resistive base-to-ground return only and is considered poor engineering practice. In this circuit, part of the RF drive voltage is rectified by the base-emitter junction and although the resistor is usually of a low value (5–10 Ω), D.C. voltage developed across the resistor adds to the RF ripple and may build up to a level higher than the BV_{EBO} rating. Figure 5-6B shows an improved base return scheme where a resistor is used to "de-Q" an inductor. The low internal resistance of the inductor L suppresses the rectification, although it is still possible for the RF voltage peaks to reach the value of BV_{EBO} in some instances. Figure 5-6C is essentially the same as B, except that the inductor L is connected to a bias source of +0.65–0.7 volts, which usually has a low enough source impedance to clamp the D.C. voltage to the level of the base bias potential.

Another failure mechanism, which is better known and easier to detect, is exceeding the collector-emitter or drain-source breakdown voltage in normal operation (BV_{CER} and BV_{DSS}). Exceeding this rating will also not usually destroy the transistor immediately but results in increased power dissipation and heat generated, which can lead to long term failure. However, it is well known that RF breakdown voltages are somewhat higher than those measured at D.C. or very low frequencies, such as the D.C. breakdown voltages specified on most data sheets. This is also the case with MOSFETs and is caused by the frequency dependency β and g_{FS}. The increase of the RF breakdown voltage with increasing frequency is usually rapid up to around the β cutoff frequency and the corresponding point for a FET, whereafter the change is more gradual. Measurements have verified increases up to 35% depending on the device and its initial value of β or g_{FS}. However, in a BJT only the BV_{CEO} and BV_{CER} are affected; BV_{CES} remains unchanged.

Although similar measurements have not been made regarding other circuit configurations, it is suspected that the same is true for common base, but with a different frequency-voltage profile. The fact that increased die temperature lowers breakdown voltages (second breakdown)[7] tends to balance the increase in

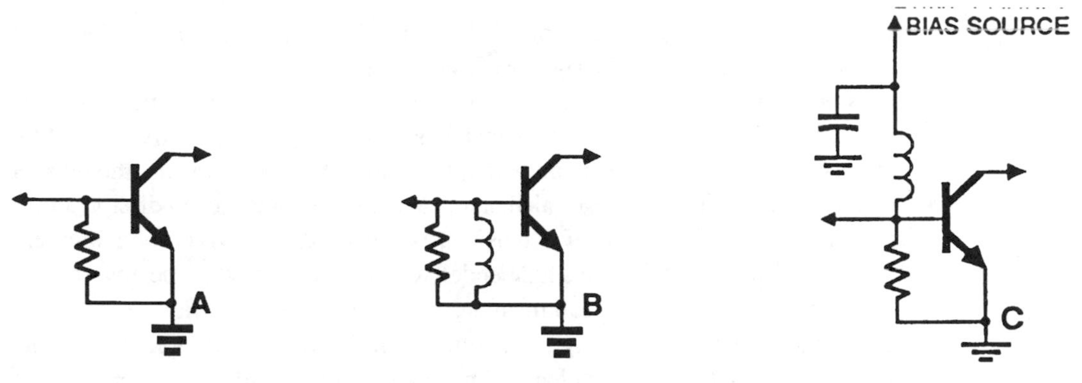

FIGURE 5-6
Base-to-ground return configurations for bipolar transistors. **A** may lead to exceeding the BV_{EBO} rating and is not recommended. **B** is recommended for Class C amplifiers, and **C** for forward biasing in Classes A and AB.

breakdown voltage with frequency and the actual breakdown voltages under operating conditions may remain more or less constant for most applications. Exceeding the BV_{CER} or BV_{DSS} is difficult to prevent in certain applications, such as power oscillators. Any kind of suppressors or clamping devices usually do not work since they would have to handle high RF currents and considerable amounts of power would be dissipated in them. However, the collector-emitter (or drain-source) breakdown problem can be minimized by proper circuit design, particularly output matching. The actual RF voltage swings at the output of a transistor depend on the loadline presented to the transistor. High loadlines lead to high RF voltage swings; low loadlines create low RF voltage swings. However, high loadlines are needed to achieve high efficiencies while low loadlines tend to give poor efficiencies. Thus it is necessary to compromise the circuit design in those instances where breakdown is an issue.

Most transistor failures are usually overdissipation related, which can occur due to overdrive, thermal runaway, self oscillation, or load mismatches. Overdrive is probably the least likely cause of failure, except during a circuit's initial design phase or in those instances where large amounts of drive power are available, such as in the case of add-on booster amplifiers. Overdrive is most likely caused by an operator error, whereas thermal runaway may be the result of an inadequately designed cooling system, improper transistor mounting to a heat sink or selection of a wrong device for a specific application. Unwanted self oscillations can easily destroy a transistor as the result of excessive currents that may flow at each frequency of oscillation. Circuit design and particularly circuit layout are critical to prevent oscillations at frequencies that may be either above or below the frequency of operation.

Transistor failure caused by a mismatched load is without question the most frequent source of device failure in using high power RF transistors.[1,2] Load mismatches will vary with application. Some can be prevented, but others cannot be prevented in normal use. It is extremely important to consider the amount of load mismatch anticipated in selecting a transistor for a specific application. As explained in Chapter 1, transistors are available in the industry for a wide variety of load mismatch conditions. Thus, this failure mechanism can frequently be avoided by choosing the right transistor for the job.

A push-pull amplifier tends to be more vulnerable to transistor failures from overdissipation caused by load mismatches than a single ended one. Even if the transistors are well matched and balanced in all of their parameters, the balance can be disturbed under a load mismatch, causing one transistor to dissipate more power than the other. Thus, if a transistor is specified to survive, for example, with a 10:1 load VSWR in a single ended test circuit, it may not be able to withstand the same amount of load mismatch in a push–pull amplifier. There are techniques to prevent this, e.g., fast acting shut down circuitry or an automatic level control (ALC) loop, which will be examined in detail in Chapter 9.

MOSFETs have another failure mode, which is unique to them and fairly common. It is rupture of the gate oxide by overvoltage, discussed briefly in Chapter 3 in the context of comparing parameters. Only a pulse on the order of a few nanoseconds in width is required to rupture the gate. The greatest danger is when the gate is open and the FET is being handled by mounting it in a circuit, etc.[6] During such operation, electrostatic discharge (ESD) can damage the gate unless proper grounding procedures are exercised. These include antistatic workbenches and floor as well as grounded tools and personnel. There are other situations in addition to ESD, which can subject the gate to a high voltage transient:

1) A transient can be initiated by the signal drive source being switched on and off. It is then amplified by the driver chain and reaches the final gate at a high amplitude.
2) A transient can be generated when the power supply is switched on or off. The transient may be originated by the power supply or it can be generated by charging and discharging capacitances in the circuit itself, including the drivers.
3) A transient can be reflected back to the gate from the output through the feedback capacitance (C_{RSS}). This is a typical mode of device failure in amplifiers driving inductive loads, but is possible in any RF amplifier under certain load mismatch conditions.

How does the device and/or circuit designer prevent transients from damaging the gates of FETs? Some of the low frequency MOSFETs have built-in zener diodes between the gate and the source, but their junction capacitances would be excessive at VHF and higher frequencies. A combination of zener and signal diodes as shown in Figure 5–7 has been used successfully. The zener diodes, which should have their zener voltages 2–3 volts below the FET's maximum V_{GS} rating, allow biasing of the FET to its V_{GS}(th) level and the signal diodes with their low capacitances eliminate the zener capacitance effect. In Figure 5-7, diodes *a* and *b* rectify the gate RF drive voltage and a D.C. potential close to the peak-to-peak value of the RF voltage appears across the junctions of *a* and *c* (positive) and b and d (negative); however, diode string *ac* will not limit the negative half cycle, and *bd* will not limit the positive half cycle of the RF drive. It is important that inductances associated with series

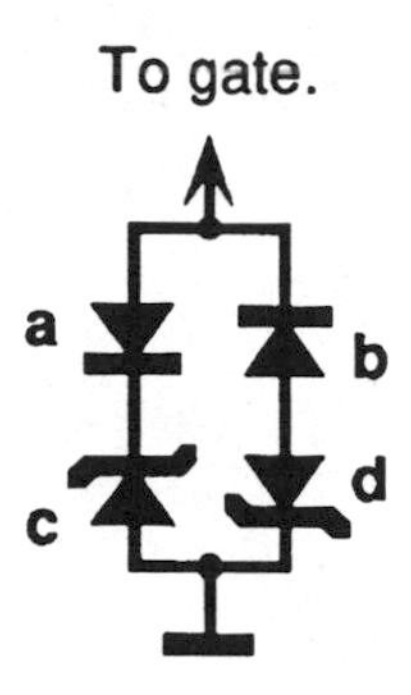

FIGURE 5-7
Low capacitance gate protection circuit for MOSFETs. The upper frequency is limited by the series inductance of the diodes and the type of diodes used for *a* and *b*.

diodes be kept extremely low in order for the circuit to function properly. It may even be advisable to employ leadless pills or other surface mountable packages. Depending on the type of diodes used, charging of the zener capacitance in pulsed operation may affect the pulse shape, but this matter has not been investigated in any detail. Pulse shape would also be a function of the pulse width in question etc.

If gas discharge devices, such as surge arrestors, were available with low enough breakdown voltages, they would make ideal MOSFET gate protectors. They are fast, they can carry high currents after the gas has ionized and their electrode capacitance can be < 0.5 pF. There are gases with ionization potentials of around 20 volts, but there are other factors that also determine the breakdown voltage of a gas discharge device, and the industry probably has not seen sufficient demand for a low voltage unit.

References

[1] Joe Johnson, Editor, *Solid Circuits*, Communications Transistor Company, 301 Industrial Way, San Carlos, CA, 1973.

[2] Various Applications Notes, *The Acrian Handbook*, Acrian Power Solutions, 490 Race Street, San Jose, CA,1987, pp. 622-674.

[3] *Transistor Manual*, Technical Series SC-12, RCA, Electronic Components and Devices, Harrison, NJ, 1966.

[4] Howard Bartlow, "Thermal Resistance Related to Flange Package Mounting," *RF Design*, April, 1980.

[5] Robert J. Johnsen, "Thermal Rating of RF Power Transistors," Application Note AN-790, Motorola Semiconductor Sector, Phoenix, AZ.

[6] All RF MOSFET Data Sheets, Motorola Semiconductor Sector, Phoenix, AZ.

[7] *High Reliability Devices*, Solid State Databook Series, SSD-207B, RCA Solid State, Somerville, NJ, 1974.

[8] A. B. Phillips, *Transistor Engineering*, New York: McGraw-Hill Book Co., Inc., 1966.

[9] James R. Black, "Electromigration Failure Modes in Aluminum Metallization for Semiconductor Devices," *Proceedings of IEEE*, Volume 57, p. 1587.

6

Construction Techniques

TYPES OF PACKAGES

Parasitic reactances, material losses, and—for higher power devices—thermal limitations combine to make package selection for RF transistors a technically challenging undertaking. It has been said the best RF package is NO package. While this is understandable, it is not practical. Thus the job for the semiconductor manufacturer is to design a package that protects the RF die, heat sinks it, and makes connections to the "outside world" with minimal deleterious effects.

Primary characteristics of packages suitable for use with high power devices at RF frequencies are shown below. Thermally the package must allow the user to maintain die temperature below a prescribed maximum, generally 150°C for plastic packages (low power) and 200°C for metal-ceramic type packages (typically used for high power devices). Low power devices have their die mounted on the collector portion of the package lead frame encapsulated in plastic and for power dissipations less than 250 mW no difficulties are typically encountered. Packages with thicker lead frames (an example is the Motorola Power Macro package) have been created to increase power dissipation limits to over 1 watt. The desirable characteristics of good RF power packages are:

1. Good thermal properties
2. Low interelectrode capacitance
3. Low parasitic inductance
4. High electrical conductivity
5. Reliable
6. Low cost
7. Form factor suitable for customer application

Low power packages come in a variety of choices for the circuit designer. The choices extend from metal can hermetic packages such as the TO-39, to plastic encapsulated packages (such as the TO-92 and the Macro-X), to plastic, surface mount packages of various sizes and power dissipation ratings, and finally to metal-ceramic, hermetic packages for the most severe environmental requirements. A collection of plastic, surface mount and metal-hermetic packages for low power is shown in Figure 6-1. Typically, however, when the power

FIGURE 6-1
Various types of packages for low power transistors.

level exceeds 1 watt, the RF die is mounted in a metal ceramic package which is constructed with a ceramic material that has excellent thermal characteristics such as BeO.

Several electrical requirements are imposed on an RF package in addition to the thermal requirements. Objectively, the semiconductor manufacturer wishes to make contact with the die while keeping parasitic capacitance and inductance along with conductivity losses at a minimum. The package design —location of external leads, thickness of the ceramic material, choice of plating materials, and the path length from die to external lead connections—allows the device manufacturer to approach these desired results. Another key objective is to keep lead length from die to the external circuit as short as possible. Leads (including current paths on the ceramic) must be sufficiently wide to prevent excess inductance. And, most importantly, the plating on both ceramic and leads must be low loss and sufficiently thick in skin depths to minimize series resistance to RF current flow.

Other factors in addition to thermal and electrical enter into the package design. These involve such matters as reliability, cost and customer convenience. Customer convenience includes items such as type of heat sink (stud or flange) and form factor (surface mount, machine insertable, microstrip compatible, etc.). Reliability covers die attach, wire attach, hermeticity, and lead solderability. Both die and wire attach are dependent on plating. High power die attach must have void free, low thermal resistance bonds (typically silicon-gold eutectic) to prevent thermal hot spots, and these bonds can be achieved most readily with an adequate amount of smooth, pure gold in the die attach area of the package.

Most modern-day transistors are constructed with silicon nitride die passivation. Many also use gold (Au) "top" metal and Au wire which alleviates the need to keep moisture and foreign material from coming in contact with the die and wire bonds. However, if packages are subjected to contaminants (such as those found in vapor phase soldering and subsequent flux removal solutions), gross leak hermeticity is important to prevent particles from getting inside the transistor, where long term chemical action could result in premature device failure. Most RF power packages used today include an epoxy seal between the ceramic cap and the package "header," which includes the ceramic substrate and heat sink.

In the early 1960s, as transistors began to deliver watts of power at frequencies greater than 50 MHz, a new RF power package evolved suitable for microstrip circuit applications (see Figure 6-2). It was called Stripline Opposed Emitter (SOE), after its planar lead construction. The package contains two opposing leads tied to the common element, which is the emitter in transistors intended for use as common emitter amplifiers. A raised bridge over the collector metalization ties the two common element leads together and allows the shortest possible wires from the die to the common element leads. Today, this package, or variations of same, is the basic package used for almost ALL high power RF discrete transistors.

One of the first variations which Motorola pioneered in the early 1970s was to add an additional metalization path on the ceramic substrate between common

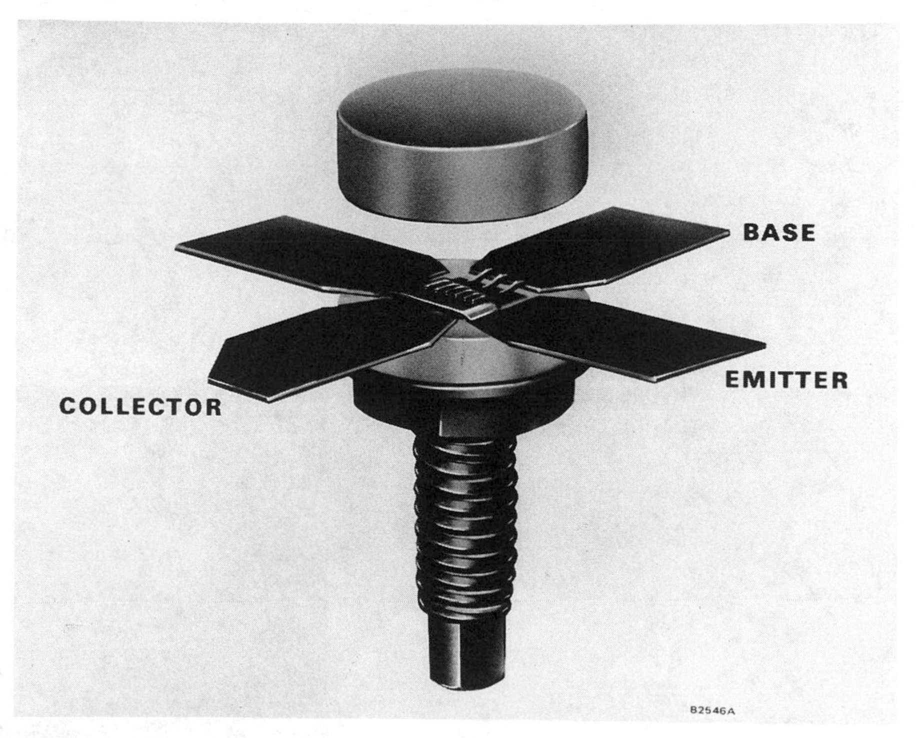

FIGURE 6-2
Stripline Opposed Emitter (SOE) RF power transistor package.

element leads. Together with the raised bridge, this path provided a means of "dual bonding" the common element wires and thereby reducing common element inductance. The package, shown in Figure 6-3, is called a Dual Emitter Bond package (DEB). The ultimate reduction of common element inductance was achieved by the "isolated collector package" originated by Microwave Semiconductor Corporation in the early 1970s. Here, as shown in Figure 6-4, the die (which is located on the collector "island") is surrounded by common element metal. Dual emitter bonds (or base bonds) can be attached to this metalization and the metal can be "wrapped around" or in some other manner attached to the bottom side of the ceramic which is generally at ground potential.

An interesting variation of the "isolated collector package" is one using internal matching for the input and/or output. One of the first internally matched package types (without use of the isolated collector) is shown in Figure 6-5. Internal matching utilizes the inductance of wire bonds along with metal over semiconductor (MOS capacitors to form low-pass "matching" networks that ideally transform input (and/or output) impedances to higher real values and lower imaginary values (which means lower Q's). Internally matched parts are sometimes referred to as "j0" or "CQ" parts, names that reflect the intent of the internal matching process.

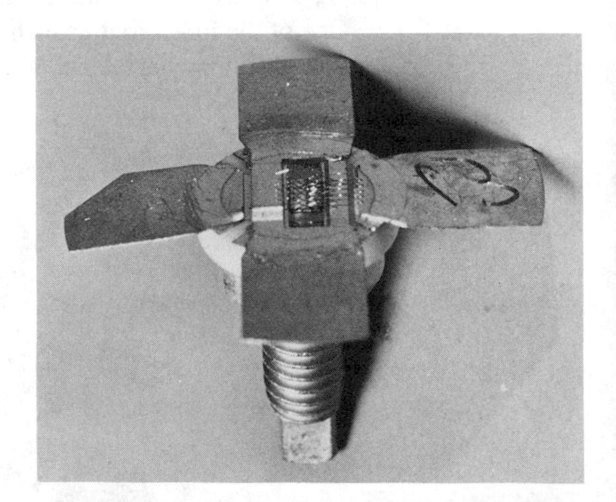

FIGURE 6-3
Dual Emitter Bond (DEB) RF power transistor package.

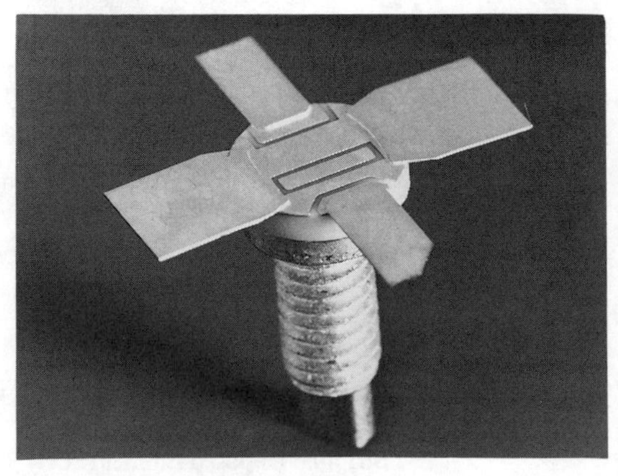

FIGURE 6-4
Isolated collector package.

FIGURE 6-5
A variation of the DEB package using internal matching for the input.

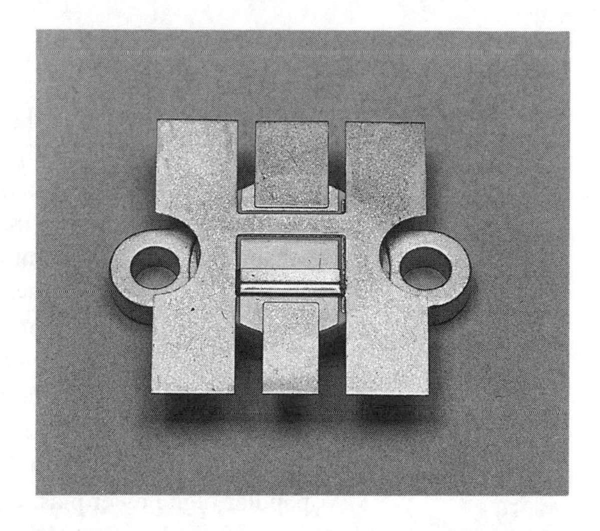

Packages with isolated collectors are ideal for internal matching because the common element metalization allows the placement of MOS capacitors on both the input and output sides of the die. Wires can then be used to attach the input and output leads to the die and at the same time serve as part of the matching networks. Note that packages with six leads (such as that shown in Figure 6-5) are really SOEs with the common element leads on each side of the package split and re-directed toward the input and output. Also note that it is not uncommon to refer to the package as SOE even when it is used for common base BJTs or even FETs.

With the advent of high frequency, high power parts and their use in broad band circuits, a company called Communication Transistor Corporation (CTC) created a unique variation of the SOE package by placing two such packages on a common flange (see Figure 6-6). They called the package "Gemini," which means twins. An evolution of the original package resulted in a single piece of ceramic containing metalization for two separate transistors. Because the transistors are used in "push-pull" circuits, the packages are commonly referred to as "push-pull" packages. The primary advantage of "push-pull" packages is the characteristic of minimum inductance between common elements of the two

FIGURE 6-6
The "Gemini" package developed by Communication Transistor Corporation.

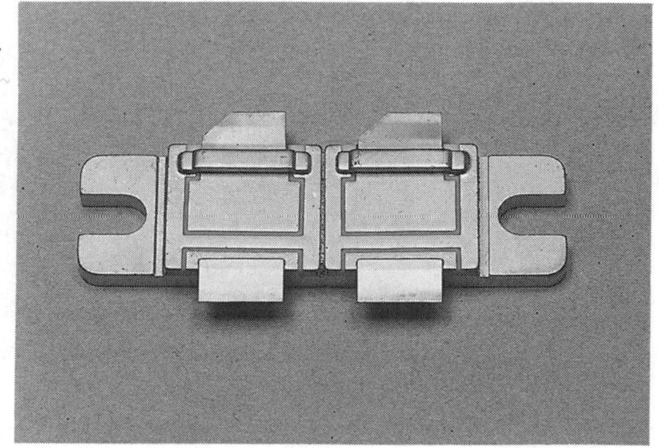

transistors (see the next section for a more complete discussion of common element inductance).

Cost considerations led to the addition of ceramic (BeO) to a conventional TO-39 package as shown in Figure 6-7. This package, called a common-emitter (CE) TO-39, allows BJT die to be isolated from the TO-39 metal and, simultaneously, allows the emitter of the die to be "tied" to the header. A similar use of the package for vertical structure MOSFETs could result in a common-source TO-39. What the CETO-39 package did was to make a significant reduction in package costs for parts in the medium power range of 1 to 4 watts for VHF and 1 to 3 watts for UHF. The power dissipation capabilities of the package in addition to the multiple wire bonds that would be required for larger die restrict its use to the power levels listed above. Package parasitics make this kind of package undesirable for use at frequencies higher than 500 MHz.

A popular high power plastic package called the TO-220 is used at low frequencies and can be adapted for "common emitter" operation by the addition of ceramic in a manner similar to the TO-39. Such a package, called the CETO-220, has never achieved widespread use at RF because of the increased complexity (and higher costs) of multiple wire bonds required for large size RF die. The added costs of assembly offset to a large extent the lower package costs (compared with SOE's). Also the configuration of the leads and internal wire bonding result in excessive package parasitics and high-Q impedances in both the input and output of the transistor.

Hermetic packages for power transistors have requirements contrary to those of good RF packages. The lead length through the hermetic seal is usually

FIGURE 6-7
The common-emitter (CE) TO-39 transistor package.

longer and more lossy than the length required for similar non-hermetic packages. In most other respects, hermetic packages are identical to conventional packages. Today, hermeticity is seldom warranted for commercial applications. Cost of hermetic packages is prohibitive except for those special applications, typically military and space, where severe environments are encountered and ultra-reliability is essential.

THE EMITTER/SOURCE INDUCTANCE

For simplicity we will concentrate on the common emitter/source amplifier configuration. It should be realized that in a common base circuit, the base-to-ground inductance is equally critical but for a different reason (see Chapter 3). To obtain the maximum power gain of a given device, the emitter/source-to-ground inductance must be kept as low as possible. Minimizing inductance is most critical in low impedance devices which means high power transistors operating at low voltages. Inside the package, two factors affect emitter/source inductance. One is the die design, and the other is the package construction.

Small signal transistor die are small (die size is 20 mils, or 0.5 mm square or less) and generally involve a single bond pad for attaching a wire to the emitter/source portion of the transistor. An example is shown in Figure 6-8. A power transistor die is, in reality, created by combining large numbers of "small signal" transistor cells on a single die and interconnecting these cells by means of a metal "feed structure" as shown in Figure 6-9. Note that several bond pads are made available for attaching wires to the emitter/source portions of the transistor die.

Bond pad sizes in RF die are made relatively small to minimize parasitic capacitances, which results in the largest wire sizes used in RF transistors to be

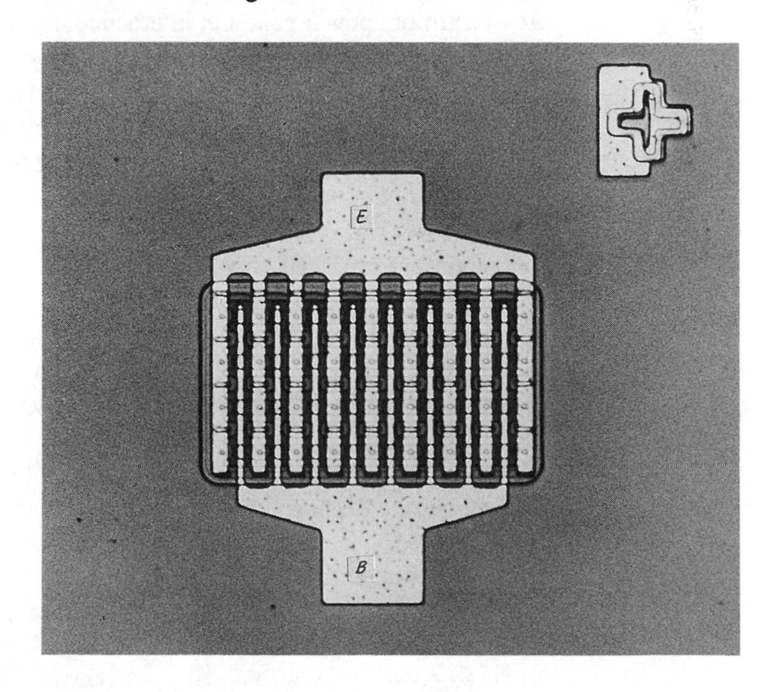

FIGURE 6-8
Low power RF transistor die.

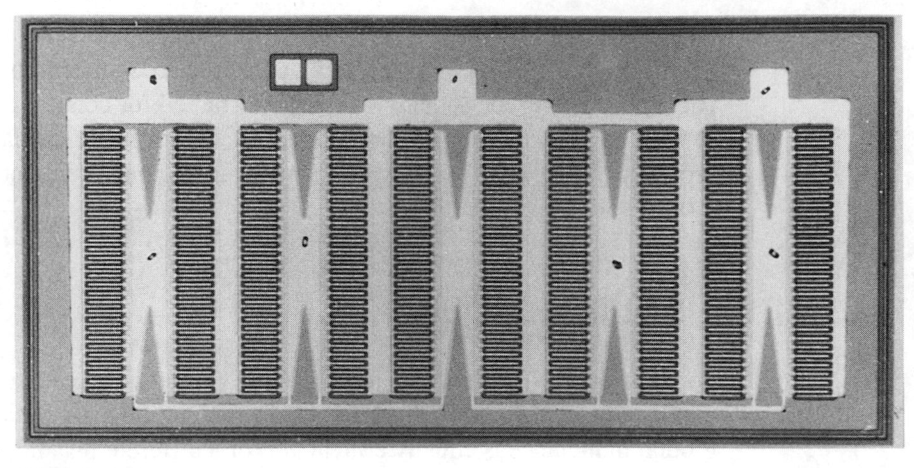

FIGURE 6-9
High power RF transistor die.

approximately 2 mils (50 μm). Many higher frequency transistors designed to use 1½ mil (37 μm) or even 1 mil (25 μm) wire. A 2 mil wire over a ground plane has an inductance associated with it of approximately 20 nanohenries per inch of length. Thus, it is essential that both die and package design be coordinated to result in minimum inductance from bond wires. In die for high power transistors, this is achieved by making the die long and narrow with a large number of bond pads. Types of packages were discussed in the previous section of this chapter.

The emitter/source inductance outside the transistor consists of the transistor lead inductance to ground and the inductance of the circuit board ground plane. Thickness of the circuit board copper clad foil could have an effect in determining maximum power gain, but in practice it has been proven to be negligible, except in special cases such as very high power and low voltage applications. In most professional designs a double-foil-sided circuit board is employed, which provides a continuous ground plane at the bottom side of the board. It is electrically accessible by plated through holes or feed-through eyelets around the transistor mount opening, near the emitter/source area. If a designer lacks the facilities to obtain plated-through holes, the emitter/source areas in the transistor mount opening can be wrapped around with straps of metal foil, connecting these areas on the top of the board to the ground plane (see Figure 6-10). Ground feed-throughs, in addition to being achieved with plated through holes and feed-through eyelets, can also be created by using small lengths of hook-up wire placed through holes drilled in the printed circuit board and then soldered to each side.

In some cases—such as low frequency 3–4 octave bandwith linear designs which include biasing circuitry and feedback networks, etc.—the ground plane must be the top layer of the circuit board, thus eliminating the need for feed-through connections for the emitter/source grounding. When RF transistors are characterized by semiconductor manufacturers, great care is taken to minimize emitter/source inductance because higher inductance, in addition to causing reduced power gain, would also lead to errors in input impedance values.

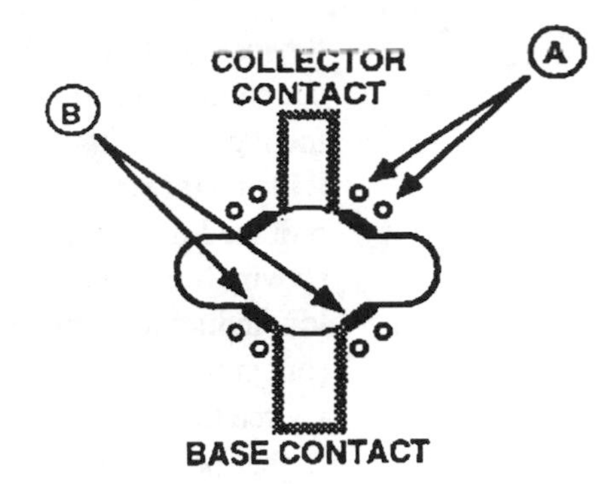

FIGURE 6-10
Emitter grounding method for a transistor in a "CQ" (controlled Q) package. (A) designates plated-through holes or feed-through eyelets to the bottom ground plane of the board. Alternatively, wrap-around metal foil straps (B) can be used.

In order to minimize lead inductance, the transistor mount opening in the circuit board, which is necessary to allow the device to be attached to a heat sink, should not be made larger than necessary for a given package type.[1, 2, 4] The emitter/source inductance affects the device's power gain to the same extent as that created by an unbypassed resistance having a value equal to the inductive reactance in question, with the only differences being that the inductive reactance does not generate a D.C. voltage drop and its effective value is more frequency dependent. If the lead inductance is converted to reactance at the frequency of operation, its effect can be compared to that of an equal value resistance between the emitter/source and ground. For small signal amplifiers the voltage gain is simply defined as $R_L/R_{E(S)}$, where R_L = load impedance and $R_{E(S)}$ = external emitter/source resistance or reactance. However with power devices at high frequencies, the situation becomes more complex because of phase errors, generally greater reactance values and lower device impedances. The effect of emitter/source inductance on the power gain of an RF power transistor can be calculated using S-parameters for example, if available. A detailed model such as that shown in Chapter 7 would be helpful in computer aided designs.

The transistor wire bond (for a specified number of bond pads) and lead frame inductances are usually fixed by package dimensions, and can only be reduced by selecting the physically smallest package in which the die can be mounted. However, if the package is made too small, it is possible to increase thermal resistance.[3] (See Chapter 2). Sometimes the same transistor die is housed in various package styles—for example, the standard 0.380 SOE, 0.500 SOE, or plastic TO-220. Out of these three, it would be possible to achieve the highest power gain with the 0.380 style since the internal package inductance is lower than in the two other case styles. The stud mounted packages, although not as good thermally as the flange types, allow closer access to the ground plane, since no openings for the flange ears are required. However, many devices are not available in these packages, nor could they be used in designs where the rear side of the heat sink is not accessible.

In a push-pull circuit configuration, the emitter/source-to-ground inductance becomes less important and the ground path only provides the D.C. supply to the devices. Analyzing the push-pull operation reveals that the RF current is now flowing from emitter-to-emitter (or source-to-source). For this reason, the devices should be physically mounted as close to each other as possible. If this cannot be done due to existing circuit layout or other reasons, some improvement can be obtained by connecting all the emitters or sources together with a wide metal strip placed over the transistor caps (see Figure 6-11). With flange mounted devices, each emitter can be connected to the flange using solder lugs or wire loops under the mounting screws, enabling the heat sink to provide a low inductance connection between emitter leads. Since the emitter-to-emitter (or source-to-source) inductance can in practice be made lower than emitter/source to ground inductance, it is obvious that a push-pull circuit will exhibit a higher power gain than a single-ended one using the same devices.[2, 4] For push-pull operation at VHF and UHF, special packages have been developed where two transistor die are attached next to each other, thus limiting the emitter-to-emitter inductance to that of the bonding wires. This is probably the only practical approach to using push-pull techniques at UHF frequencies, particularly at higher power levels.

LAYING OUT A CIRCUIT BOARD

Depending on frequency, power level, and voltage of operation, different requirements are dictated for RF circuit layouts. Practically all solid state RF circuits today use some type of laminate of dielectric material and metal foil (usually copper). This circuit board laminate is referred to in the industry as copper clad laminate and is available either with the foil attached to only one side or both sides. The foil thickness is measured in ounces per square foot. Half ounce laminate is the thinnest standard and converts to a foil thickness of 0.0007″ or 0.018 mm. Similarly, one ounce would be 0.0014″ or 0.036 mm, etc. For small signal circuit layouts the half ounce laminate is sufficient, and in some instances can be used at higher power levels at UHF to microwaves, where the skin depth is shallow. As most readers know, A.C. current is concen-

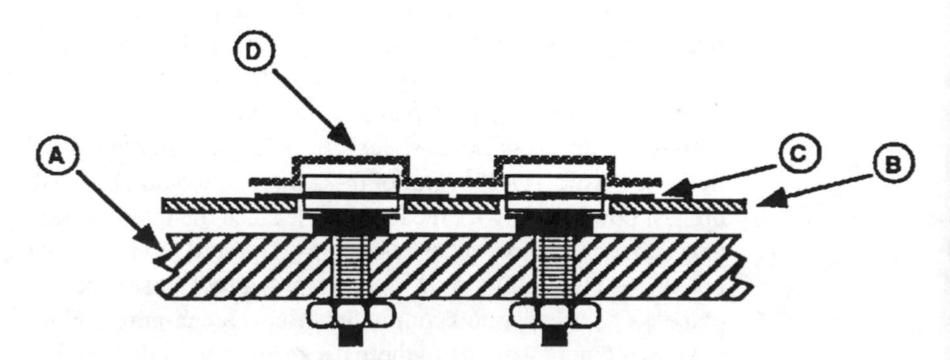

FIGURE 6-11

An example of how emitter-to-emitter inductance can be reduced in a push-pull circuit. (A) represents the heat sink, (B) the circuit board, and (C) points to one of the emitter leads. (D) is a metal strap of 0.25–0.3″ in width (usually copper), formed to allow the ends and the center part to be soldered on top of the emitter or source leads.

trated in the surface layer of a conductor, which is the result of the phenomenon known as "skin effect." The lower the frequency, the deeper this layer extends (or the thicker it is), and vice versa. However part of the current also passes below the top layer and thus one can think of a conductor as having a number of layers, each one of the thickness of one skin depth.

By definition, a skin depth is the distance in inches (or meters) in which the current decreases to a value of 1/e or 36.8% of its initial value.[13] It is analogous to sigma in a Gaussian distribution. That is to say, all current for practical purposes is contained in about five or six layers each having the thickness of a skin depth. The skin depth (δ) of a copper conductor is approximately 0.00035" or 0.009 mm at 100 MHz. Using the skin depth of copper at a frequency of 100 MHz as a reference, the skin depth versus frequency can then, for practical purposes, be figured as:

$$\delta = 0.00035 \times \sqrt{\frac{1}{f/100}}$$

where δ = the skin depth in inches or mm, f = the actual frequency in MHz, and 100 = the reference frequency in MHz. From the formula above we can conclude that low frequencies require heavier foil thicknesses than high frequencies do. However, there are also other factors to be considered in assessing total circuit losses. Some of these are dielectric losses, IR losses in the RF circuitry, and IR losses in the D.C. carrying conductors.

The single sided circuit board is used primarily in circuits designed for very low frequencies where the ground plane inductance is not critical. It can also be used at UHF and microwave circuits where the so called coplanar striplines (waveguides) are used for impedance matching or other functions. Since there is no ground plane, the coplanar waveguide is most practical where relatively high impedances are involved.

A two sided circuit board laminate is the most commonly used material.[5, 6] It is useful in applications requiring a ground plane on the top of the board with other circuitry underneath it (method #1). Openings to the top ground plane can be made in locations requiring clearances for component lead feed-throughs. Probably the most common circuits requiring two sided laminate board layouts are the ones designed for higher frequencies (VHF to microwave), with a continuous ground plane on the bottom side of the circuit board (method #2). This may be required to provide solid grounding points for the components on the top by means of feed-throughs, or to establish a specific impedance for a microstrip. (Microstrip is commonly referred to as stripline, although stripline to be technically correct is the name given to a transmission line consisting of a current-carrying conductor with a ground plane above and below.[14] Stripline has its major applications in constructing filters, hybrid couplers and other passive components.) Sometimes it may be advantageous to combine the two grounding methods described above and have a partial ground plane on each side of the board (see Figure 6-12). In lower frequency applications two sided circuit boards are used to reduce the emitter to ground inductance for increased power gain (see preceding section).

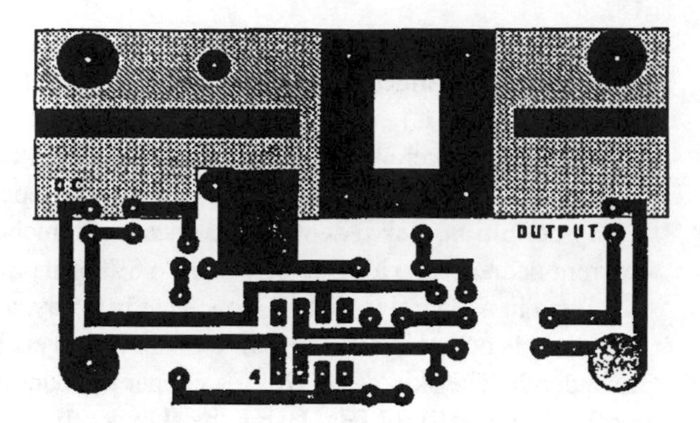

FIGURE 6-12
An example of a partial ground plane on each side of a circuit board. The black rectangle (bottom) is connected to the top ground plane (shaded area) via feed-throughs.

There is a variety of dielectric materials employed in making printed circuit board (PCB). In a single sided laminate the quality of the dielectric and how well the foil adheres to it at elevated temperatures are more important than the material's relative dielectric constant ($\in_r$), except in designs with high impedance lines or relatively high Q resonating elements [relative dielectric constant of a material is defined as the material's actual dielectric constant referred to the dielectric constant of a vacuum, ($\in_0$)] . The dielectric constants most commonly used for insulating mediums in circuit board laminates range from 2 to 7, although aluminum oxide with its $\in_r$ of 9.5 has gained popularity due to advances in ceramic technology, laser machining and metal deposition on such substrates. The higher $\in_r$ of Al_2O_3 results in more compact circuit designs and circuits capable of withstanding high temperatures without changes in performance compared to ones using organic based dielectrics.

The dielectric constant of a material is, in reality, a complex number consisting of a real and imaginary part. The imaginary part, sometimes referred to as $\in''$, when divided by the real part, sometimes referred to as $\in'$, is called the *loss tangent* of the material. The term $\in''$ is referred to as the "loss factor," which is a measure of the "lossiness" of the material.[19] In most cases the real part $\in'$ is normalized to the dielectric constant of a vacuum ($\in_0$) and is referred to as $\in_r$. In all insulating materials, $\in_r$ is temperature and frequency dependent. The changes are largest in high loss materials like phenolic and epoxy fiberglass and lowest in materials such as Teflon™ (TFE) and Al_2O_3. This is true for both the real and imaginary parts of $\in_r$, which results in phenolic materials being usable only at audio or ultrasonic frequencies, or perhaps to frequencies of 10 to 20 MHz, particularly in small signal applications. Epoxy fiberglass is usable to about 200 MHz, but beyond that frequency area, glass TFE, Duroid™, or plain TFE should be used. Figure 6-13[2] shows the relationship of $\in_r$ to the characteristic impedance of microstrip. There are rather complex formulas to calculate microstrip impedance, when $\in$ and H (the height—see Figure 6-14) are known. As a matter of fact, the graph of Figure 6-13 has been plotted from data obtained by such calculations and should be accurate enough for most applications. The term "H" in the expression W/H is confusing to many, but it actually refers to

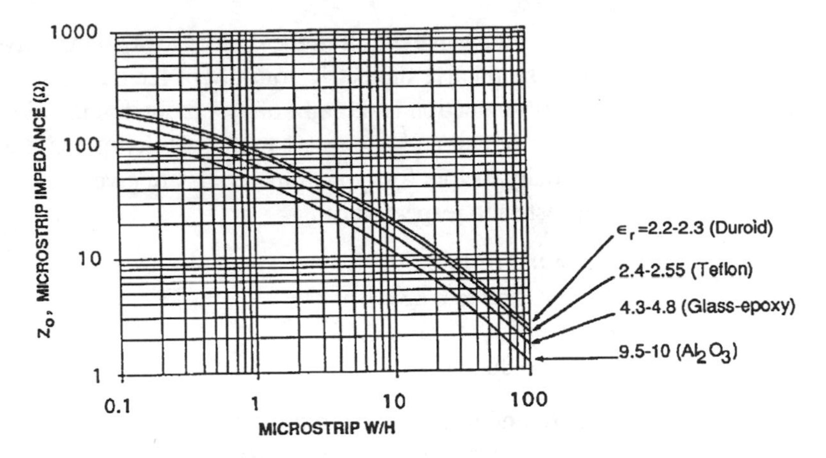

FIGURE 6-13
Microstrip impedance vs. width/height for some of the most popular dielectric materials.

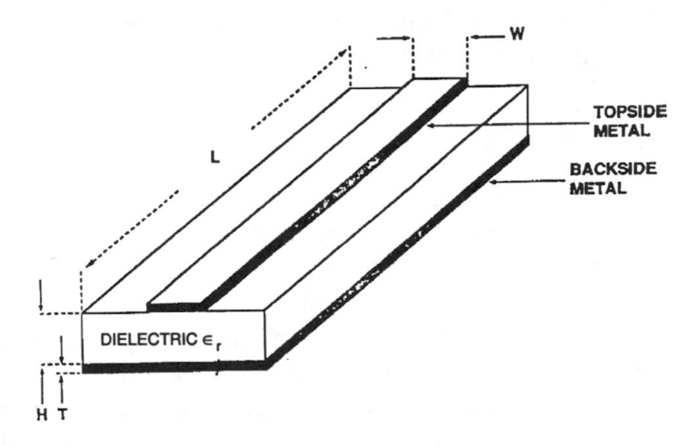

FIGURE 6-14
Cross section of a microstrip. Although not shown here, in practical circuits the backside metal and the dielectric must extend a minimum of two linewidths to each side.

the thickness of the dielectric medium (Figure 6-14). It can be noticed that for a given characteristic impedance the line gets wider with a dielectric medium of low ϵ_r and vice versa. As a point of comparison, air is assumed to have an ϵ_r of 1.0. Materials with their ϵ_r's in the range of 100 and higher have been used for dielectrics in capacitors for many years. However, only recently high dielectric ceramics (most of them barium-totanium oxide based substances) have been developed, which are sufficiently stable with temperature to be used for microstrip substrates. Also it should be noted for high values of ϵ_r, the line width decreases and, thus, more accurate dimensioning is required.[5, 7]

Electrical wavelength in transmission lines that have a relative dielectric constant greater than one is reduced by a factor that is related to the square root of the dielectric constant. That is to say,

$$\lambda_m = \frac{\lambda_0}{\sqrt{\epsilon_r}}$$

where λ_m = the wavelength in the dielectric, λ_o = the wavelength in free space and ϵ_r is the relative dielectric constant of the transmission line. Sometimes this wavelength reduction is thought of as a reduction in velocity of the propagated wave (which it is) and one refers to a "velocity factor" for the dielectric material. Velocity factors for typical materials are given below from the lowest to the highest ratio respectively:

Duroid: 0.68 to 0.66
Teflon: 0.65 to 0.63
Glass-epoxy: 0.52 to 0.49
Al_2O_3: 0.36 to 0.32.

In UHF and microwave circuit layouts using coplanar waveguides, striplines or microstriplines, sharp corners (as shown in Figure 6-15A) in folded lines should be avoided because they create standing waves in these areas and result in the line having an irregular impedance along its length. A common practice to avoid this discontinuity is illustrated in Figure 6-15B. In case of a folded microstrip, a length correction factor must be entered as

$$L - \frac{Wn}{\sqrt{2}}$$

where L = the calculated line length, W = line width, and n = number of folds.

As recently as 1980 circuit board layouts had to be made with black adhesive tape laid down on a transparent or opaque Mylar foil, or for more demanding applications by cutting and peeling of a red "ruby" coating from its Mylar backing. Since the "ruby" peel-off layer is very thin, more precise layouts can be realized with this method than with the adhesive tape. The tape and "ruby" layouts are typically done oversize (×2 to ×8) and reduced to normal size by photolithography. This process yields very accurate dimensioning required for microwave and other microstrip designs. The "ruby" printed circuit (PC) artwork technique is still used for the most demanding circuit designs.

Today most circuit board layouts are made using computers. There is special software available from a variety of sources designed especially for this purpose.[15, 16, 17, 18] Depending on the hardware, accurate dimensioning is possible, especially if the initial artwork is made in an expanded scale. The image

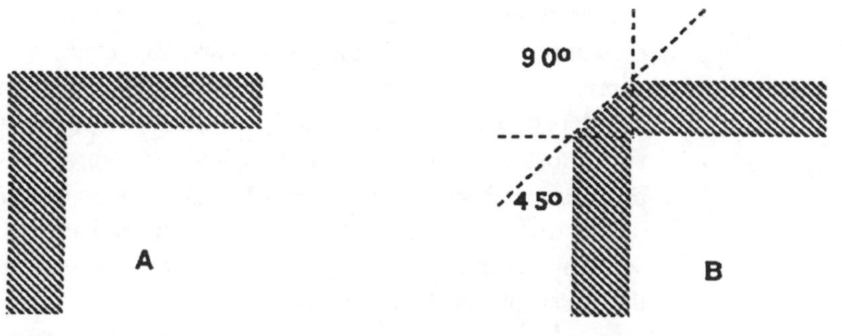

FIGURE 6-15
Making a fold in a microstrip incorrectly is shown in (A), whereas (B) represents a correct procedure that provides a constant characteristic impedance along the line.

can be printed with a suitable printer to produce a black and white artwork, which can photographed and reduced for the final layout film. The computer image can also be directly transferred to a PC board manufacturing facility via a modem, which in some cases speeds up the procedure but does not guarantee accuracy. Note that printers such as a "laserwriter" can change the artwork dimensions typically in a negative direction by 1–2%, thus requiring a correction factor to be used.

Low frequency and high power amplifier circuits are frequently realized in push-pull configurations with wide band transformers as matching elements, even if only single frequency operation is desired. Such designs are much easier to implement and more reliable than either a push-pull or a single-ended one with narrow band LC matching networks.[2, 4, 6] In the latter case, the L's would reach very low values and the C's would be required to carry high RF currents, resulting in excessive heating of the components and possibly even heating the solder to its melting point. Transistor manufacturers frequently show single-ended narrow band circuits as test circuits in the device data sheets, but not many of the circuits (for very high power devices) are suitable for continuous operation, although they may permit achieving the rated peak performance of the device. In the characterization of a 600 watt transistor, where narrow band test circuits were required for three frequencies (3 to 30 MHz), problems similar to those described above were encountered. These circuits barely held intact during a 2–3 minute tune-up operation, even when forced air cooling on the passive components was provided.

It should be remembered that even at low frequencies (such as 2–30 MHz) when one is designing a high power amplifier (such as 300 watts), ground planes and circuit layout are just as important as they are at higher frequencies (such as 500 MHz) for lower power amplifiers (say 50 watts). This is true because impedance levels in these two instances are comparable, both being relatively low. And if wideband transformers are used, attention must be paid to minimize especially the inductance between the device's output terminal and the transformer's connection points. Excessive series inductance here would result in loss of output power along with early saturation and deterioration in wide band performance. The latter would also be affected by excessive series inductance on the input side in the form of possible resonances within the passband. However, it is not necessary to have exact 50 Ω lines at the inputs and outputs of amplifiers below VHF because the line lengths are only a fraction of the wavelength and, therefore, the discontinuities would not noticeably affect circuit performance.

In addition to the RF carrying conductors, attention must be paid to the D.C. circuitry. At a power level of 300 W, for example, and with a 28 volt supply, typical D.C. currents are approximately 16 A, assuming a 50% efficiency. In pulsed operation, over twice the power output of CW can be obtained from the same device under certain conditions, meaning that the peak D.C. current can be as high as 40 A. Especially in pulsed operation, any IR voltage drops would in addition to reducing the peak power output also deform the shape of the pulse. Thus it may be necessary to specify for the RF circuitry a laminate foil thickness heavier than normally required.

TIPS FOR SYSTEMATIC PC LAYOUT DESIGN

Assuming the frequency or frequency range of operation, power output level desired, and supply voltage as well as the circuit details are known:

1) Determine whether a double sided design is necessary and if partial ground plane on each side would benefit the design (for example, by saving board space).
2) Select the laminate with proper dielectric characteristics and foil thickness according to the specifications above.
3) Make a rough sketch to see how much area is required. If there is a size limit, some parts of the layout, such as the widths of the RF and D.C. carrying foil runners, may have to be compromised.
4) Remember, in intermittent operation such as two way communications or in pulse operation, much less stress is placed on the passive components than in applications that tend to have longer "on" durations like TV or FM broadcast. Thus in applications intended for intermittent operation, many components (such as capacitors) can have lower RF current ratings, which allow them to be smaller in size.
5) Provide as many grounding feed-throughs as possible within practical limits. Too many is always better than too few!
6) In a push-pull circuit, locate the transistors as close to each other as physically possible. This will give the most optimum power gain and broad band performance.
7) Most of the laminates with organic dielectric materials have a much higher temperature expansion coefficient than that of the transistor. If the circuit board becomes very large, it is recommended that it be divided into two separate sections, one for the input and one for the output.[12] By splitting the input and output boards, the expansion will minimize the stresses on the transistor leads and their solder joints. This procedure is especially useful in high power push-pull designs.

MOUNTING RF DEVICES

Power Transistors

RF power transistors are reliable devices, capable of operating in excess of 100,000 hours without a failure when proper mounting techniques and electrical specifications are observed. Excessive torque in mounting both the stud mounted and the flange mounted devices has been noted to be a relatively common cause of premature transistor failures. Measurements have shown that overtorque especially with the flange mounted devices makes the thermal contact to the heatsink worse than undertorque. Overtorque prevents the flange from expanding longitudinally with heat, resulting in upwards bowing and separation of its center area from the heatsink. The maximum recommended screw torques are as follows: 6.5 and 11.0 in.-lbs for the 8-32 and 10-32 stud mounted packages respectively and 5.0 in.-lbs for flange mounted packages. In all cases split lock washers and flat washers are recommended, of which the latter should be in immediate contact with the flange's top surface or with the bottom of the heatsink when using stud packages. Some equipment manufacturers use the so-called "Belleville" washers along with flat washers. The "Belleville" washer is one made of thin steel sheet which is "bowed" 10–30 degrees. It is available in a variety of torque ratings and, thus, probably provides the most constant and controlled pressure for the flange mounted devices.

There have also been controversies about the metric system hardware with its finer threads when compared with the British standard hardware (#4 = 3mm, #6 = 4mm). According to experts in thermal studies, a difference could be noticeable only if proper mounting torque and selection of other mounting hardware are not observed.[21]

Certain transistor packages are less critical in their mounting procedures than others. An example is one having the flange made of mechanically hard material such as a copper-tungsten mixture, commonly referred to as Elkonite™.[22] This material, although not as good a heat conductor as copper, is often used because its thermal expansion coefficient is a closer match to the expansion coefficient of ceramics than is that of pure copper.

On the other hand, large push-pull packages (Gemini), which can be up to 1.5″ (40 mm) in length and have flange thicknesses of only 0.06″ (1.5 mm), are more critical in mounting than normal flange packages. With these headers excessive torque of the mounting screws and an excessive amount of thermal compound can make the flange bow upwards in the center, and in addition to resulting in a bad thermal interface, may cause fractures in the BeO insulators and the dice.[10]

Good deburring of the mounting holes, whether a stud or flange mounted device, is of utmost importance. One must insure that the transistor makes a good physical contact to the heat sink instead of "sitting" on burrs that may surround the mounting holes. Also flatness required of the heat sink depends on the transistor package type. Most flange mounted transistors have an unpublished flatness standard of ± 1 mil or ± 0.25 μm. This means that the heat sink surface at the mounting area should have at least the same flatness, which is not difficult to achieve.

In addition to hardware mounted transistor packages, some devices come in "pill" headers, which are conventional headers with standard lead configurations without the mounting stud or the flange. Pills are mounted to the heat sink by soldering, using heat conductive epoxy or pressing the "pill" against the heat sink by some mechanical means with a thermal compound interface. The soldering technique yields the best thermal interface of all mounting methods and is, therefore, common practice in most high reliability equipment. Figure 6-16 shows two common mounting methods of RF power transistors. Both are equally good and used in the industry, except in Figure 6-16B the heat sink surface is thinner by the amount of the depth of the recess. If the ratio of the recess to the total thickness is 1:4 or more, there will not be an appreciable difference in thermal transfer.[8, 9]

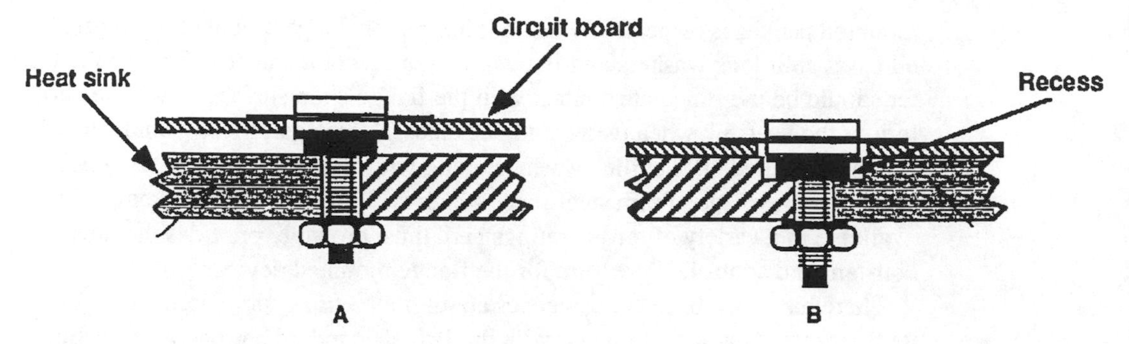

FIGURE 6-16

Two possible mounting methods for RF transistors. Stud mounted devices are shown for simplicity. In (A), the transistor is mounted directly on the top surface of the heat sink, with the circuit board spaced an appropriate distance to be at the same level as the package leads. In (B), the circuit board is flush mounted to the heat sink, into which a recess has been cast or machined. Note that for stud packages the upper portion of the stud is shaped like the letter "D" to provide a self-locating "key."

In order to improve the reliability and to minimize the lead inductance, the seating plane of the transistor lead frame should be close to that of the circuit board. Figure 6-17C shows a correct relationship for the two. In Figure 6-17C the lead ends have been bent upwards to aid the soldering process and to make removal of the device easier in case of a failure. It should be noted that if the circuit board is very large and the soldered area of leads is small, a long term failure mechanism may develop. Depending on the temperature excursions, a difference in thermal expansion coefficients may result in eventual separation of the electrical connection due to the phenomenon of solder fatigue. Gold plating on the transistor leads is approximately 50–100 μ″ or 1.25–2.5 μm in thickness and is essential in manufacturing of RF power transistors since they are exposed to temperatures of 425–450°C during die bonding. This temperature would oxidize or liquify most metals which have good properties of solderability. However, gold forms a very brittle intermetallic compound with tin (present in most solders) thereby making the immediate interface of the leads to the solder a more likely candidate for the solder fatigue problem. Use of a large amount of solder helps to some extent, but in military projects the transistor leads are generally pre-tinned in order to "leach" the gold out of the solder joint before mounting. If formation of intermetallics turns out to be a serious problem, it is recommended that any tin based solders be switched, for example, to ones having an indium base.

Some military specifications call for strain reliefs in the transistor lead frame in the form of small loops formed in the leads, close to the transistor housing; however, these are generally difficult to implement effectively. Some improper transistor mounting procedures are shown in Figure 6-17A and B. Both may lead to long term reliability problems in addition to resulting in excessive lead inductance. The procedure shown in Figure 6-17B should especially be avoided since the joint between ceramic cap and the BeO ceramic disc is composed of a plastic sealing material which loses strength above 175°C. While the strength of

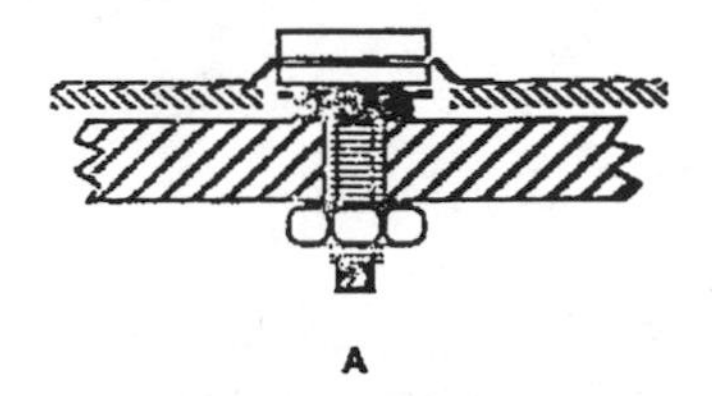

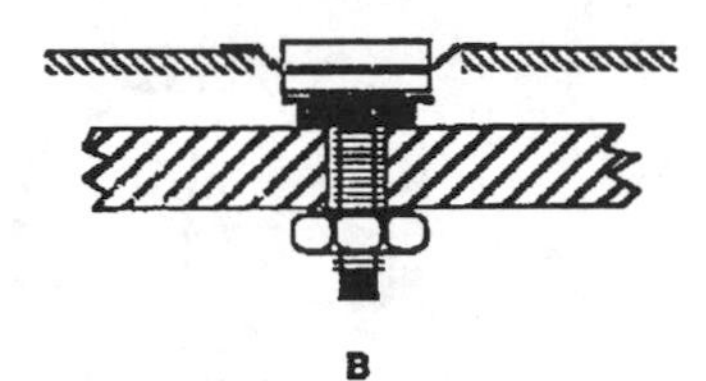

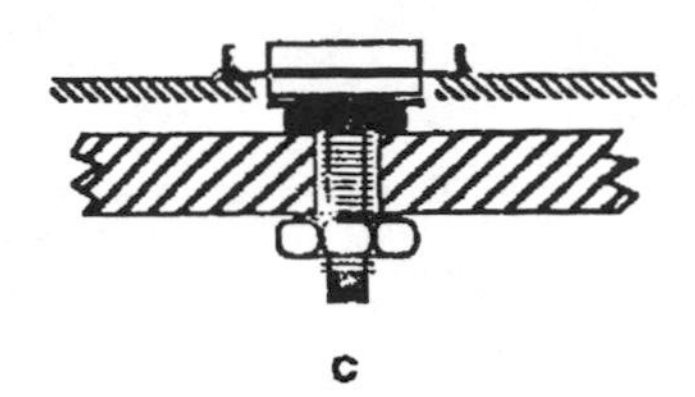

FIGURE 6-17
(A) and (B) show improper mounting of RF power transistors. Both have increased lead inductance, and (B) may have reliability problems due to breaking the epoxy seal of the cap. (C) is a proper mounting method in which board and lead heights are equal. The lead ends are bent upwards to make lead soldering and removal of the device easier.

the material returns upon cooling, any force applied to the cap at high temperature may result in failure of the cap-to-BeO joint.

RF power transistors should always be fastened to the heat sink prior to soldering leads to the circuit board, although in some cases the opposite procedure may make automated assembly easier in production. If the leads are soldered first, it is almost impossible to achieve tolerances tight enough to prevent tensions that push the leads up, down or sideways[8, 9], and in such cases semiconductor manufacturers usually limit their responsibility for failed devices.

The leads of RF semiconductors are made of material such as Kovar or other nickel based alloys, which have their thermal expansion coefficients close to those of ceramics. These alloys are hard, have low ductility, and they harden easily. Thus the leads need only a few sharp bends in a given area to break. Breakage usually occurs in the lead-to-ceramic interface. If repeated bends cannot be avoided, they should be done as far from the transistor housing as possible.

A unique method, which is solderless, can be used for mounting flanged devices. An example of such an arrangement is shown in Figure 6-18 for a 0.5″ SOE package. The flange is first fastened to the heat sink with spacers. The transistor leads are pressed against contacts on the circuit board surface by a Teflon™ ring, followed by silicone rubber and aluminum rings. It is important that the circuit board contact areas are clean. Solder plating is acceptable but gold plating would be best in the areas of contact. The figure itself is self explanatory regarding the mounting process. Depending on the rigidity of the circuit board and the locations of its mounting points to the heat sink, additional support may be required to prevent the board from bending downwards from pressure. An advantage with this type transistor mounting technique is that it allows for various degrees of expansion of the heat sink, circuit board and the transistor with temperature.

As discussed earlier, the more standard method of mounting which involves soldered lead connections does not allow for any movement due to expansions caused by temperature. And this results in mechanical tensions in the transistor leads. Breakage of the leads or separation of the solder joints may occur, which can lead to long term failures. Since there is no guarantee that the transistor leads make a positive contact to the circuit board surface at uniform distances from the transistor housing, the solderless method of transistor mounting just

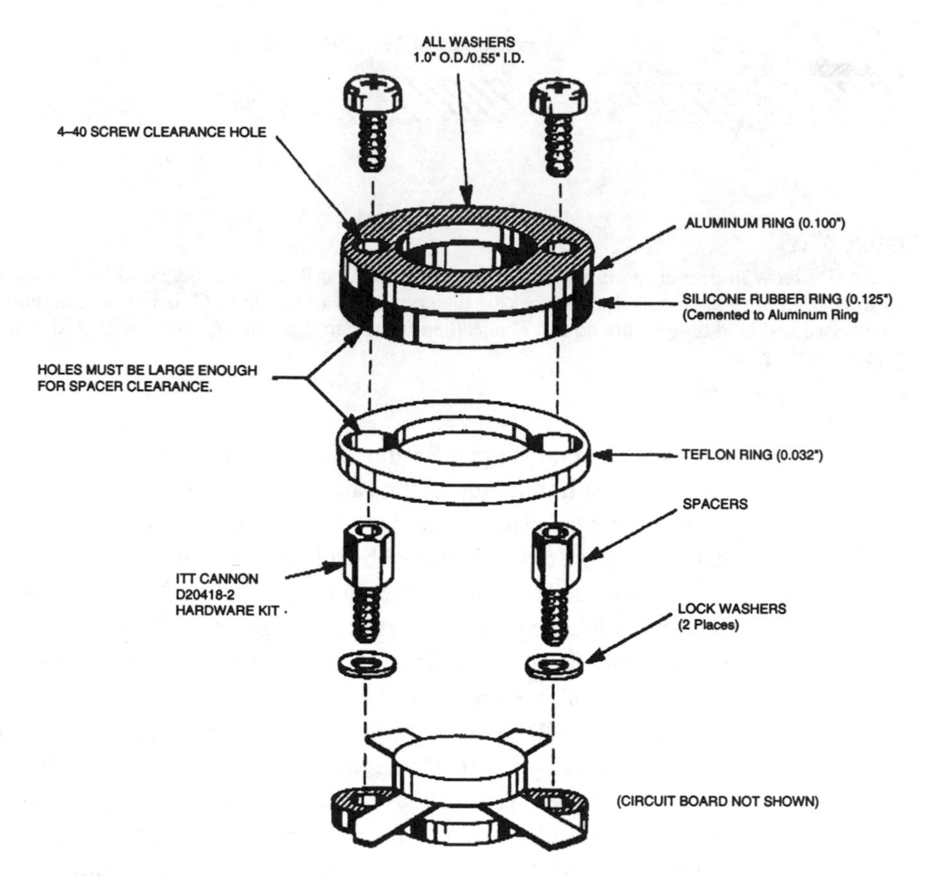

FIGURE 6-18
Solderless method of mounting flanged devices.

described is not recommended for use at frequencies higher than 100–150 MHz because the circuit series inductances of the leads become increasingly critical. Mechanically clamping transistors has been tested in a 2 kW prototype RF amplifier (Figure 6-19) for a period of ten years with over 5000 hours of operation without failures.

It is common practice with stud mounted transistors to tighten the mounting nut while holding the device by the leads to prevent it from turning. This should never be permitted! Stud mounted transistors generally have a "wrench flat" at the end of the stud which is provided to prevent the stud from turning while tightening the mounting nut. An even worse practice is to fasten the transistor to the heat sink initially with a low torque, then solder the leads and finally tighten the mounting nut to its full torque. Such a method of mounting will always leave some twisting tension in the leadframe.

A Brief Summary: Do's and Don'ts of RF Power Transistor Mounting

1) Control the torque of mounting screws and observe the manufacturer's specifications. Avoid overtorque, which usually results in worse thermal interface than undertorque. Use proper mounting hardware.

FIGURE 6-19
Mechanically clamping transistors in an RF power amplifier

2) Avoid applying an excessive amount of thermal compound. This results in a poor thermal interface and may deform the flange in some device types.
3) When soldering down the leads, apply solder abundantly. This will leach most of the gold from the leads and prevent formation of a brittle joint.
4) The device should never be mounted in such a manner as to apply force on the leads in a vertical direction towards the cap.
5) The device should never be mounted in such a manner as to place ceramic-to-metal joints in tension.
6) Make sure that device mounting holes are deburred.
7) Always fasten the device to the heat sink prior to soldering the leads.
8) Avoid bending the leads repeatedly to prevent their breakage.
9) When fastening stud mounted devices, do not hold the leads to prevent device rotation while tightening the nut. A wrench flat is provided for this purpose.
10) Leave sufficient clearance between the circuit board opening and the device body. 0.05" or 1.25 mm is considered adequate.

Low Power Transistors

Low power transistors are much less critical in mounting because of their low power dissipation. In many cases, the transistor is suspended in air with its leads attached to the RF circuitry. While in most cases thermal considerations are unimportant, it is essential that lead length be kept to a minimum to prevent excessive parasitic inductance.

Some moderate power packages such as common-emitter TO-39s should have their case soldered to a heat sink not only to take advantage of the potentially higher thermal rating of these packages but also to achieve the lower common element inductance which leads to higher device gain.

Plastic packages such as Macro-X and surface mount types like SOT-23 and SOT-143 have the die mounted on the lead frame. If power dissipation levels approach the thermal rating of the packages, care must be taken in providing adequate thermal dissipation area at the point in the circuit to which the collector lead is attached.[11]

RF MODULES

Two kinds of modules are common in today's RF world. One is a linear amplifier type used primarily for overcoming line losses in cable television systems. The second type is a power amplifier generally used in the transmitter portion of two-way radios. The former module requires little care in mounting other than to minimize lead inductances provided that the temperature of the heat sink portion of the amplifier (generally a block of aluminum) is maintained at less than the maximum value specified on the data sheet. The amount of power to be dissipated can be predicted by the amount of D.C. power required by the amplifier since the actual RF output power of these amplifiers is generally on the order of a few milliwatts.

Mounting power modules can be similar in complexity to that of flange-type power transistors. Two types of substrates are currently used in the industry. If the substrate is ceramic, then flatness of the heat sink over the mounting surface is more critical than if the substrate is PCB. Ceramic substrate modules usually require heat sink flatnesses on the order of ±0.003 inches while PCB substrate modules are at least two to three times less critical.

When mounting power modules, care must be taken to prevent "bending" the flange by tightening the mounting screw at one end of the module before tightening the screw at the opposite end of the module. This same consideration must also be taken when mounting flange-type power transistors. The proper sequence in applying torque is to alternately "snub down" each end of the module to a "finger-tight" condition and then torque each mounting screw to its proper specification limit.

Power modules generally have their maximum case temperature specified and do not specify thermal resistances from die (inside the module)-to-case as is done for discrete transistors. The manufacturer has taken thermal resistances into his design considerations and has arrived at maximum case temperatures such that device temperatures will always remain at safe levels provided case temperature limits are not exceeded. Here again to determine power dissipation it is necessary to consider ALL input powers—both RF and D.C., and then subtract RF output power.

The RF circuitry in both linear and power modules is made available to the "outside world" by means of round or flat leads (pins). These leads have a typical inductance of up to 20 nanohenries per inch, which explains why keeping the lead length at a minimum is essential to satisfactory RF performance, particularly at higher and higher frequencies.

Soldering the leads of linear or power modules must be done carefully to prevent melting the solder connections inside the module. Earlier comments about soldering that were made when discussing power transistors apply with modules also (comments about "embrittlement"). However, unlike discrete transistors that are constructed at very high temperatures (die bonds over 400°C and package leads attached by brazing at temperatures over 800°C), modules are typically constructed using solders that have a melting point close to 200°C. Therefore, soldering leads must be done quickly to prevent raising the temperature of the substrate to any value over 180°C.

References

[1] John G. Tatum, "VHF/UHF Power Transistor Amplifier Design," Application Note AN-1-1, ITT Semiconductors, 1965.

[2] Joe Johnson, Editor, *Solid Circuits*, Communications Transistor Company, 301 Industrial Way, San Carlos, CA, 1973.

[3] "Power Circuits—DC to Microwave," Technical Series SP-51, RCA Solid State, Somerville, NJ, 1969.

[4] Various Applications Notes, *The Acrian Handbook*, Acrian Power Solutions, 490 Race Street, San Jose, CA,1987, pp 622-674.

[5] *RF/Microwave Devices Databook*, Series SSD-205C, RCA Solid State, Somerville, NJ, 1975.

[6] H. O. Granberg, "Mounting Procedures for Very High Power RF Transistors," Application Note AN-1041, Motorola Semiconductor Sector, Phoenix, AZ.

[7] *Transistor Manual*, Technical Series SC-12, RCA, Electronics Components and Devices, Harrison, NJ, 1966.

[8] Lou Danley, "Mounting Stripline-Opposed-Emitter (SOE) Transistors," Application Note AN-555, Motorola Semiconductor Sector, Phoenix, AZ.

[9] Bill Roehr, "Mounting Considerations for Power Semiconductors," Application Note AN-1040, Motorola Semiconductor Sector, Phoenix, AZ.

[10] H. O. Granberg, "Good RF Construction Practices and Techniques," *RF Design*, September/October, 1980.

[11] Harry Swanson, "Mounting Techniques for Powermacro Transistor," Application Note AN-938, Motorola Semiconductor Sector, Phoenix, AZ.

[12] H. O. Granberg, "Building Push-Pull, Multioctave, VHF Power Amplifiers," *Microwaves & RF*, November, 1987.

[13] Frederick E. Terman, *Electronic and Radio Engineering*, New York: Mc Graw-Hill, Inc., 1955, pp. 21–24.

[14] W. Alan Davis, *Microwave Semiconductor Circuit Design*, New York: Van Nostrand Reinhold Co. Chapter 5, 1984.

[15] Design CAD, American Small Business Computer, 327 S. Mill St., Pryor, OK, 918-825-4844.

[16] Easy CAD2, Evaluation Computing, 437 S. 48th St., Ste. 106, Tempe, AZ 85281, 602-967-8633.

[17] Design Workshop 2000, 4226 St. John's, Suite 400, DDO Quebec, Canada, H9G1X5, 514-696-4753.

[18] MINICAD by Graphsoft, Inc., 8370 Court Ave., Ste. 202, Ellicott City, MD 21043, Phone # 301-461-9488.

[19] Arthur R. Von Hippel, *Dielectrics and Waves*, New York: John Wiley and Sons, Inc., 1954.

[20] Norman E. Dye, "Packaging Considerations for RF Transistors," *Proceedings of RF Technology Expo 86*, 1986, p. 261.

[21] J.G.A. Scholten, "Modeling RF Transistors When the Heat's On," *Microwaves & RF,* February, 1984.

[22] Data sheet for 10W3 Bases, Contacts Metal Welding, Inc., 70 S. Gray St., Indianapolis, IN 46206.

7

Power Amplifier Design

SINGLE-ENDED, PARALLEL, OR PUSH-PULL

Each of the major power amplifier configurations—single-ended, parallel, or push-pull—has its own application with regards to frequency spectrum, bandwidth and power level. A single-ended narrow band amplifier usually produces optimum performance of a device. These circuits are employed when power gain or other information must be compiled for a specific application, or if an amplifier for single frequency operation is required. Lumped constant matching networks can be used up to about 500 MHz and stripline designs are common at 100 to 200 MHz and higher, and in fact are the most practical design concepts at UHF and microwave. At VHF and low UHF, etched air line inductors—which resemble lumped constant elements—may be the best solution for inductance, particularly with respect to production repeatability. With proper techniques, it is possible to achieve bandwidths of an octave or more.[1,2]

Paralleling transistors is usually employed when higher power levels than are attainable from single transistors are desired. The paralleling technique is in fact widely used at microwaves, where push-pull designs for reaching higher power levels become too critical. Many problems can be encountered in paralleling transistors (such as extremely low impedance levels and uneven power sharing if the devices are not closely matched) and it is because of these problems that paralleling is not usually recommended.[2,3] However, it can be done successfully by following the special guidelines to be given.

At low band and up to UHF, push-pull circuits are common because they offer certain advantages over the other two circuit configurations. The most important of these is probably the feature of suppressing the even order harmonics, but its effectiveness depends on the matching of the two devices. Other advantages are wider bandwidths, higher input/output impedances, and less critical bypassing, especially in the output circuitry.[1]

SINGLE-ENDED RF AMPLIFIER DESIGNS: LUMPED CIRCUIT REALIZATION

If a single frequency or relatively narrow band RF amplifier will fulfill a required application, one designed with lumped constant LC elements is probably the most economical and easiest to design, especially since the capacitances and in some cases the inductances can be made variable. Suitable matching networks for these circuits are discussed later in this chapter. The circuits, using lumped constant elements for impedance matching from the device's input and output to 50 Ω, are widely used for transistor test circuits up to about 300 MHz (or up to 900–1000 MHz for low power designs) since the variable elements allow for adjustments to achieve optimum performance and compensate for transistor parameter tolerances which occur from unit to unit.[1, 2] An example of such a circuit is shown in Figure 7-1A. Although shown as a Class C configuration, it can be biased to Classes A, AB, or B as well with proper biasing arrangements (see Chapter 4).

Good emitter grounding is essential, and it is recommended that a lower ground plane be provided at least in the immediate area where the transistor is mounted. Depending on the exact circuit layout, it may be a good idea to provide a continuous ground plane just to give low inductance grounding points for the C's through feed-throughs to the top of the circuit board. Foil pads are usually provided in appropriate locations for element interconnections to be soldered down. Since all L's and C's are surface mounted, there is a danger that at higher power levels and at continuous operation some circuit elements will heat up to temperatures high enough to melt the solder. This is a disadvantage of the lumped constant approach in an RF power amplifier design, although it may be lessened by providing some air flow to these elements.

The advantages and disadvantages of the lumped element approach to single-ended amplifier designs can be summarized as follows:

Advantages: Adjustable components permit achieving a transistor's best performance at a specific frequency; no special components required; relatively inexpensive.

Disadvantages: Not suitable for continuous operation at high power levels; limited power output capability; limited frequency range; poor repeatability for mass production.

DISTRIBUTED CIRCUIT REALIZATION

Figure 7-1B represents a typical UHF or microwave common base amplifier circuit. The impedance matching is done completely with microstrip transmission lines. In designing these circuits (and common emitter circuits also) the exact dielectric constant (ϵ_r) of the substrate material must be known. The maximum ripple tolerable within a specified bandwidth determines the number of reactive elements (n) required in matching the input and output to 50 Ω. Inductors are formed by lengths of stripline of specified widths and lengths and capacitors by open stubs at specific points on the lines.[3, 4] There is a 3–4 dB difference in the ripple between $n = 1$ and $n = 4$, but after that only ½ dB to $n = \infty$. Manually the reactive elements can be designed as Chebyscheff lumped constant matching networks, which are then converted to microstrip format.

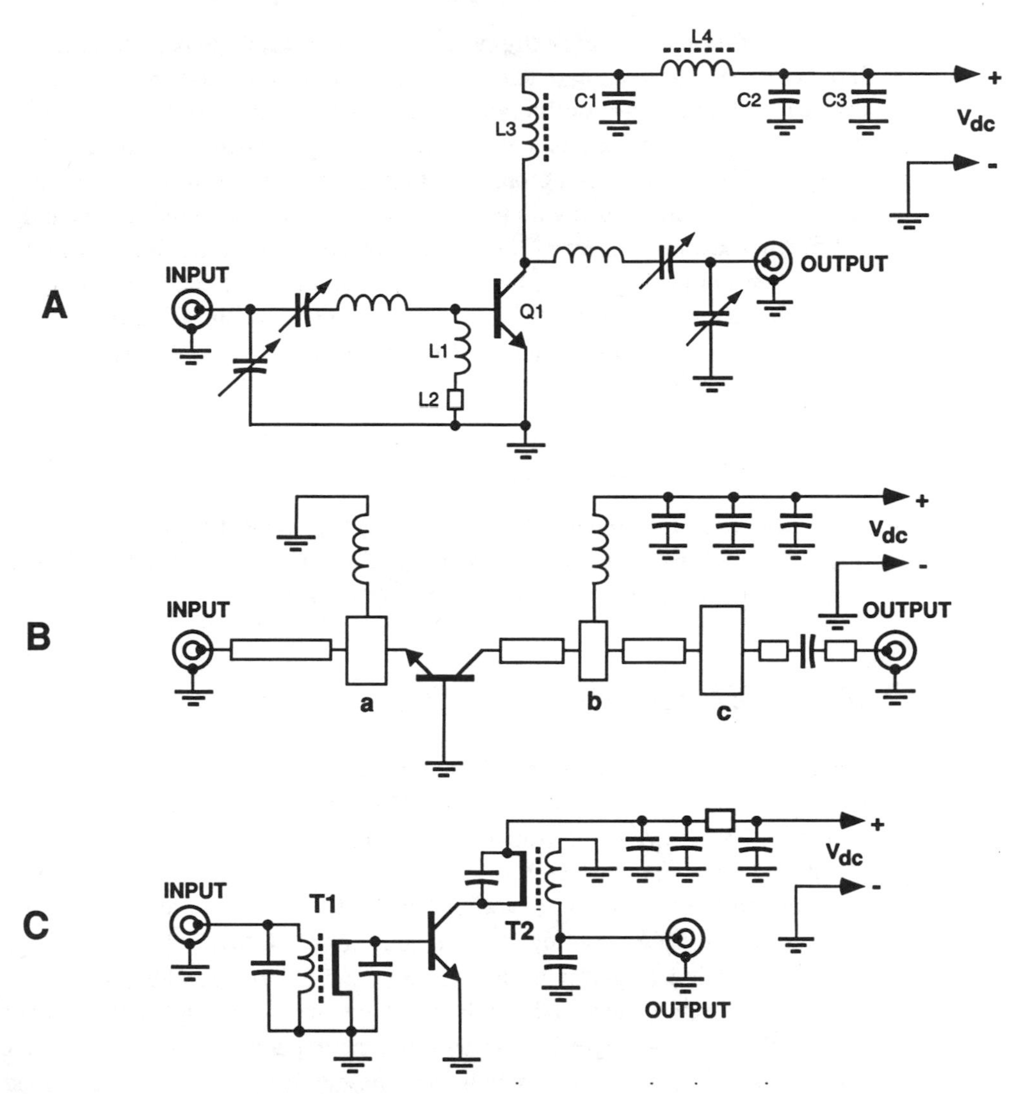

FIGURE 7-1

Examples of single-ended RF amplifier circuit configurations. In (A), the lumped constant matching technique limits its use to relatively narrow band applications and frequencies up to VHF or low UHF. (B) is a typical UHF or microwave common base circuit where the impedance matching is done completely with microstrips. (C) represents a wideband amplifier circuit where transformers are used for impedance matching. It is usable up to UHF in small signal designs.

Today there are numerous computer programs available which calculate the line and stub dimensions directly. For prototype development, the line and stub dimensions can be modified by "cutting off" metal or adding copper foil with conductive adhesive backing specially developed for these kinds of applications. There is no need for feed-throughs to the circuit board bottom ground plane as was required in lumped circuit, lower frequency designs since the capacitors are formed with stubs. The stubs are shown in the schematic (Figure 7-1B) as "a," "b," and "c." Exceptions are shorted stubs and ground returns for the input RF choke and collector supply by-pass capacitors.

For stability reasons it is extremely important to keep the base-to-ground inductance at its minimum (Chapter 6). This is relatively easy with modern transistors if a package configuration is chosen with base leads connected directly to the mounting flange which, in turn, is grounded to the heat sink along with the circuit board ground. Common-emitter amplifier configurations also are usually concerned with minimum emitter-to-ground inductance but for a different reason—to prevent loss of gain. Again dual-emitter packages (such as SOE) and/or packages with "wrap around" emitter metalization (no external emitter leads) will minimize the effect of common element inductance.

Here are some advantages and disadvantages of the stripline design approach for single-ended amplifiers:

Advantages: Easy design procedure; low number of components; good repeatability for mass production.

Disadvantages: Limited to a narrow frequency spectrum without redesign; expensive substrate material; power handling capability limited at CW (used mostly for pulsed operation at higher power levels).

QUASI-LUMPED ELEMENT REALIZATION

The circuit in Figure 7-1C is mostly suited for low frequency operation up to power levels of 50 W, or perhaps up to 500 MHz in small signal use (up to 100–200 mW) where the impedance levels are high. This circuit uses conventional, wide band transformers, (T1 and T2) which are limited in bandwidth compared to transmission line types (Chapter 10). However, this is not the main problem. 50 to 75 W amplifiers up to frequencies of 30 MHz have been designed using the circuit configuration shown in Figure 7-1C, but good bypassing of the transformer ground return is difficult to achieve even if multiple capacitors of mixed values are employed in parallel since the impedance levels are extremely low and the RF currents extremely high at these points. If the circuit is biased to any class requiring a positive base voltage, the same problems would exist on the input side as the input transformer ground returns cannot be D.C. coupled and would have to be similarly by-passed to ground. Good quality chip capacitors will improve the situation, but then connections to a solid ground become even more important than in other circuits, such as the circuit shown in Figure 7-1A.

The advantages and disadvantages of the wide-band transformer approach to realizing single-ended amplifier designs are listed below:

Advantages: Broad band performance (3–5 octaves); inexpensive; good repeatability for mass production; can be designed to achieve peak performance from the transistors.

Disadvantages: Limited power output capability; critical circuit layout for optimum performance (large signal); limited frequency range; requirement for special components.

PARALLEL TRANSISTOR AMPLIFIERS: BIPOLAR TRANSISTORS

The purpose of paralleling transistors in RF power amplifiers is to achieve a higher power output than what can be obtained from a single transistor. Paralleling is usually done with the highest power devices available for a given application; otherwise, it would be more economical and simpler merely to select a higher power single device for the job. For this reason, the impedance levels (especially at the input) would become extremely low if the devices were directly paralleled. To avoid creating such low impedances, which would make matching networks lossy and make their design difficult into 50 Ω, it is customary first to perform an impedance transformation for each device to an intermediate level such as 10–25 Ω. These intermediate impedance points are then paralleled and the resultant is transformed to 50 Ω by additional matching networks. (See Figure 7-2A).[1, 2, 3] Paralleling more than two devices is rarely attempted.

The larger the number of transistors paralleled, the more impractical the situation gets. In addition to the intermediate impedance getting lower, all transistors must be closely matched in power gain and output capacitance. In addition for Class A or AB, the V_{BE} (forward) and h_{FE} must be matched unless the devices are individually biased. The intermediate impedances for each device must also be identical, which is difficult to achieve except in microstrip designs.

Paralleling many transistors in low power applications can be feasible, for example, if the desired power output is moderately low (say 2–5 watts) and the designer wishes to use inexpensive 1 watt devices in a TO-39 or similar header. The transistor paralleling technique described can be used from low-band to microwaves and is commonly seen in L-band radar equipment.

The advantages and disadvantages of paralleling bipolar devices are as follows:

Advantages: High power output is possible by using two or more devices; circuit repeatability is good in microstrip designs; does not require 180° phase shift in input and output.

Disadvantages: Requires closely matched devices and tight passive component tolerances; no even order harmonic suppression; at lower frequencies the design is more critical than push-pull.

MOSFETs

Many designers who have tried to parallel MOSFETs have, to their surprise, experienced some unusual and seemingly unexplainable behavior. Devices can "blow" when biased to a low idle current, or if not biased, when RF drive is applied. There is an explanation for this—the parallel configuration forms an oscillator comparable to the emitter coupled multivibrator known from bipolar circuit technology. MOSFETs have a high enough unity gain frequency that the inductance formed by the gate/source bonding wires, the leads, and their exter-

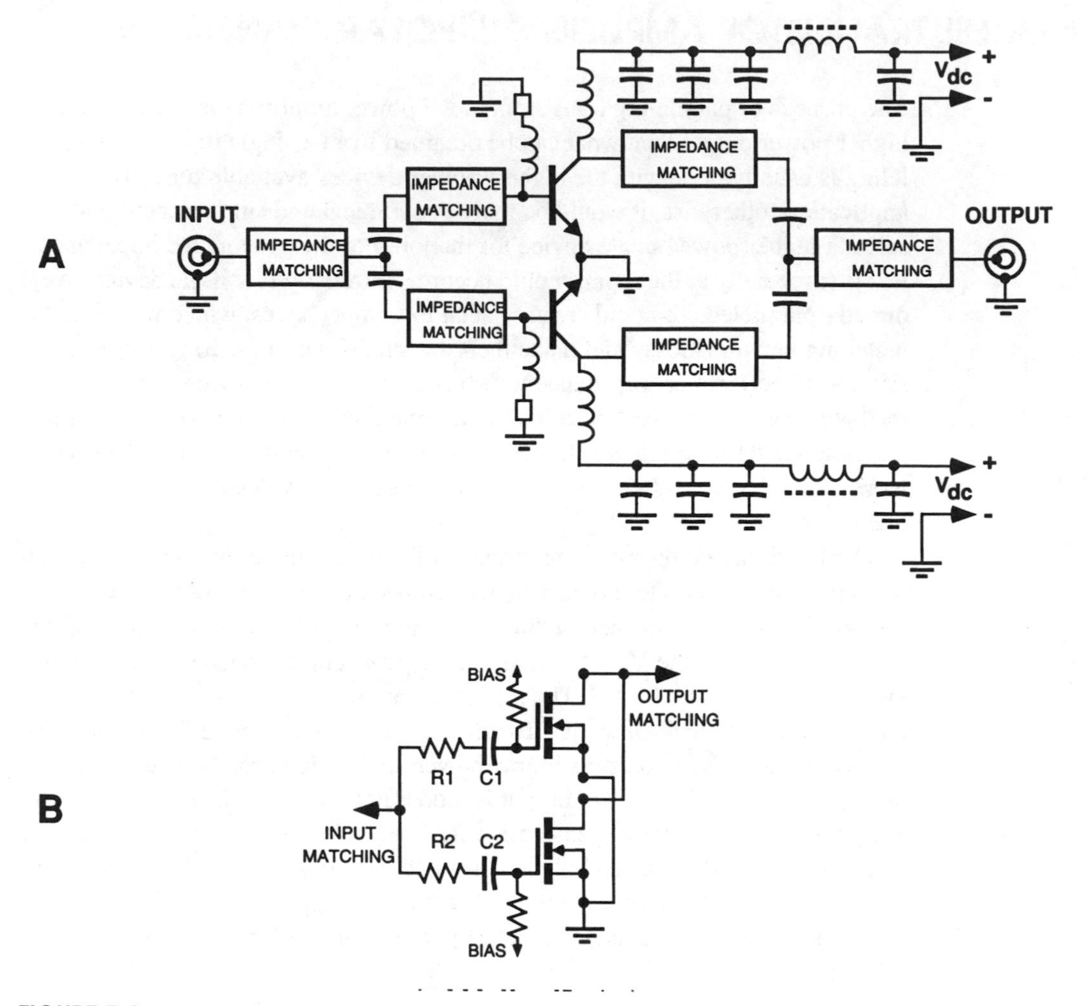

FIGURE 7-2

RF power solid state amplifier circuits using paralleled transistor configurations. In (A), part of the impedance matching is done separately for each branch. The final matching is then done from an intermediate impedance level to 50 Ω. The circuit in (B) demonstrates a paralleling technique for MOSFETs. Note the presence of gate isolation resistors (R1 and R2) to prevent high frequency spurious oscillations.

nal connection together with the device's internal capacitances form a resonant circuit which permits oscillations to occur. These oscillations are usually at a frequency beyond the passband of the intended amplifier. The resonant frequency can be as high as 400–500 MHz for higher power devices and as high as 1000–1500 MHz for lower power ones. High currents can flow at the oscillating frequency resulting in the destruction of the device. Unless the designer accidentally detects the oscillations (usually with a spectrum analyzer) and takes corrective action, many devices may be lost and headaches experienced.

MOSFETs *can be paralleled* but their gates must be isolated and the Q value of the resonant circuit lowered by some means. This can be done with resistors as shown in Figure 7-2B or comparable values of low Q inductive reactances. It is obvious that either method affects the device's high frequency performance since an RC or LC low-pass filter is actually formed between the outside input terminal and the gates because the C is the device's C_{iss}. This limits the frequency of operation of the configuration to VHF at best, where the input impedance levels are still relatively high even with the isolation components added. Because FETs have higher input/output impedance values, it is not necessary to employ the impedance matching procedure described earlier for bipolar transistors (shown in Figure 7-2A). However, it may be possible for the intermediate matching technique to provide the gate isolation necessary, but it has not been pursued by the authors.

As mentioned earlier, the isolation scheme limits the high frequency performance of an amplifier. For example with devices rated for a power output of 150 watts, a resistance or comparable value of low Q inductive reactance of 3–5 Ω would be required at the gate of each transistor and this would limit the maximum frequency of operation to below 100 MHz. With smaller devices (30–40 W), these resistance or reactance values would be on the order of 10–20 Ω. The above discussion indicates that paralleled MOSFETs are suitable only for applications up to low VHF. The gate isolation is also applicable to push-pull circuits, which will be discussed later in this section.

The advantages and disadvantages of paralleled MOSFETs are:

Advantages: High power output is possible by using two or more devices; no intermediate impedance matching required; excellent power sharing even with poorly matched devices.

Disadvantages: Limited high frequency operation due to necessary gate isolation.

PUSH-PULL AMPLIFIERS

A push-pull circuit configuration offers certain advantages over single-ended and parallel transistor designs. A push-pull circuit can be designed as a narrow band system using lumped constant elements or using some microstrip techniques at higher frequencies. These circuits are rather critical, however, requiring extreme symmetry between each side. Wide band designs using transformers for impedance matching are much more tolerable in this respect due to a "floating" center tap that can be provided.

A "floating" center tap, whether in the input or output transformer, means that no physical center tap is provided. A 180° phase shift across the transformer winding exists in either case. In a center tapped design the ground reference is well defined, but any imbalance between the two winding halves is reflected to the transistors. The imbalance would result in an amplitude difference in the drive signals to each side of the balanced circuit or in unequal loads to the transistors in the output.

In the input of a push-pull amplifier, a transformer with a floating (or physically nonexistent) center tap provides a much more balanced drive to the two transistor inputs. The return ground path for the "on" transistor, when using a floating transformer, is created by the input capacitance of the "off" unit. Assuming that the input capacitances of both devices are equal, and since the RF voltage amplitude across the whole winding is twice that from one side to a center tap, amplitudes to both the "on" and "off" transistors are equal in each case. No change in the input return loss should occur either. In the output the same conditions exist, except we do not rely on the output capacitance of the "off" transistor, which is at the power supply voltage potential (D.C.) while the "on" unit is at "ground."

Thus, there is always a voltage close to the D.C. supply superimposed by the RF voltage swing across the output transformer primary. Peak voltages as high as five times the D.C. supply across the transformer winding are common. This voltage (collector-collector or drain-drain) is twice the peak RF voltage that exists from the collector /drain of a single device to ground, which represents a 4:1 difference in impedance. In the output matching of a push-pull circuit, the symmetry is more important than in the input matching. In addition to the well known suppression of the even harmonics, the balance affects the amplifier's stability, efficiency, and susceptibility against mismatched loads. One of the best ways to reach a good balanced condition in wide band transistor output matching is to employ a separate collector/drain stricture as shown in reference 1 at the end of this chapter.

Push-pull circuits with only lumped constant elements are not really considered feasible since creating the exact 180° phase shift becomes too critical and every unit would have to be individually adjusted in production. A hybrid design as shown in Figure 7-3A is a much better choice. The initial matching to an intermediate impedance is done with LC networks (as in Figure 7-2A) while the 180° phase shift is realized with simple and well performing 4:1 and 1:4 transmission line transformers. The intermediate matching networks can also be microstrips, which designs are common in UHF amplifiers. Another possibility is to bring the impedances of each device directly to 50 Ω, in which case only a 1:1 balun would be required to provide the phase shift.

Figure 7-3B shows an amplifier circuit best suited for low frequency applications up to 50–100 MHz. The upper frequency limit is determined by the exact types of transformers used. If they are the so called conventional type, the upper frequency limit is usually 30–50 MHz. There are some conventional type RF transformers that will perform up to 200–300 MHz, but these will be discussed in Chapter 10. Both circuits shown in Figure 7-3A and 3B are for Class C. With proper base forward biasing they can be converted to linear amplifiers (Class A or AB). If transmission line transformers were used in the design of Figure 7-3B to extend the bandwidth, the circuit would become fairly complex since impedance ratios such as 16:1 and 25:1 would be required, especially in high power and low voltage applications.

Figure 7-3C is a typical push-pull amplifier designed with MOSFETs.[1, 4] Since the impedances of MOSFETs are higher than those of BJTs in general (at least up to UHF), their impedance matching is easier. As shown, the circuit configuration of Figure 7-3C is directly adaptable for a 200 watt VHF amplifier design, although the compensation capacitors for the transformer's leakage

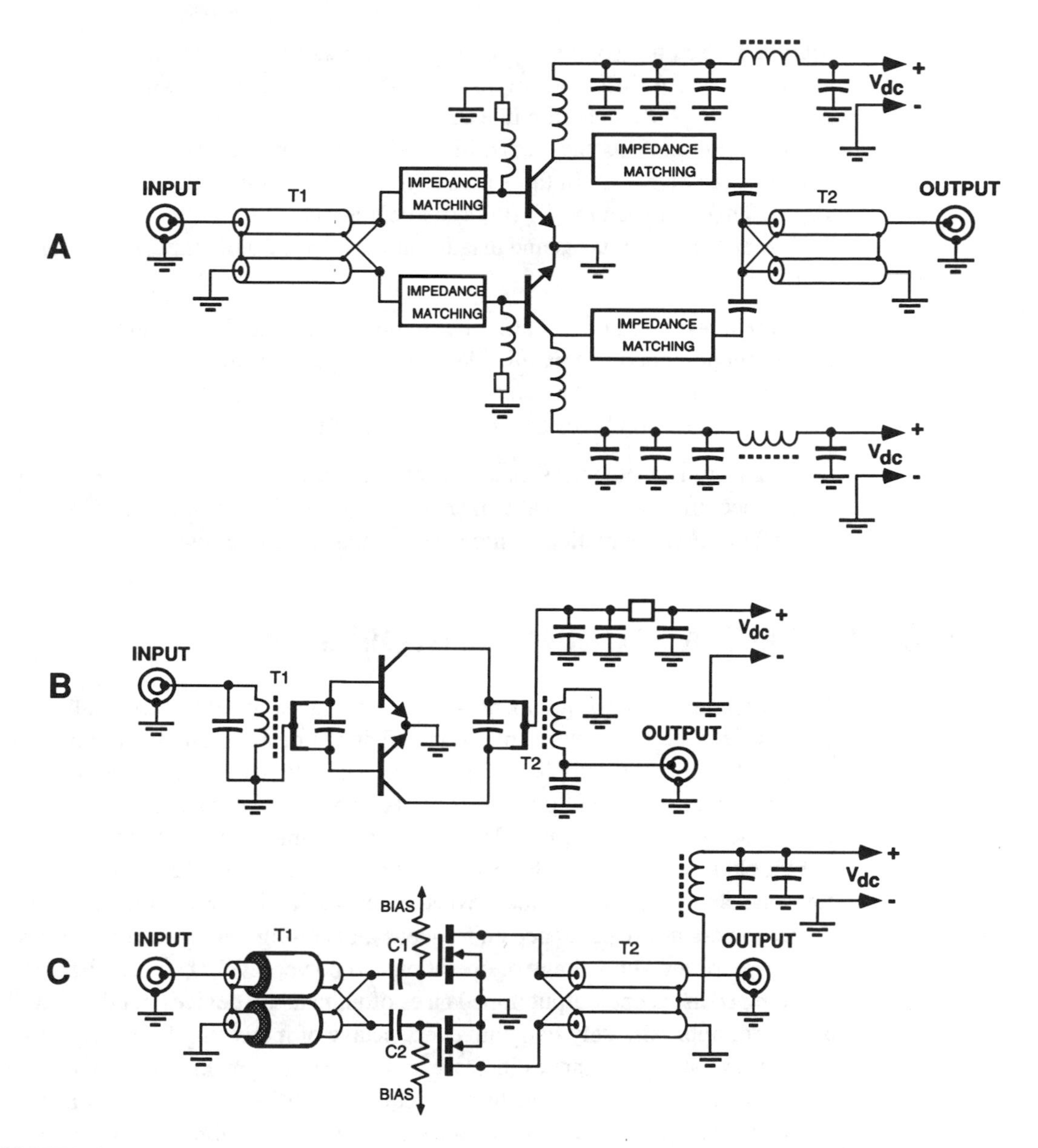

FIGURE 7-3
Push-pull circuit configurations. In (A), part of the impedance matching is done with individual matching networks such as microstrips. Transformers are used for further impedance matching and generation of a 180° phase shift. (B) represents a straightforward push-pull circuit using conventional transformers to provide the phase shift and input/output impedance matching. 4:1 and 1:4 transmission line transformers are used in C to provide the functions above. With the impedance ratios shown, the circuit is useful for operation from high supply voltages where the output impedance is relatively high.

inductance have been omitted. Even if low frequency operation is not required, it is recommended that T1 be loaded with suitable magnetic material to provide the isolation necessary between the FET gates (see parallel operation of MOSFETs above). In all push-pull circuits the h_{FE}/g_{FS} should be matched. In

MOSFET circuits also, the V_{gs}(th)'s must be matched if biased from a single voltage source. A difference of 50 mV is acceptable for devices with their g_{FS}'s of 4–6 mhos. Since the drain idle current is directly related to the g_{FS} versus V_{gs}(th), matching becomes less critical with lower power devices (30–40 W) where values of g_{FS} are in the 1 mho range. In such cases differences of 100–150 mV between the V_{gs}(th)'s can be tolerated.

Here are the advantages and disadvantages of push-pull amplifier configurations:

Advantages: Even order harmonic suppression; easier input/output matching over single-ended and parallel designs due to higher impedance levels; emitter/source grounding and collector/drain by-passing less critical; automatically combines the power outputs of two devices.

Disadvantages: Requires matched devices; creating the required 180° phase shift becomes more critical at increasing frequencies, which makes the configuration impractical at high frequencies, e.g., microwaves.

IMPEDANCES AND MATCHING NETWORKS

Many designers of RF equipment with vacuum tubes or solid state small signal circuits are not familiar with solid state RF power designs, and the importance of many factors in developing suitable hardware. We will concentrate on RF power because the design guidelines are far more critical than those for low power and small signal design. It is true that the same rules apply in each case, but the physical layout of RF power circuits is much more critical due to the low input and output impedance levels involved. The important factors are the device impedance dependence on frequency, operating voltage and power level. As a rule, for a given voltage of operation and power level, the normalized input impedances and output impedances of unmatched devices get divided by a factor of approximately two with every octave of increasing frequency. However, the inductive reactances increase at the same rate, making the impedance matching more complex at the higher frequencies. This applies to both BJTs and FETs, except that the input impedance of the FET is higher by an order of magnitude and at very low frequencies approaches infinity since the gate represents only a pure capacitance.

On the other hand, a BJT has a base-emitter diode junction which must be forward biased to turn the transistor "on." Thus the base input impedance, even at low frequencies, depends on the conduction angle and base forward bias. When RF transistors are characterized by the manufacturer (also described in Chapter 1), e.g., the impedance values are measured along with other parameters such as power gain, linearity and efficiency, the unit is usually inserted into an optimized test circuit. The connections to the transistor are made with as short lead lengths as possible and a clamping structure is used to temporarily mount the device. This allows a number of units to be tested with easy insertion and removal. The test circuit should have the necessary elements for fine adjustment of the input return loss to its maximum and the power output to a specified level while maintaining a desired efficiency.

After the test and adjustments the transistor is removed from the test circuit and resistive terminations are connected to the circuit's input and output. A special probe consisting of a lead frame of the same type the transistor has is clamped in the circuit in place of the transistor. Connections from the probe to a network analyzer are made with short lengths of precision coaxial cable, such as the semirigid type. The numbers obtained from the network analyzer are the conjugates of those given in data sheets (see Chapter 1). A number of devices are usually tested and measured in the manner described to ensure consistency of the parameters. Since the values measured may be extremely low, errors in the form of stray inductances, etc., may limit the accuracy to about ± 20% and in most cases are only guidelines for a designer.

Although there are other methods to determine transistor impedances, the one described here, known as the indirect method, is by far the most common. (A similar description including figures is also given in Chapter 1.) In some instances, especially at UHF and microwaves, so called "load pull" contours are plotted on a Smith chart to indicate the device's behavior at multiple frequency points. All Smith chart data of impedances given in the data sheets is in serial form but in some cases (especially for low frequency designs) it is advantageous to convert it to parallel form to determine the actual resistive and reactive components. This can be done with formulas:

$$R_p = R_s[1+(X_s / R_s)^2] \text{ and } X_p = \frac{R_p}{X_s / R_s}$$

where R is the resistive component and X is the reactive component. In most cases, if the value of X is not very large compared to the value of R, a fairly accurate composite impedance (Z) can be obtained as:

$$\sqrt{R_s^2 + X_s^2} \text{ or } \sqrt{R_p^2 + X_p^2}$$

S- parameters are standard with small signal Class A devices and sometimes with Class A power devices. However, power devices are seldom characterized with S-parameters because most experts question their accuracy and usefulness under large signal conditions except for Class A and stability calculations. SPICE parameter modeling, which is a newer approach to describe RF transistor behavior by a model suitable for use with computer aided design (CAD) programs, is claimed to give more accurate results. Again, this is more likely to be true for linear operation rather than for non-linear operation. The so-called "Gummel-Poon" model of a bipolar transistor (used in Berkeley SPICE) is a linear model and it would not be applicable to large signal, non-linear bipolar transistors. It also does not include package parasitics. A "macro-model" can be created for high power non-linear parts but the problem is determining a model that would apply for more than one set of operating conditions.

Figure 7-4 shows a MOSFET model. The model consists of data involving

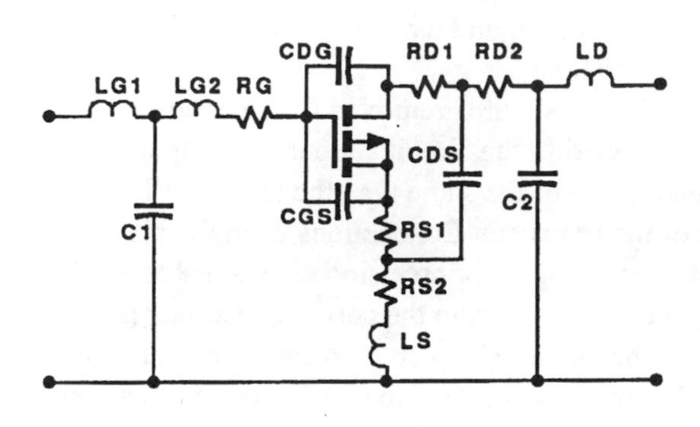

LG1 - Package lead inductance.
LG2 - Wire bond inductance.
C1 - Package lead capacitance.
RG - Poly gate sheet resistance plus gate metal runners to bond pads.
RS1 - Metal finger and material sheet resistance.
RS2 - Wire bond pad and metal runner resistance.
CGS - Source plus channel capacitance under poly gate.
CDG - Gate metal to drain MOS capacitance.
CDS - Drain - source and intrinsic diode capacitance.
LS - Source wire bond and lead inductance.
RD1 - Resistance of epitaxial layer.
RD2 - Resistance of substrate material.
C2 - Package and lead capacitance.
LD - Package lead inductance.

FIGURE 7-4
A typical model of a MOSFET that can be used, for example, for SPICE parameter extraction. The model would be very similar for bipolar transistors. Most of the information above must be obtained from the transistor manufacturer and must be derived according to the geometry and process profile.

die parameters, package stray inductances and capacitances and wire bond inductances. This data is generated by the device die designers in conjunction with applications engineers who characterize the device. Building the model is a rather time consuming operation and its accuracy for multiple applications is questionable. These are reasons why such data is not presently included in most device data sheets.

For the output, the impedance levels are more or less dictated by the supply voltage and the level of power output. The output impedances with each type device are capacitive at lower frequencies, but turn inductive when the wire bond inductances become dominant, which is mainly determined by the device's output capacitance. Output impedance matching into 50 Ω is usually easier than input impedance matching due to its higher level in most cases. Output impedance also remains capacitive up to higher frequencies than input impedance. At low frequencies the output impedance can be determined with a fair accuracy as

$$\frac{(V_{CC} - V_{SAT})^2}{P_{OUT}} \quad \text{or} \quad \frac{[V_{DD} - V_{DS}(ON)]^2}{P_{OUT}}$$

but beyond 100 MHz or so—depending on the device's electrical size—the complex impedance values must be taken into account.[1, 7] The nature of the output impedance and its matching is more critical than the input impedance since it also determines the overall efficiency of operation, whereas input matching only relates to the input return loss.

One of the problems facing a circuit designer is the design of high frequency matching networks. Developing networks that will accomplish the required matching, harmonic suppression, bandwidth, etc., and consist of components having realizable values can result in many hours of design time unless the design engineer has access to a computer aided design capability (see Chapter 8). The design of matching networks involves an infinite number of possibilities, and any kind of tabulation of all possible network solutions would be virtually impossible.[5, 6, 8]

Some commonly used matching networks are shown in Figure 7-5. These

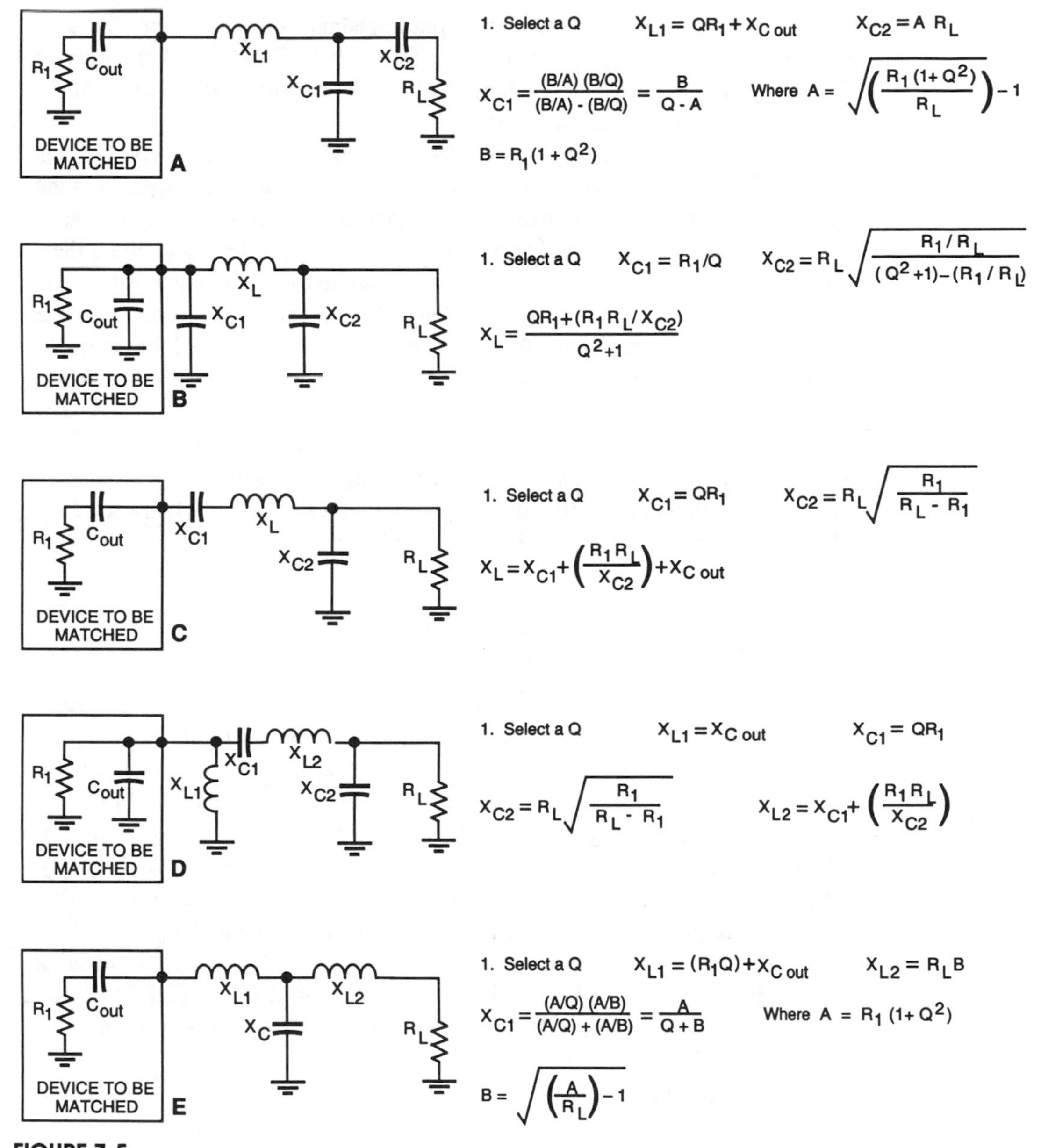

FIGURE 7-5

Narrow band LC matching networks applicable up to UHF. Although shown for output matching only, all configurations can be reversed. They can also be used for interstage matching depending in the impedances involved. In certain cases, the component values may be physically or electrically impractical.

networks can be used for matching in transistor RF power amplifier circuits that have a wide variety of source and load impedances. All networks are reversible, but the R_L side is in general more suitable for matching into a higher impedance. They can be used for interstage matching as well, but with modifications since R_L becomes a complex function.[5] An alternate approach is to utilize intermediate impedance values as discussed earlier in this section. Finally, one must determine whether the component values remain within practical limits. R_1 and

C_{out} represent the complex input or output impedance of a transistor. These complex impedances are shown in series form for A, C and E and in parallel form for B and D. However, each network can be converted to either form if more convenient for a particular application.

In Figure 7-5, network **A** is applicable only when R_1 is less than 50 Ω. When R_1 approaches 50 Ω, the reactance of C_1 approaches infinity. Network **B** is the PI network widely used for matching at higher impedance levels. It may be impractical for use where R_1 is small. For values of R_1 of less than 50 Ω, the inductance of L becomes impractically small while the capacitances of both C_1 and C_2 become very large. Networks **C** and **D** have very similar characteristics. In both, R_1 must be less than 50 Ω. However, it must be stressed that these network configurations often yield the most practical component values where low values of R_1 must be matched. Network **E** is a "T" network. This network is useful in matching impedances of greater than 50 Ω and, therefore, is especially applicable to small signal circuit design. It has also been observed in laboratory tests that this network yields very high efficiencies when used for output matching in transistor RF power amplifiers.

A PRACTICAL DESIGN EXAMPLE

Assume an amplifier of P_{out} = 125 W at 100 MHz is to be designed. The supply voltage is 28 V, and the power gain required 40 dB. Browsing through various device data books, we can notice that the 40 dB gain requirement at 100 MHz can be met with two stages if MOSFETs are used. For example, the MRF174 has an indicated power gain of 14 dB at f = 100 MHz, meaning that 5 watts are required to drive it to a P_{out} of 125 W. As a driver, the MRF134 will do nicely with its 27 dB power gain at 100 MHz. These two stages should thus satisfy the 40 dB gain requirement. Next, we need to establish which matching network configuration to use. From the data sheets we find the MRF134 output impedance as 20.1–j46.7 Ω (P_{out} = 5 W) and input impedance of MRF174 as 1.33–j2.98 Ω at 100 MHz. These converted to parallel form are 130–j55 (Z_p = 140) and 8.0–j3.6 (Z_p = 8.8) Ω respectively.

Since the output impedance of the driver is relatively high, the most suitable networks are **B** and **E**. In this case **B** may provide a more versatile function since C_1 and C_2 can both be made variable elements, whereas low value L's are difficult to make adjustable except with a very limited range. The Q is defined as R_1/X_{C1} and since R_1 = 130 Ω, C_1 would need to become extremely small in value if the Q is to be kept low. On the other hand, if the Q is high instability problems may be encountered and the bandwidth would get narrow. A Q of 5 will give X_{C1} as 26 Ω or C_1 as 60 pF and as established earlier, X_{Cout} = 55 Ω (28 pF), R_L = 8.8 Ω. Then:

$$X_{C2} = 8.8\sqrt{\frac{130/8.8}{(5^2+1)-(130/8.8)}} = 10.1\Omega \text{ or } 158\text{pF}$$

and

$$X_{L1} = \frac{5 \times 130 + (130 \times 8.8 / 10.1)}{26} = 29.4\,\Omega \text{ or } 47\text{nH}.$$

We now have the values for all three elements as:

$C_1(\text{actual}) = C_1 - C_{out}$ or $60 - 28 = 32$ pF.
$C_2 = 158$ pF.
$L_1 = 47$ nH.

COMPONENT CONSIDERATIONS

Each of the matching networks presented here has its own limitations. Although the network configuration is normally up to the discretion of the designer, it is sometimes necessary to use one configuration in preference to another in order to obtain component values that are more realistic from a practical viewpoint. Component selection in the VHF and UHF frequency ranges often becomes a major problem, and the network configurations to obtain realistic component values are of vital importance to a design engineer. Design calculations for matching networks can become completely meaningless unless the network components are measured at the operating frequency.

For example, a 100 pF silver mica capacitor that meets all specifications at 1 MHz can have as much capacitance as 300 pF at 100 MHz due to its series inductance. At some frequency, the capacitor's series inductance will finally tune out the capacitance, thus leaving the capacitor with a net inductive reactance. Values of inductance in the low nanohenry range are also difficult to achieve, since the inductance of a one inch (25 mm) straight piece of AWG #20 solid copper wire is approximately 20 nH. Component tolerances have no meaning at VHF frequencies and above unless they are specified at the operating frequency. It cannot be emphasized enough that components *must* be measured at their operating frequency.

The unencapsulated mica capacitors (known under such names as "Unelco," "Underwood," "Standex," "Elmenco," "Semco," etc.) widely used in RF designs from low band to UHF, are more rugged than ceramic chip capacitors but have higher series inductances. "Unelco" is a common name in the industry for these capacitors, which come in two basic physical sizes shown in Figure 7-6. When used at VHF or UHF, their real values must be adjusted according to the frequency of operation. Parasitic inductances for the "Unelco" and "Miniunelco" are 1.5-2 nH and 1-1.2 nH respectively. The following equation has been proven to have sufficient accuracy for determining the required low frequency value when the effective value and frequency are known:

$$C_{nom} = \frac{C}{1 + ((2\pi)^2 LC)10^{-9}}$$

where

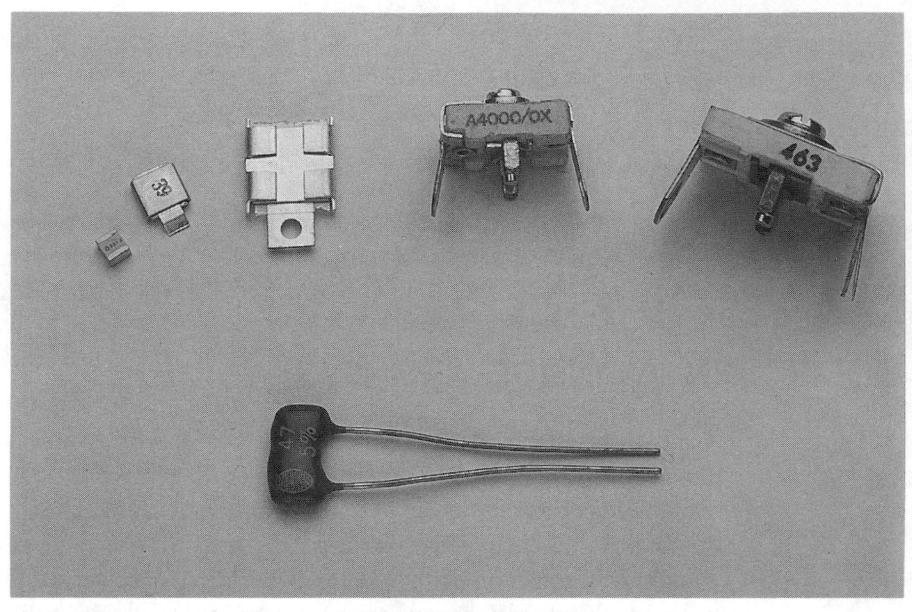

FIGURE 7-6

Various types of capacitors used in RF power circuits. From left to right (upper): a multilayer ceramic chip, "Miniunelco," standard "Unelco," and two types of compression mica variables. Lower center is a so-called dipped mica or silver mica suitable for use up to VHF with very short leads.

C = effective capacitance required in pF,
L = Parasitic series inductance in nH,
f = Frequency in MHz.

Assume a capacitance of 100 pF is required at 400 MHz and we wish to use a "Miniunelco." Substituting the values:

$$C_{actual} = \frac{100}{1+((2512)^2 1.0 * 100 * 10^{-9}} = 61.3 \text{ pF}.$$

From the above calculations, we notice that the actual low frequency value of the capacitor required for a 100 pF effective value is almost 40% lower. The nominal value at 150 MHz would be 91.8 pF (as a comparison), which is well within the standard 10% tolerance limits for these components. The nominal value of any type capacitor can be calculated in a similar manner using the equation above as long as its parasitic inductance is known.

STABILITY CONSIDERATIONS

Instability in solid state amplifiers may well be the most difficult problem a designer must face. We talk about unconditional stability, which means that—no matter what the amplifier load is—it does not exhibit spurious oscillations even with drive levels and supply voltages outside their nominal values. In reality, such conditions rarely exist with RF amplifiers except possibly in low power Class A designs. Instabilities in RF amplifiers can be observed in several ways. Without instruments such as a spectrum analyzer or an oscilloscope, one may notice evidence of erratic tuning (if tuning elements are provided) or current being drawn when the drive power is removed. Some types of instabilities may be too low in amplitude to be detected with anything but a spectrum analyzer or they appear only outside the nominal level of drive power and supply voltage. Thus we have at least three variables that affect an amplifier stability: The load (R and X), drive level, and supply voltage.[9, 10]

In mobile communications the nominal supply voltage is 12.5 V, but can vary to a value as low as 10.5 V or as high as 16 V. With low level amplitude modulation and SSB (see Chapter 4), the effective drive power varies with the modulation. With high level AM (also Chapter 4), the collector/drain voltage varies between zero and the maximum with the modulation. Considering the two latter variables. plus the R and phase angle (X) of the load, we must realize how difficult a task it is to design a stable RF amplifier operating under these conditions. Many designers have spent sleepless nights and hundreds of hours trying to get an amplifier to meet a stability specification even at 3:1 load mismatch.

Forgetting the drive level and supply voltage variables for now, it is relatively easy to reach stability in an amplifier operating into a resistive (usually 50 Ω) load. In real life, however, there is always some type of a load mismatch. In communications, for example, the load is an antenna connected to the amplifier output through a harmonic filter. In industrial and medical applications the load can consist of various types of matching networks presenting at least momentarily an undefinable load to the amplifier. In amplifiers to be designed for frequency modulated communications, there is one less variable because the power input remains constant. This should make the design of stable amplifiers for FM somewhat easier than for AM.

There are several types of instability in RF amplifiers. Some of them are circuit or layout oriented and some are device oriented or may be a combination of both.[2, 9, 10] Many of the modes of oscillation depend strongly on nonlinear effects. That makes them very difficult to analyze compared with small signal feedback type oscillations, which are adaptable to linear circuit analysis. In a power transistor, the ratio of feedback capacitance to the input impedance (feedback capacitance to the input capacitance in a MOSFET) determines much of the stability criteria. The higher the ratio is, the more possibility for stability the device has. Normally the feedback capacitance would introduce negative feedback, reducing the power gain, but at certain frequencies the feedback will turn positive due to phase delays, etc. Thus it is evident that devices with higher ratios of feedback capacitance to input impedance exhibit the most stability.

Transistors processed for low voltage operation in general have lower ratios of these capacitances, making stability in 12.5 volt systems more difficult to achieve than it would be in, for example, a 50 volt design. For bipolar transistors, the feedback capacitance is not given in data sheets because it is not easy to measure. It is a function of many parameters such as the device geometry, the types and values of emitter ballast resistors and the silicon material resistivity. The ratio of input impedance/capacitance to feedback capacitance is somewhat higher with MOSFETs than BJTs, concluding that the MOSFETs are more stable in this respect.[4, 11]

One of the most common instabilities occurs at low frequencies (1–10 MHz), where the device power gain (assuming use of VHF or higher frequency transistors) can be as high as 30–40 dB. This oscillation can be strong, but noticeable only by its mixing products with the fundamental (f_o) as seen in Figure 7-7. In professional circles this display is referred to as a "Christmas tree," the width of its skirts depending on the amplifier bandwidth. The frequency of oscillation (f_1, Figure 7-7) although high in amplitude, may not be detected on a spectrum analyzer due to bandwidth limitations of the circuit. In some instances this low frequency oscillation can be strong enough to cause a transistor to exceed its dissipation limits and destroy itself.[2, 11] This mode of instability is almost completely circuit oriented and is mostly preventable by controlling the low frequency power gain of the amplifier. It is helpful to select a transistor with low h_{FE}, which controls its low frequency gain, but has a minimal effect at high frequencies. Conversely, the emitter/source inductance and resistance have a larger effect in the device's high frequency gain and a lesser effect at low frequencies. Thus, it is advisable to keep these values at their minimum regarding the amplifier's high frequency performance.[9, 10]

Although the available gain of the device itself remains unchanged, the low frequency gain of the amplifier can be lowered by certain simple design practices. In Figure 7-8, L1 and L3 are the most critical DC feed elements. Their values should be selected to block the low frequencies, where the device gain is considerably higher than it is at the frequency to be amplified. Their values should be as low as possible without resulting in any loss in power gain or efficiency. To be on the safe side, their reactances should not exceed 5-10 times the impedances at the base and collector.

The Q values of L1 and L3 must also be controllable. If this cannot be done with lossy ferrite beads (L2 for example), parallel resistances can be used. It is

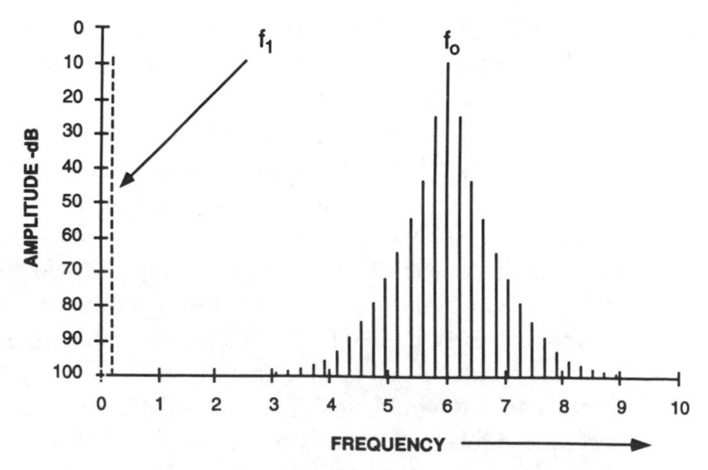

FIGURE 7-7

A display of low frequency instability. Frequency f_1 is mixed with the carrier (f_o) which produces a series of sidebands around f_o.

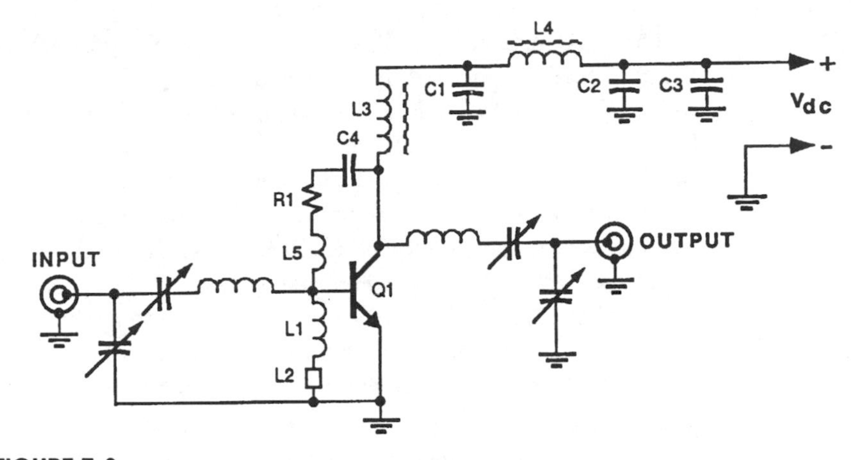

FIGURE 7-8
Schematic of a typical RF amplifier showing the appropriate D.C. feed structures and a negative feedback network.

a common practice to wind these chokes (L1 and L3) over low valued (10–50 Ω) non-inductive resistors. This practice should be applied especially to the collector feed choke (L3), where a lossy ferrite bead may face excessive heating due to the high power level. C1 (Figure 7-8) must have a large enough value to bypass these low frequencies to ground. To avoid possible resonances, multiple capacitors of different values (0.01 and 0.1 μF for example) are sometimes paralleled. L4 (Figure 7-8) has a high enough reactance to pass a minimum of the low frequencies. In fact, it should be as large in value as possible up to the point where IR losses start producing an excessive D.C. voltage drop. C2 and C3 again are two different values and paralleled to avoid resonances, and their values should be large enough to bypass all low frequencies to ground. The purpose of the L4/C2/C3 network is to prevent any RF from feeding back to the DC power source.[2, 4, 9]

Another, and more effective means, of reducing the low frequency gain of an RF amplifier is to introduce negative feedback.[2] In Figure 7-8, this is accomplished by the network C4/R1/L5. C4 is merely a D.C. blocking capacitor and its value is not critical, except the value must be large enough to provide a low reactance at the low frequencies. The feedback slope is controlled by L5. The function it plays is based on its increasing reactance with frequency. A value of L5 is determined to produce minimum feedback at the operating frequency and the maximum feedback at low frequencies, where its reactance is low. R1 (Figure 7-8) is used to control the overall level of the feedback. Its value is normally very low, except in cases where R1 is used in conjunction with L5 to control the gain slope.

Another cause for low frequency instabilities can be the physical layout of the circuit.[2, 4, 9] The most important point in an RF amplifier layout is to provide a good and solid ground plane. It will minimize the possibilities of generating RF ground loops that can feed RF energy back to the input in a suitable phase to make an oscillator out of the amplifier. In most cases this problem cannot be fixed except by making a new circuit layout, which generally proves to be costly. Also, excessively high Q's of the matching networks and high Q's of the D.C. feed networks (if the input and output networks happen to resonate) can result in self-oscillations at some intermediate frequency. However, these high Q's can be prevented by following proper design guidelines for RF amplifiers. Such guidelines were discussed in Chapter 6.

An instability which can be prevented by following the proper RF design guidelines is an inductively induced feedback.[10] It is more common in HF and VHF amplifiers using lumped constant matching elements than, e.g., in microstrip designs. Enough RF energy can be coupled to the inductor(s) of the input matching network from the output to trigger an oscillation. The oscillation occurs at a frequency where the input-output phases approach 360°. It may not be of a level high enough to destroy the transistor, and when input drive is applied the oscillation usually disappears and "snaps" to the driven frequency. Helpful hints to prevent this type of instability are to locate the input and output matching networks as far apart from each other as physically possible and orient the inductors of the input and output networks in 90° angles. One might also try electrostatic shielding, although it has only been proven effective in small-signal designs.

A completely different type of instability is caused by the so-called varactor effect.[2, 9, 10] It is known that a varactor multiplier in addition to multiplication can generate sub-frequencies as well if a selective circuit is provided for those frequencies. This is what is known as the varactor effect instability. Varactor effect instability would imply also that there are $2xf_o$ and $3xf_o$ products present, but they would fall on the harmonics and would be hard to distinguish. Their amplitudes are probably much lower than that of the $1/2\ f_o$, since the system power gain is much lower at these higher frequencies. Instead, a stronger $1/2\ f_o$ spur is generated since in most cases the bandwidth extends to those frequencies and sufficient power gain is available (Figure 7-9). The $1/2\ f_o$ oscillation is usually of a fairly low amplitude and does not affect the amplifier's performance noticeably. There is no real cure for the $1/2\ f_o$ instability, and it is most likely to occur in low gain amplifiers of Classes B and C. One possibility may be to add a half frequency band reject filter to the amplifier output, but this only works in relatively narrow band designs. In Classes AB and A, the diode junctions do not go out of forward conduction and the $1/2\ f_o$ phenomenon does not usually occur.

It would be ideal to provide a proper resistive load to the amplifier at its harmonic frequencies even if there is a load mismatch at f_o.[9] This can be accomplished with a diplexer, which is discussed in Chapter 9. Another common practice for RF amplifier stabilization is inserting a resistive attenuator between the amplifier output and the load. The attenuation need be only 1–2 dB, but there is always a power loss. Thus this technique is practical where the stability is more important than the system efficiency. An advantage of resistive loading is that in addition to isolation, the resistive load (although not 50 Ω) is provided to the amplifier at all frequencies. For 1 dB attenuation, the resistor would have a value of 440 Ω, while for 2 dB the value would become 220 Ω.

Testing an amplifier for instabilities can be accomplished using a spectrum

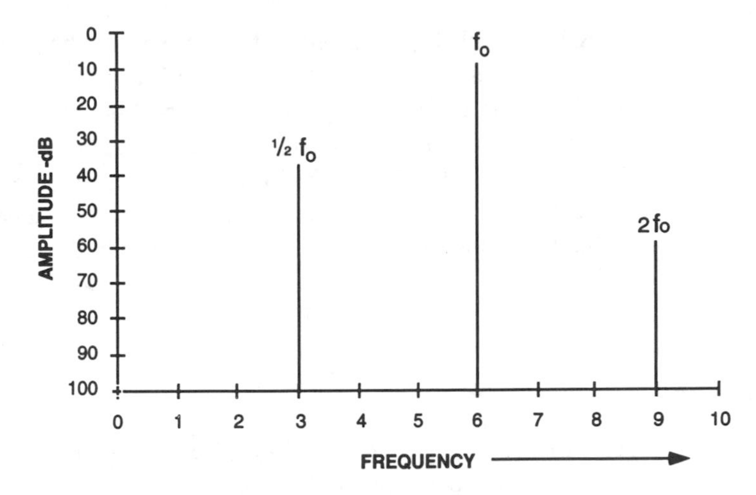

FIGURE 7-9

A display of the so-called half f_o instability, which is caused by varactor effects primarily in the base-emitter junction.

analyzer to see the spurious responses (if any) and an LC network to simulate a load mismatch having a reflection coefficient of near unity in magnitude and all possible phase angles. A description of such a network and formulas to calculate the component values of the network are given in Chapter 9. Any value of load mismatch can be realized by inserting an attenuator between the amplifier output and the "complete mismatch" simulator (Figure 7-10). This is illustrated in Figure 9-7 and formulas of the attenuator values for various amounts of load VSWR are given in the text. Generally at UHF and microwave frequencies, the desired magnitude of reflection coefficient is achieved by a transmission line attenuator terminated in a short circuit. Variation in phase angle of the load reflection coefficient is accomplished by means of a line stretcher.

It has been noted that Class C amplifiers operating at low voltages are the least stable and high voltage units of Classes A and AB exhibit the best stability. Complete stability (no spurious oscillations) of a low voltage Class C amplifier into a 3:1 load mismatch is usually considered adequate and into 5:1 would be considered excellent. The stability of an amplifier can be analyzed

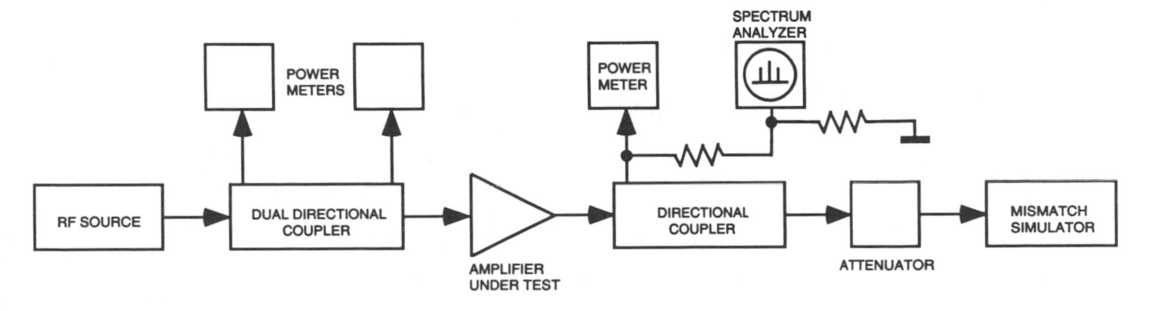

FIGURE 7-10

A test set-up for measuring the stability of RF amplifiers under various load mismatch conditions.

using large-scale S-parameters[12], but this will not help to design a stable amplifier. There are many variables regarding the stability issue and since stability is largely circuit layout and component dependent, a computer aided design can hardly guarantee a stable amplifier. If the results of stability measurements are unacceptable, there is not much to be done except "go back to the drawing board" and carefully re-examine circuit design including board layout following the many guidelines set forth in this chapter.

References

[1] *RF Device Data Book*, DL110, Rev 4, Volume II, Section 7, Motorola Semiconductor Sector, Phoenix, AZ.

[2] Joe Johnson, Editor, *Solid Circuits*, Communications Transistor Company, 301 Industrial Way, San Carlos, CA, 1973.

[3]"Power Circuits—DC to Microwave," Technical Series SP-51,RCA Solid State, Somerville, NJ,1969.

[4] Various applications notes, *The Acrian Handbook*, Acrian Power Solutions, 490 Race Street, San Jose, CA,1987, pp. 622-674.

[5] Frank Davis, "Matching Network Designs With Computer Solutions," Application Note AN267, Motorola Semiconductor Sector, Phoenix, AZ.

[6] Philip Cutler, "Passive Networks," Volume 1, *Electronic Circuit Analysis*, New York: McGraw-Hill Book Company, Inc., 1960.

[7] "VHF/UHF Power Transistor Amplifier Design," Application Note AN-1-1,ITT Semiconductors.

[8] "RF Large-Signal Transistor Power Amplifiers," Part 3-Matching Networks, Publication RS 762, ITT Semiconductors, 1965.

[9] Peter A. Kwitkowski, "RF Power Amplifier Instability—Causes and Cures," *Proceedings of RF Expo West,* 1987.

[10] Nathan O. Sokal, "Parasitic Oscillation in Solid State Power Amplifiers," *RF Design*, December, 1980.

[11] H. O. Granberg, "Good RF Construction Practices and Techniques," *RF Design*, September/October, 1980.

[12] R. Jack Frost, "Large-Scale S-Parameters Help Analyze Stability," *Electronic Design*, May, 1980.

8

Computer Aided Design Programs

GENERAL

The process of impedance matching the input and output of high power RF transistors is best performed using computer aided design programs such as EEsof's TOUCHSTONE™ or Optotek's MMICAD™ . This is particularly true when matching involves a band of frequencies instead of a single frequency. It is not the intent of the authors, however, to describe the details of a particular computer aided design (CAD) program. Detailed instruction manuals accompany any CAD program and these describe the inputs required and the sequential steps involved in achieving a particular optimized matching network. There is still a place, though, for the use of the Smith Chart as a tool in the initial phase of the matching process. A most unique program to aid the design engineer in using the Smith Chart has been created at Motorola.[1] It has been given the name Motorola Impedance Matching Program (MIMP), and it is available free of charge to anyone who desires a copy.

What the program does is to provide a simple environment for entering and analyzing impedance matching circuitry. It focuses solely on impedance transformations. A standard library of passive circuit elements is provided by MIMP, including various combinations of capacitors, inductors and transmission lines in both series and shunt configurations. It also contains a unique "distributed capacitance element" that models a capacitor distributed along a transmission line.

The real nucleus of MIMP is its computer-aided Smith Chart. It is uncommon for computer-aided design (CAD) programs to incorporate the benefits of manipulating actual impedance transformations on a Smith Chart. MIMP's Smith Chart facility provides this electronically. It also displays each circuit element's contribution to the total impedance transformations. Included in MIMP is an auxiliary data base in which are tabulated the input and output impedances for many of the RF power transistors contained in Motorola's RF Data Book.

Details of MIMP are contained later in this chapter. Copies of the actual program suitable for use on IBM compatible personal computers can be obtained from Motorola Semiconductor Products by contacting their nearest sales office and requesting DK107. Requirements to successfully use the data disk are an IBM compatible (MS-DOS) personal computer with at least 640K of RAM, a 80286 or higher processor, and a VGA graphics adapter. A "mouse" is recommended.

The following paragraphs describe the combination of MIMP and a CAD program to design matching networks for the input and output of a high power UHF transistor. MIMP was used initially to arrive at lumped circuit matching configurations. The number of matching elements was increased systematically until the desired circuit match (measured in terms of return loss) was obtained across the frequency band of interest. Then the job of optimizing the circuit was turned over to a CAD program for final circuit design, much of which used distributed circuits for element realization. The input circuit was optimized for best match while the output circuit was optimized for best gain and efficiency simultaneously.

Let's start with the MRF658 RF power transistor. This device is intended to provide 65 watts of output power from 400 to 520 MHz while operating from a 12.5 volt supply. The first step is to take the impedance data supplied by the manufacturer (re-stated in Figure 8-1) and use the impedance matching program (MIMP) to generate lumped element matching networks (shown in Figure 8-2) to provide high return loss (the goal was 20 dB) over the frequency band from 470 to 512 MHz. In this instance, the number of matching elements was determined experimentally through an iterative process of "cut and try." Filter theory can be used to predict the number of matching elements needed to obtain a "passband" response that covers the desired frequency range and has a "ripple" in the pass-band no greater than the equivalent specified amount of return loss. However, in this instance, MIMP allows you to start with a single element and add experimentally additional elements until you achieve suitable return loss (an indication of impedance match) over the desired frequency band.

The next step in the design of the amplifier is to convert the design to a microstrip transmission line configuration and add the bias circuitry. Some of this can be accomplished with MIMP and the rest through the use of transmission line equations, such as the one given in Chapter 9 that determines a microstrip equivalent for a specified value of inductance L. Figure 8-3 shows the

FREQ MHz	Z_{IN} ohms	$Z_{OL}{}^*$ ohms
400	0.62+j2.8	1.2+j2.5
440	0.72+j3.1	1.1+j2.8
470	0.79+j3.3	0.98+j3.0
490	0.84+j3.4	0.91+j3.2
512	0.88+j3.5	0.84+j3.3
520	0.90+j3.6	0.80+j3.4

FIGURE 8-1

Input and output transistor impedances for MRF658. P_{out} = 65 watts and V_{dc} = 12.5 volts.

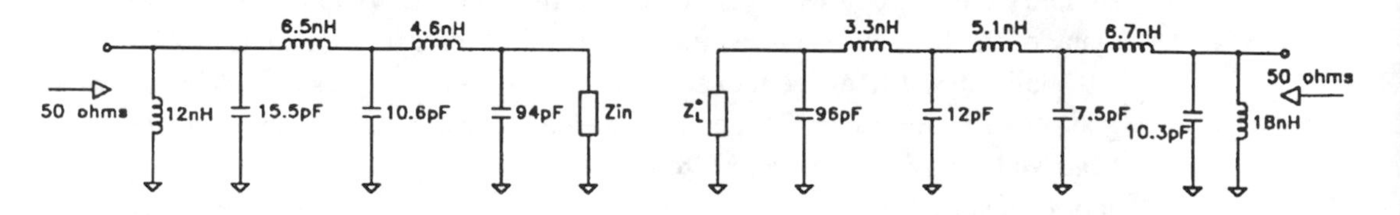

FIGURE 8-2

Lumped element matching networks from 470-512 MHz, determined experimentally by using MIMP.

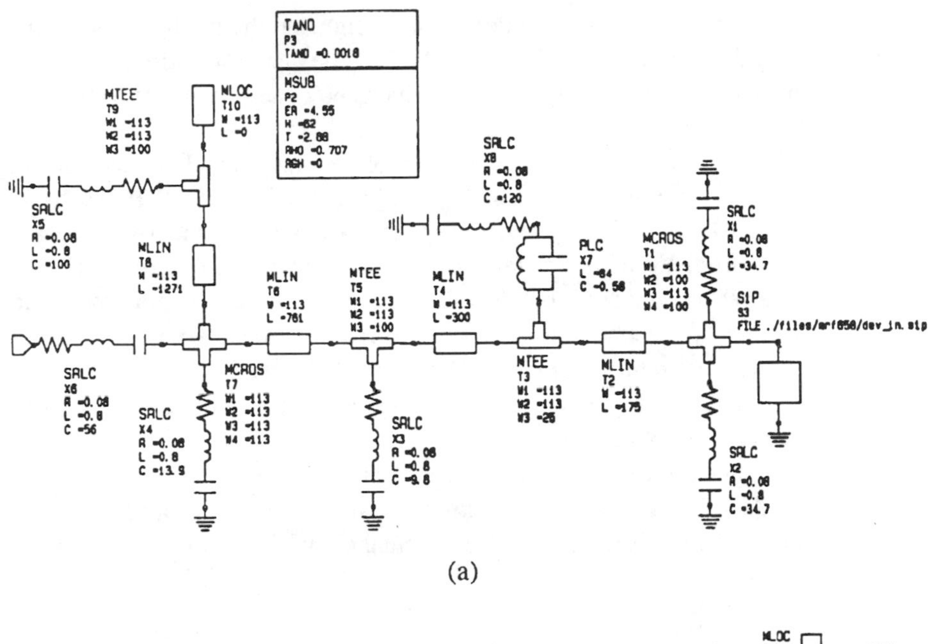

(a)

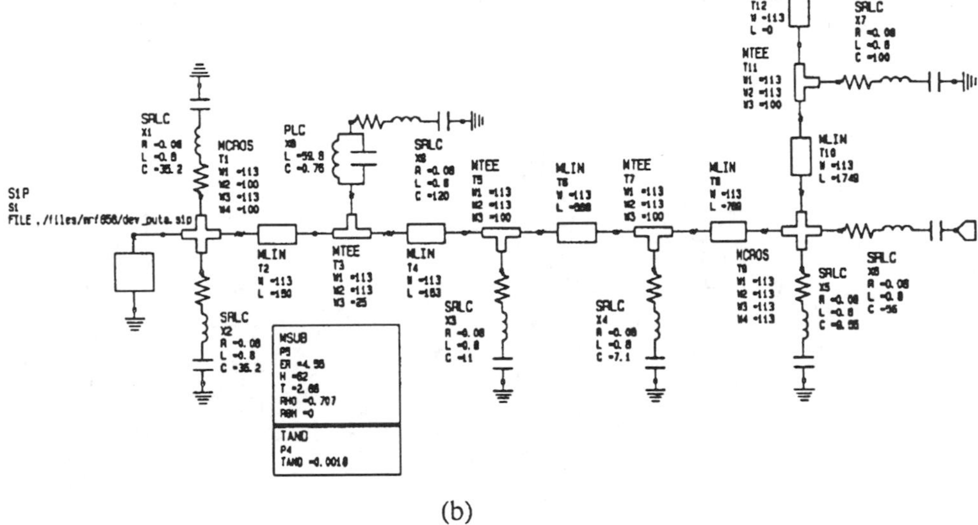

(b)

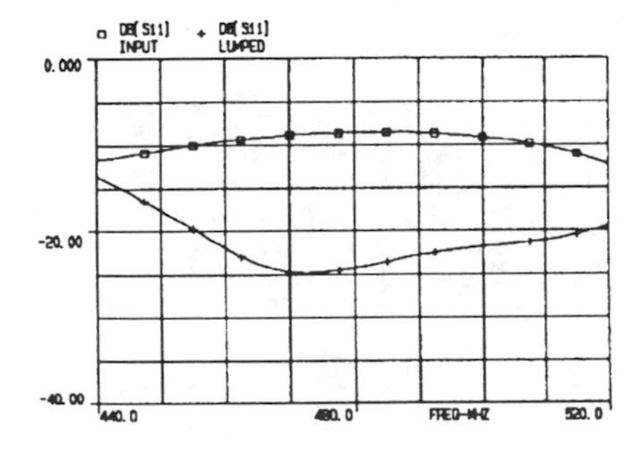

(c) Input reflection coefficient

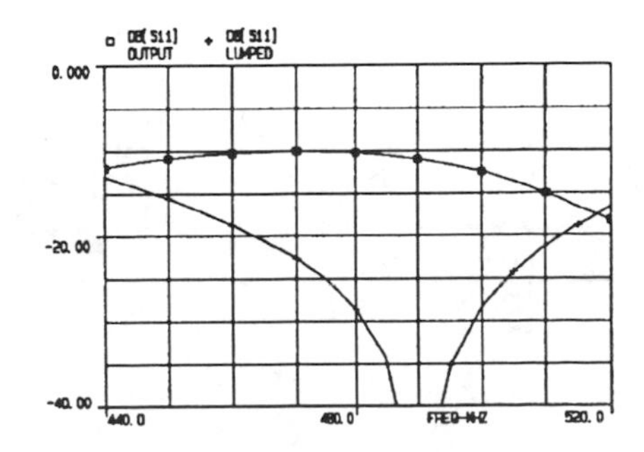

(d) Output reflection coefficient

FIGURE 8-3

Microstrip configuration of amplifier design.

initial microstrip configuration of the design and the predicted performance using the program called ACADEMY™. The need for optimization is readily apparent from Figure 8-3C and D which show the return loss to be less than desired over the frequency band of interest.

Modifications to the value of some of the matching elements is next accomplished by use of CAD optimizing programs such as TOUCHSTONE or MMICAD. ACADEMY will also optimize and automatically generate a board layout from a schematic diagram. Normally input and output matching networks are optimized separately. Usually the optimization goal for the input network can be stated in terms of return loss. In the example given, the goal was an input return loss (IRL) of at least 18 dB across the frequency band of interest. Generally the output matching network is optimized for best gain and efficiency across the frequency band.

Figures 8-4 and 8-5 show the schematic diagrams optimized by MMICAD and ACADEMY. Figure 8-6 shows the final schematic diagram of the amplifier and Figure 8-7 is a diagram of the actual circuit layout. Finally, Figure 8-8 shows the performance obtained from a series of amplifiers constructed from the ACADEMY circuit layout.[2]

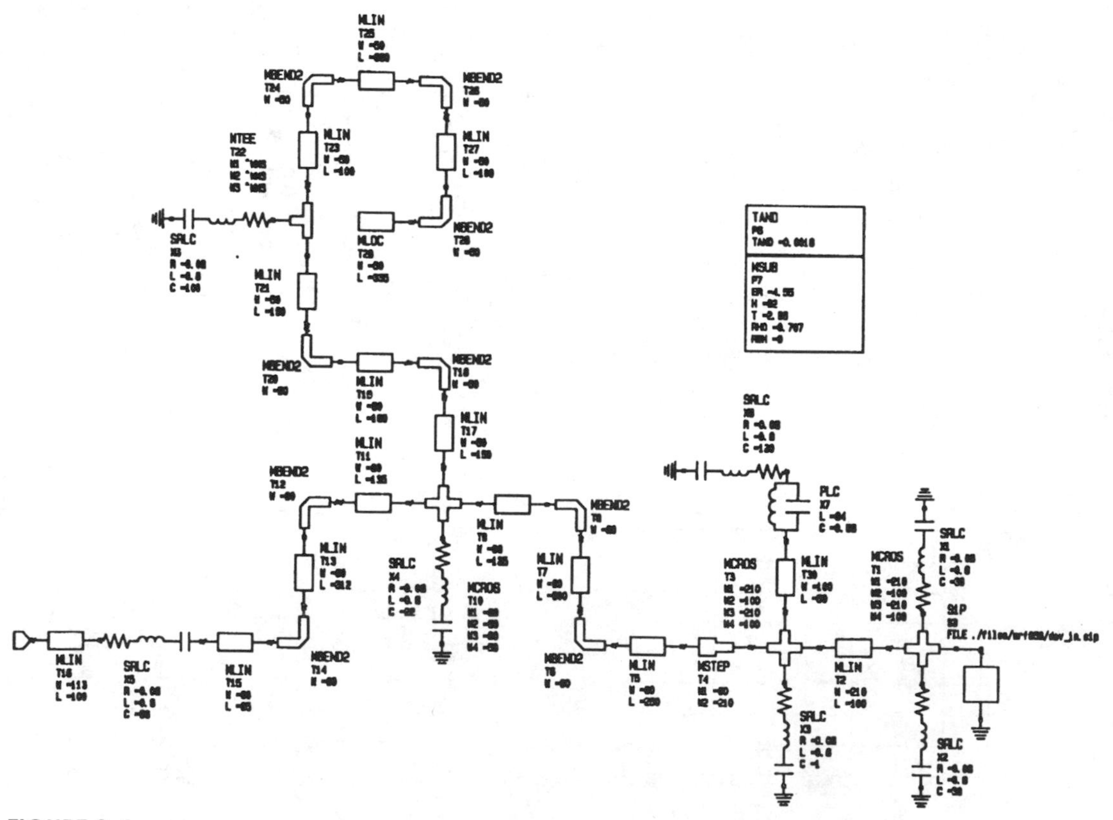

FIGURE 8-4

Optimized input matching network from ACADEMY.

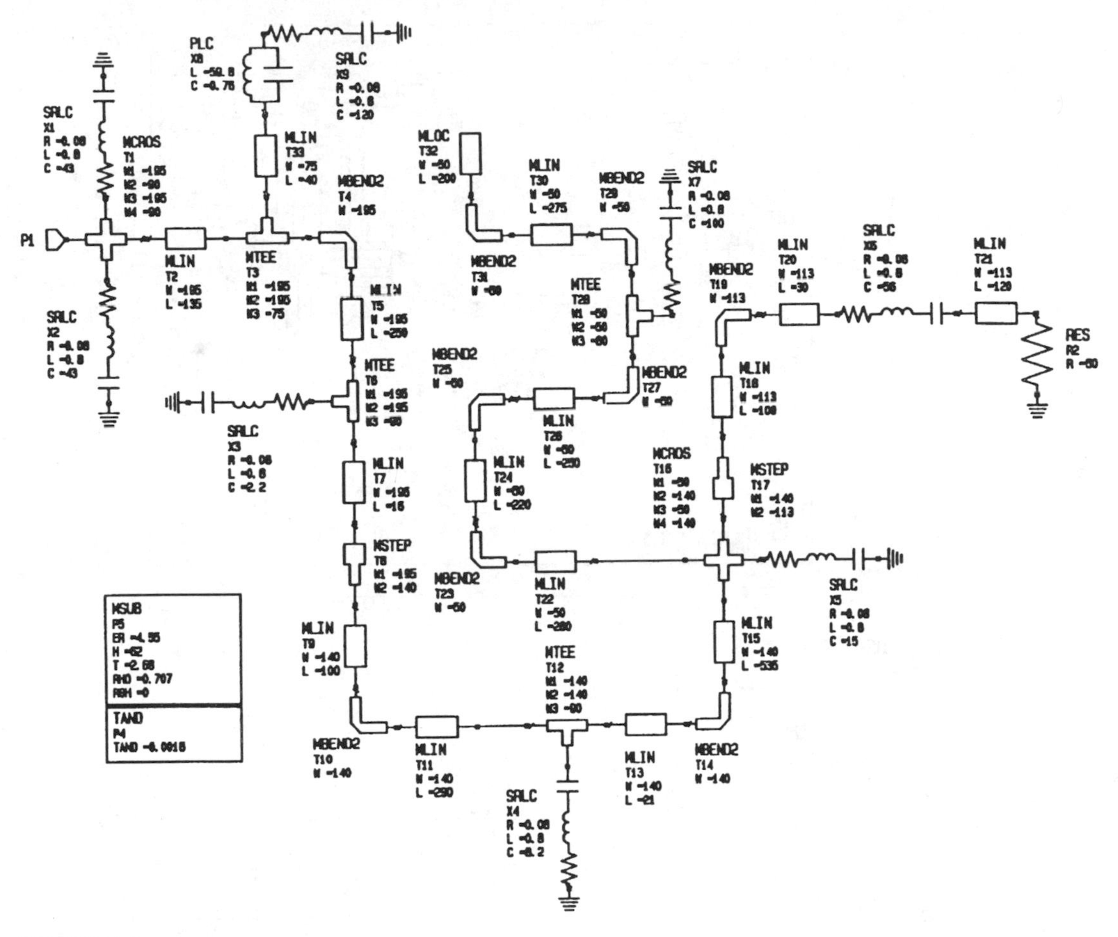

FIGURE 8-5

Optimized output matching network from ACADEMY.

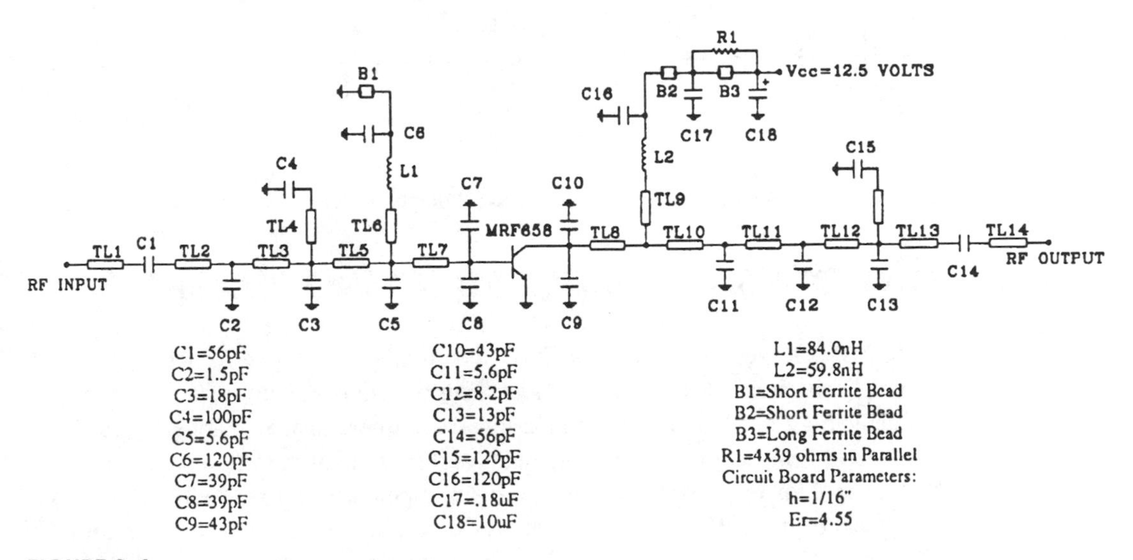

FIGURE 8-6

Schematic diagram of completed Class C UHF power amplifier.

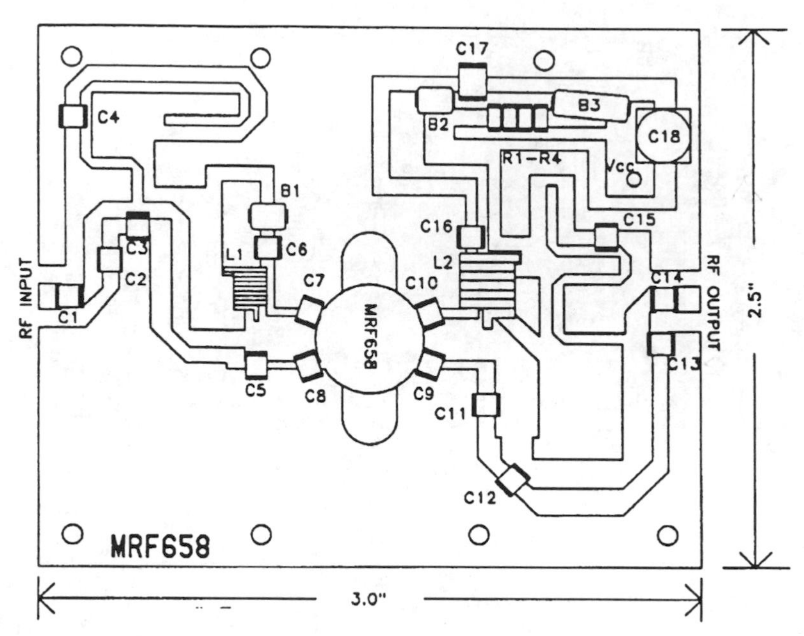

FIGURE 8-7

Diagram of assembled RF power amplifier.

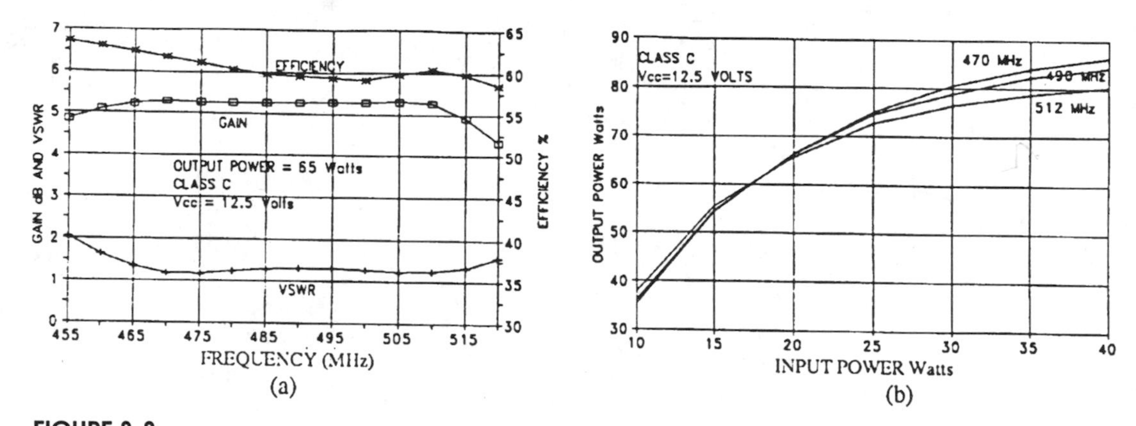

FIGURE 8-8

(a) Average gain, VSWR, and efficiency for six amplifiers, and (b) output power versus input power.

INSIDE MOTOROLA'S IMPEDANCE MATCHING PROGRAM

MIMP is a specialized form of CAD specifically developed for RF power amplifier circuit design and provides a simple environment for entering and analyzing impedance matching circuitry. Commercially available programs include a multitude of circuit elements and provide numerous analytical capabilities. However, MIMP focuses *only* on impedance transformations. This is typical of most RF power amplifier design problems, since data sheets for RF devices only present large signal impedances measured at a single combination of frequency, voltage, power level, and power dissipation. These impedances generally must be transformed to another set of impedances, such as 50 Ω or the input/output impedance of another device. To do this, MIMP provides a standard library of passive circuit elements including various combinations of capacitors, inductors, and transmission lines in both series and shunt configurations. It also contains a unique distributed capacitance element that models a capacitor distributed along a transmission line.

The real nucleus of MIMP is its computer-aided Smith Chart. It is uncommon for CAD programs to incorporate the benefits of manipulating actual impedance transformations on a Smith Chart. If commercially available programs were used merely for impedance matching, a typical final result of a computer run would be S_{11} versus frequency. To supplement this, many RF designers still keep a Smith Chart, compass, straight edge, and pencil handy so they can pictorially represent each circuit component's contribution to the total transformation. MIMP's Smith Chart facility provides this service electronically. It also displays each circuit element's contribution to the total impedance transformations.

Here are some additional advantages of the Smith Chart display function:

a) The Smith Chart can be instantly "re-normalized" to any characteristic impedance. All impedances (with interconnecting arcs) are automatically recalculated and displayed.
b) There is an option for overlaying constant return loss circles for any complex source impedance independent of the normalized characteristic impedance. (Most other programs constrain the use of constant return loss circles to the center of the Smith Chart.)
c) Multiple transmission line transformations (each with different characteristic impedances) are displayed simultaneously and in exact graphical relationships to each other independent of the Smith Chart's normalized impedance. (Drawing transmission line transformations by hand requires an iterative denormalize/renormalize/replot/redraw procedure.)
d) A tabular impedance display is provided to view the impedance at any "node."
e) Constant Q arcs can be added to the Smith Chart.
f) Real-time changes in the impedance transformation are displayed while individual circuit elements are tuned. This utility is provided to perform manual circuit optimization. A scalar display of input return loss is updated simultaneously as an additional tool for optimization.
g) If Motorola's RF power transistors are used in the design, then an auxiliary database containing the input and output impedances for many such devices can be accessed.

MIMP DESCRIPTION

MIMP is divided into three screens: the *impedance entry* screen, the *circuit entry* screen, and the *Smith Chart display*. A mouse is recommended for easy entry and manipulation of data, although there are keyboard equivalents for most of the mouse functions.

Once the program is "launched" on the computer, the screen in Figure 8-9 is displayed. This is the Impedance Entry screen, and is separated into four basic sections:

1) frequency table,
2) load impedance table,
3) source impedance table,
4) data entry keypad.

MIMP first prompts the user for the number of frequencies to be entered. The program will only accept values less than or equal to 11. (If zero is entered, the program advances to a standard device entry sequence, as will be described later.) After the number of frequencies has been supplied, each frequency should be entered sequentially starting with the lowest value. When the last frequency has been entered, the user is prompted to supply the load impedance data for each frequency. An option to specify 50 Ω is supported by pressing the ENTER key. After all the load impedance data is furnished, the source impedance data is requested. Again, there is an option to display 50 Ω by pressing ENTER. (Parallel equivalents are calculated and displayed for all impedances.) After the data is entered, the user may proceed to the Circuit Entry screen or edit any of the frequency or impedance data.

If a standard device is to be selected as the load impedance, users can select from the 2N, MRF, JO, and TP prefixes and then enter the remainder of the number. The program next prompts the user to select either the device's input or output impedance as a load. If the requested device is included in the database, the impedance information is displayed at their corresponding frequencies. The

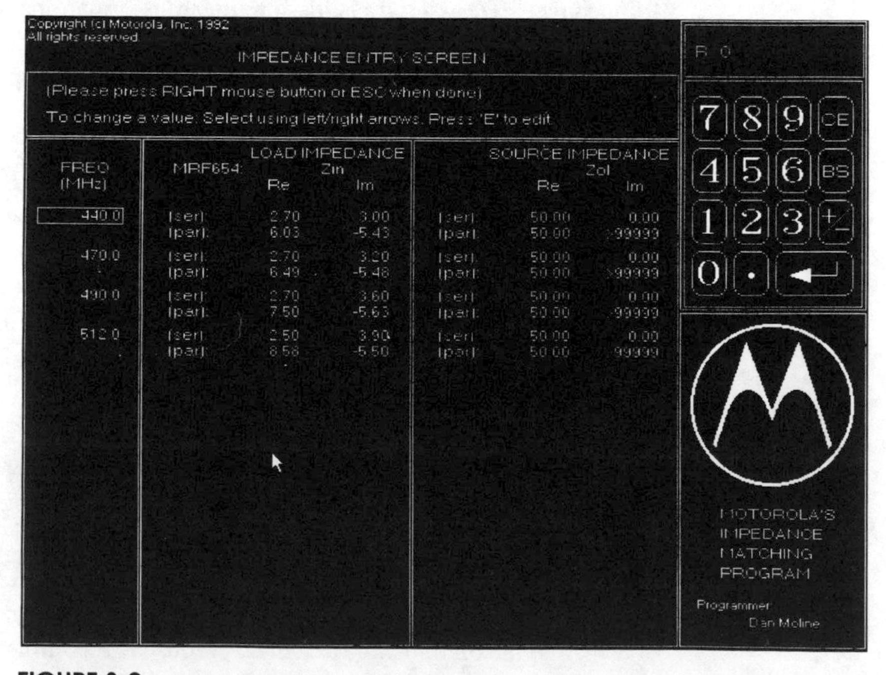

FIGURE 8-9

The Impedance Entry screen in MIMP.

source impedance is entered manually. Editing can be performed on any frequency or impedance before proceeding to the Circuit Entry screen.

Figure 8-10 shows the Circuit Entry screen. It is separated into three parts:

1) the component library,
2) data entry keypad,
3) circuit display area.

A component is selected by clicking it with the mouse or pressing the appropriate keys. Immediately after a component is selected, the numeric keypad is activated and the user is prompted to enter component values. Inductors are recognized in nanohenries and capacitors in picofarads, and inductive/capacitive reactancc can be specified.

A transmission line is defined by its characteristic impedance in ohms and its electrical length in fractions of a wavelength. The electrical length needs to be referenced to a specific frequency in MHz. Transmission lines (microstrip) may also be classified in physical terms. MIMP will prompt users for the conductor's width and length. Whenever the first transmission line is selected (whether defined in electrical or physical properties), MIMP will also request information on relative dielectric constant, dielectric thickness, and conductor thickness.[3] This information is assumed to be the same for all subsequent transmission lines and is displayed in the upper right hand corner of the display. To change these values, all existing transmission line data must be deleted.

Most CAD programs assume that a capacitor has no width; i.e., it contacts the circuit at a single point. As frequencies increase, this assumption introduces a significant error in circuit analysis, particularly at low impedances. Since a

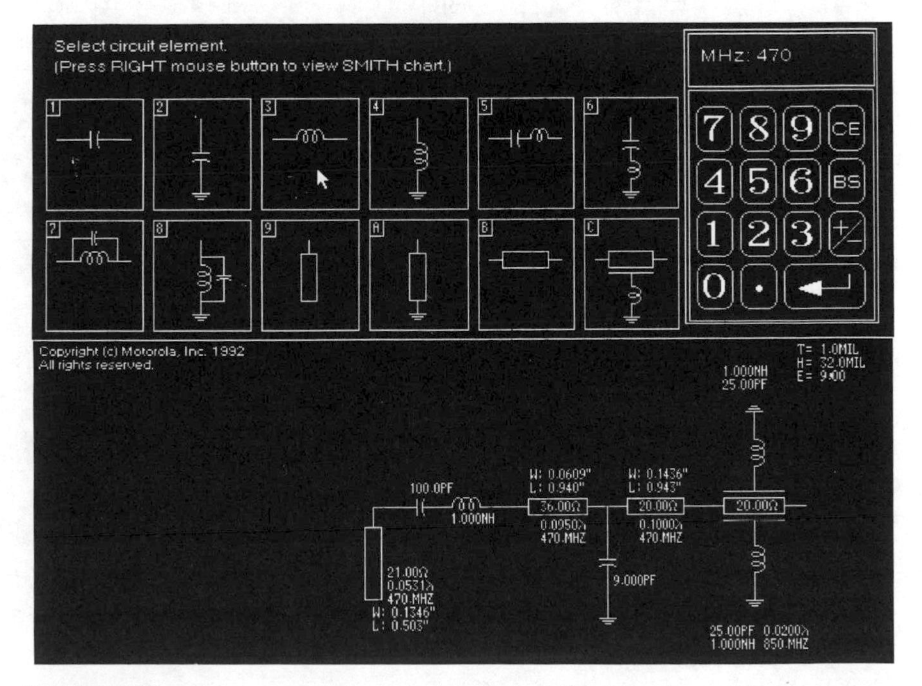

FIGURE 8-10

The Circuit Entry screen in MIMP.

capacitor is typically mounted on a transmission line, a significant phase shift can occur across its width at higher frequencies. (A 100 mil capacitor can have an electrical width of 0.02λ at 1 GHz if mounted on Al_2O_3).This error can be reduced by modeling a capacitor as a "distributed" component. On most CAD programs, this involves subdividing the capacitor and transmission line into several smaller sections to comprehend the collective capacitive effects and transmission line transformation. MIMP provides a component, called the DISTRIBUTED CAPACITOR, which first prompts the user for a capacitor value along with any accompanying series lead inductance. It next asks for the characteristic impedance of the transmission line on which the capacitor is mounted. Finally, it asks for the transmission line's electrical length (in fractional wavelengths) for that portion of the transmission line on which the capacitor is mounted. MIMP then calculates the combined effect of the two.

Figure 8-11 shows the Smith Chart display. It is divided into four sections:

1) the Smith Chart,
2) the menu bar,
3) the nodal impedance display,
4) the scalar input return loss graph.

The Smith Chart graphically displays the impedances transformed by each shunt or series element. These impedances are represented by small "x" letters. Each frequency is depicted by a different color. If there are multiple series or shunt elements, the combined effect of all elements is lumped together as one element.

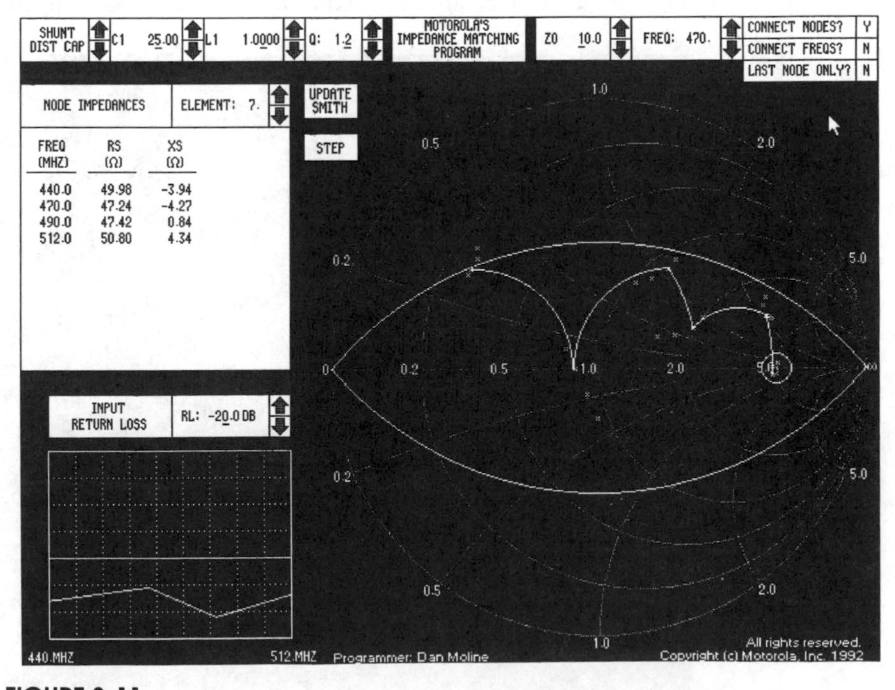

FIGURE 8-11
The Smith Chart display screen in MIMP.

There are several default conditions set whenever the Smith Chart display is entered. The conjugate of the source impedance is shown by a series of yellow "x" letters encircled by a –20 dB return loss circle (the conjugate is the desired transformed impedance). The Smith Chart is initially normalized to 10 Ω. The constant Q arcs are set to 0, and the nodal impedances are listed for the last node. The first circuit element is selected for tuning. These default conditions may be altered by making selections from the menu bar and using the mouse or keyboard to change the conditions and values. The menu bar is also used to select and tune the various circuit elements.

The nodal impedance display shows the actual transformed impedance produced by each circuit element. Multiple series or shunt elements are lumped together. (Note: these are not really nodal impedances. It is the transformed circuit impedance, starting from the load up to and including the selected element.) Different points in the circuit can be selected, and as the "node" is changed the corresponding "x" letters are highlighted.

The return loss display shows on a scalar chart how well the transformed impedance matches the source impedance. The reference return loss is indicated by the yellow line on the scalar display and the yellow circle on the Smith Chart display. If the source impedance is frequency dependent, the circles on the Smith Chart will be relocated for each frequency.

SMITH CHARTS AND MIMP

The accepted practice for plotting impedance transformations on Smith Charts requires that each circuit impedance be first normalized to its respective transmission line's characteristic impedance. Once normalized, the impedances can be plotted (and transformed) on a similarly normalized Smith Chart. Each time a new transmission line is encountered in the circuit, the impedances must then be denormalized using the old Z_0 and renormalized with the new Z_0 before additional graphical manipulations can be accomplished. The normalized Smith Chart (typically 1 Ω) remains unchanged with each new transmission line transformation. This results in a series of noncontiguous impedance transforms whose visual relationships have little or no value. See Figure 8-12.

One of the unique features of MIMP's Smith Chart display is its ability to have various transmission lines with different characteristic impedances displayed together on one Smith Chart. Instead of constantly denormalizing and renormalizing circuit impedances and plotting them on the same normalized Smith Chart, the reflection coefficient plane for each transmission line transformation is renormalized instead. The locus of points, representing the transmission line's impedance transformation, is remapped into this new plane. The origin of the reflection coefficient (RC) plane is repositioned along the real axis and the magnitude of the RC is rescaled. In effect, a second Smith Chart of an adjusted size is overlayed on the original Smith Chart. Figure 8-13 shows this. This approach permits multiple transmission lines to be displayed on the same Smith Chart in a contiguous flow while maintaining exact graphical relation-

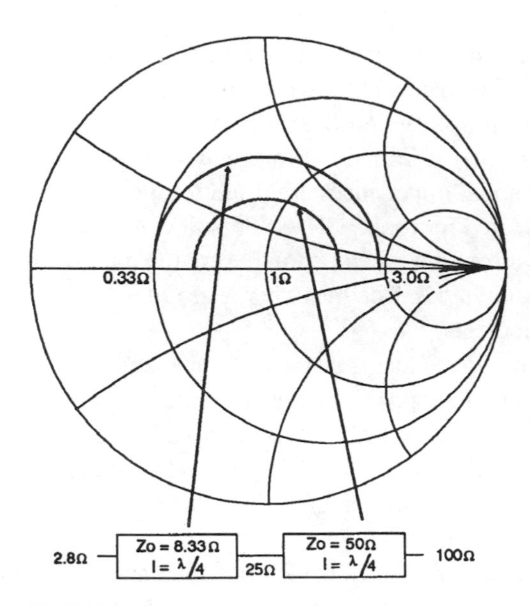

FIGURE 8-12
Conventional impedance transforms for different values of Z_o.

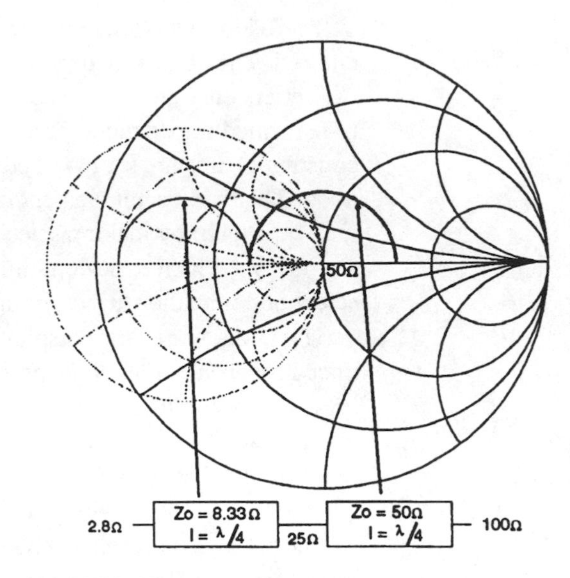

FIGURE 8-13
MIMP display for impedance transforms with different values of Z_o..

ships between the various transformations. Because of the additional calculations, this technique is obviously more applicable for CAD than by hand. Once the program is set up to handle these calculations, all impedance transformations maintain their relative positions as the relative characteristic impedance of the Smith Chart is changed. MIMP also allows the user to specify any relative Z_o for the Smith Chart.

References

[1] Dan Moline, "Impedance Matching Program," to be published in *RF Design,* January, 1993.

[2] Robert Baeten, "Computer Aided Design of A Broad-Band, Class C, 65 Watt, UHF Amplifier, to be published in *RF Design,* March, 1993.

[3] E. Hammerstadt and O. Jensen, "Accurate Models for Microstrip Computer Aided Design," *IEEE International Microwave Symposium Digest,* June, 1980.

9

After the Power Amplifier

VSWR PROTECTION OF SOLID STATE AMPLIFIERS

Most transistor failures in solid state amplifiers occur at load mismatch phase angles presenting a high current mode of operation to the output transistor(s), which results in an increase in the power dissipated by the transistor(s). Since the temperature time constant of a typical RF power transistor die is 0.5–1.0 ms, any protection system employed (including all delays in the AGC/ALC loop) must react faster than this time. Although there are a number of methods to accomplish solid state RF power amplifier protection against load mismatches, the *reflectometer* VSWR sensing is the most commonly used method. The reflectometer is usually located in series between the amplifier output and the load. A voltage proportional to the amount of output mismatch is obtained from the reflectometer, which is processed accordingly, and fed back to either the power amplifier input or one of the preceding stages in a manner to gradually reduce the power gain or to completely shut-down the amplifier.

A standard reflectometer principle, presented in numerous publications[1, 2, 3, 5, 6] over the years, is used in this design to detect the RF power amplifier output mismatch. It is also commonly known as a *VSWR bridge*, and its use can be extended to microwave frequencies (greater than 1 GHz) with proper mechanical design. The UHF and microwave designs generally employ microstrip transmission line techniques, whereas lower frequency circuits favor lumped constant implementations. In fact, up to UHF the lumped constant concept is probably the most practical means to approach the coupling coefficient required between the current line (amplifier output) and the sample line in order to produce an output voltage of moderate amplitude. A tight coupling in a lumped constant system is achieved by passing the current line through a multiturn pick-up coil, thus forming a transformer, where the current line is the primary and the multiturn coil the secondary. The multiturn winding is usually in the form of a toroid, which allows magnetic material to be used as the core to increase the low frequency response. The inductive reactance of the multiturn winding must be greater than the load impedance of the current line at the lowest frequency of operation. However, the ports are usually terminated into low reactance dummy loads of a value equal to the characteristic impedance of the system, which in most cases is 50 Ω. The high frequency limit of the VSWR bridge is determined by leakage inductances and the physical length of the multiturn winding. There will be resonances whenever the length of the secondary winding (sample line) equals a wavelength divided by 2^n, where n is an integer 1, 2, 3, etc. However the amplitudes of the resonances diminish as n increases. If the length of the sample line is kept shorter than $\lambda/16$, the amplitudes of any resonances are negligible.[1, 2]

The principle of operation is as follows: the voltage across the multiturn secondary of the transformer is proportional to the current passing through the current line and the number of turns in the sample line. When the amplifier has a perfect load (at J2, Figure 9-1) the RMS RF voltage measured across the forward port (terminated with 50 Ω) would be V_{rms}/R_L decreased by the amount of coupling (Cp) between the current line and the sample line. Similarly, the RF voltage at the reflected port is related to the RF power reflected at the output port (J2). When the load at the output port is totally mismatched, (i.e., open or shorted) the voltage at the reflected port will equal the voltage at the forward port.

In order to produce a voltage of a practical level in either the forward or reflected port, the coupling coefficient (Cp) should not be higher than 30 to 40 dB for power levels in the main line of 100 to 1000 watts. For example, if Cp = 30 dB and we have a 100 W amplifier, the power appearing at the forward port of the sample line will be 100 mW and the voltage will be $\sqrt{PRL} = \sqrt{0.1 \times 50} = 2.24$ V. The coupling coefficient (Cp) can be figured as:

$$C_p = 20\log\left(\frac{1+(1/(2N^2))}{1/N}\right)$$

where N = number of turns in the sample line. Reversing the equation: $N = 10^{Cp/20}$, where Cp = port coupling coefficient required in dB. Then if Cp = 30 dB: $N = 10^{30/20} = 32$ turns (for 1 turn primary).

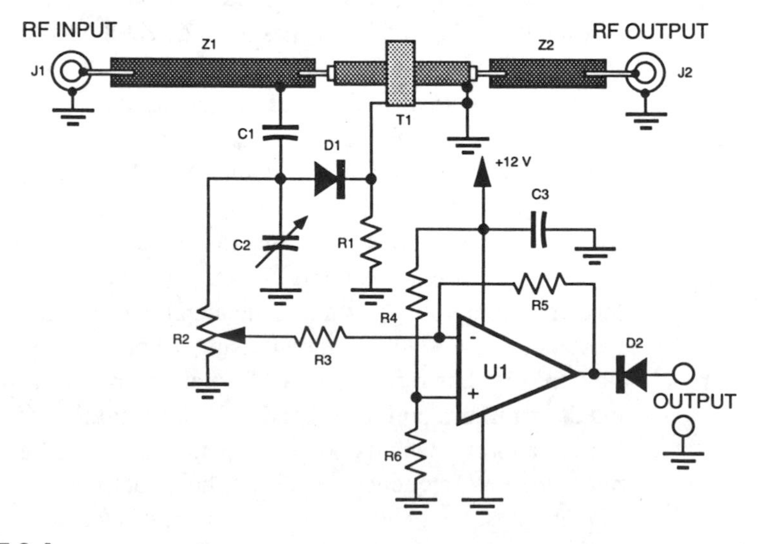

FIGURE 9-1

Schematic diagram of VSWR protection circuit described. See the text for details. Referred to in subsequent figures as "VSWR sensor."

$$\text{Input - output insertion loss} = 20 \log\left(1 + \frac{1}{2N^2}\right) = 0.0042 \text{ dB}$$

$$\text{Input return loss} = 20 \log\left(2N^2 + 1\right) = 66 \text{ dB}$$

In addition to the voltage derived from the secondary of the toroidal transformer, a voltage sample is taken from the current line by means of a capacitive divider C1-C2 (Figure 9-1). Half wave rectified voltages are created at the junction of C1 and C2 by the transformer secondary. These voltages are 180° out of phase in case of a nonmismatched load. The amplitudes are adjusted equal with C2, which reduces the voltage to near zero at the junction of C1 and C2 until a mismatch in the load is present, which results in the phase shift deviating from 180°.

There are mechanical restrictions that clearly place a limit to the bandwidth of the circuit. It is difficult to design extremely wideband and high power systems, since for high frequencies the toroidal pick-up coil should be as small as possible, and for low frequencies it should be large enough for the minimum reactance required. High permeability ($\mu = 100$ and higher) ferrites in the toroid are usually too lossy at high frequencies and will heat up even at moderate power levels. For example at 150 MHz, materials with μ of 15 or less have been found acceptable.

A Faraday shield is usually employed between the current line and toroidal winding to prevent capacitive coupling between the two. It can be best accomplished with a length of coaxial cable of proper characteristic impedance, where the inner conductor forms the current line and the outer conductor the Faraday shield. Normally, only one end of the Faraday shield is grounded to prevent formation of a shorted RF loop. However, if length of the Faraday shield is considerably smaller than the ground loop, the shield can be grounded at both ends (if desirable for mechanical reasons for example). The circuit details of the unit described here are shown in Figure 9-1. The forward port has been omitted since the system is not intended for forward power measurements. For good high frequency performance, a solid ground plane in the "current line" and "sample line" area is extremely important. Otherwise the resulting ground loops may reduce the frequency response of the circuit or result in uneven response characteristics as a function of frequency.

The circuit shown in Figure 9-1 has been tested simulating a load mismatch of 5:1 at a power level of 1 kW at 30 MHz and up to 200 W at 220 MHz. Some of the data is shown in Figures 9-2 and 3. If a fast operational amplifier such as MC34071 (which has a 13V/μs slew rate) is used, the output switching can be accomplished in 2 ms, which is fast enough to protect most RF power amplifiers. Most operational amplifiers are capable of sinking currents up to 20 mA. This output is sufficient to be used directly to turn off the bias voltage of an enhancement mode MOSFET for example, or an emitter follower can be added for higher current requirements. The output of the operational amplifier can in fact be made the main bias source to provide the MOSFET gate bias voltage. Controlling the gate voltage of a MOSFET for a gradual gain reduction would not be possible in linear operation since a steady idle current is required. In

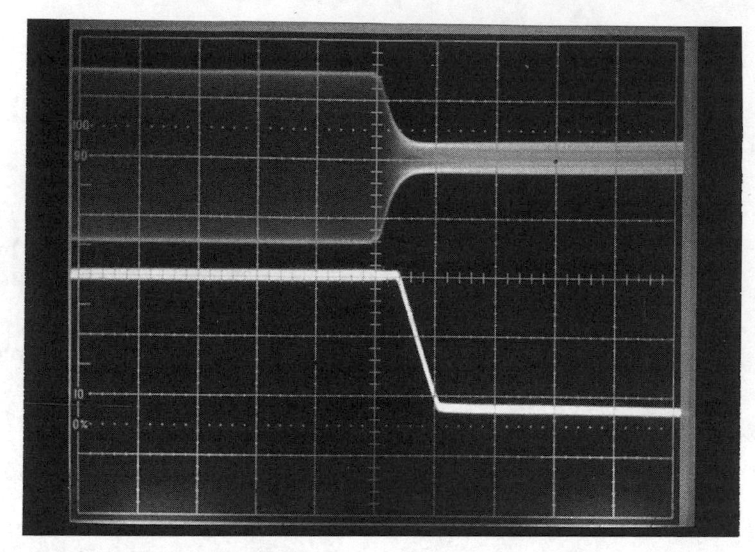

FIGURE 9-2
RF envelope and amplifier output of VSWR sensor. Horizontal scale: 2µs/division. Vertical scale: 5.0 V/division.

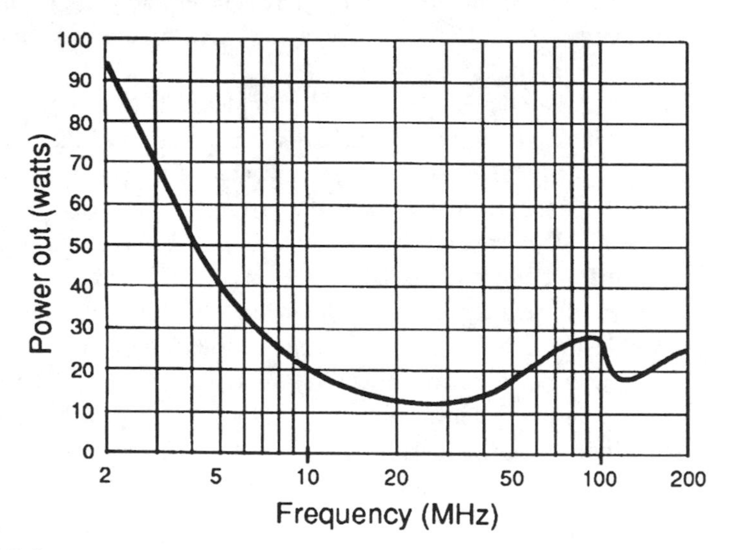

FIGURE 9-3
Sensitivity vs. frequency response of the VSWR sensor at 5:1 load mismatch.

such cases some type of voltage or current controlled RF attenuator must be used, preferably in the low level pre-stages (usually operating in class A), which are insensitive to variations in the output load. Using attenuators would be the only way to control the power gain of bipolar transistor amplifiers since the AGC function of MOSFETs is not available. One possibility is a PIN diode attenuator shown in Figure 9-6. Depending on the attenuator characteristics and the power level, the power output can be adjusted as desired for a given output

FIGURE 9-4
A typical application for the VSWR sensor (shown in Figure 9-1), where its output directly controls a MOSFET gate bias voltage at a low level stage.

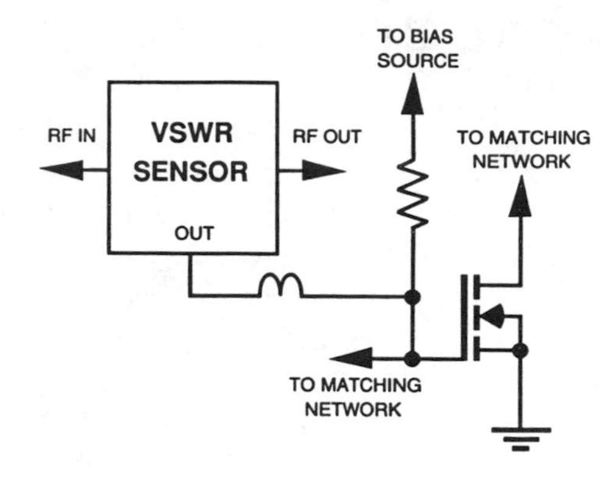

mismatch with the combination of R2 and R5 (see Figure 9-1). For this, as well as the circuit in Figure 9-5, D2 (in Figure 9-1) must be shorted in order to employ the output of IC1 for a voltage "pull-up" function. If only a fast shut down of the amplifier is desired without linearity requirements, circuits in Figures 9-4 and 5 are not only adequate but also simple. It is recommended that an early stage in the amplifier chain be controlled since low power MOSFETs have low gate input capacitances, which speeds up the shut off. In Figure 9-4 the FET bias is supplied by an external source, whereas in Figure 9-5 the bias source is the operational-amplifier output of the VSWR sensor.

FIGURE 9-5
Similar to Figure 9-4, but in this circuit the VSWR sensor acts as the gate bias voltage source for the FET in addition to providing a shut-off function.

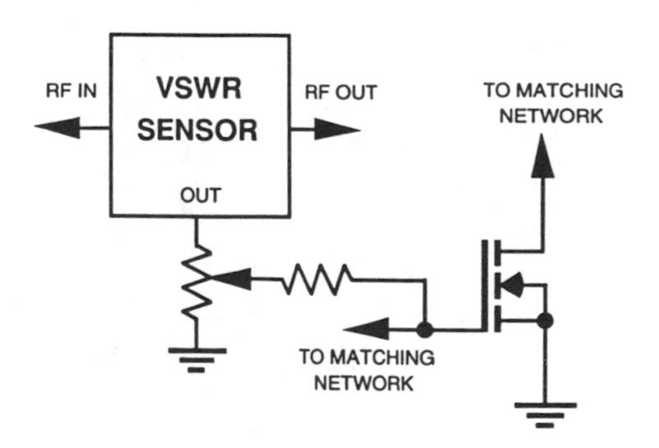

FIGURE 9-6
A PIN diode switch is used here and is adaptable with either MOSFET or BJT amplifiers. A current boost for the diodes may be necessary to drive them into full conduction, depending on the type of diodes used and the signal level.

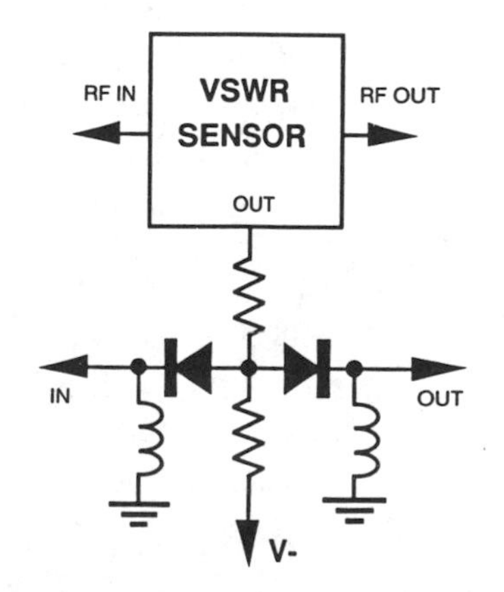

TESTING THE CIRCUIT

Specific amounts of load mismatch must be developed for testing a system such as the one described here. For example, in some applications a fold-back may be desired at 3:1 or 5:1 output VSWR, whereas in others a load VSWR of 10:1 may be tolerable. Since an infinite mismatch cannot be reached due to component losses, a value of 30:1 has been adapted as a standard for "infinite mismatch" by the industry. As the scale is logarithmic, there is not much practical difference between 30:1 (or even 20:1) and infinite. A 30:1 mismatch covering all phase angles and the R from nearly zero to open circuit can be simulated with an LC network shown in Figure 9-7. C consists of two similar variable (air) capacitors, their voltage ratings depending on the RF power level. C can also be a butterfly dual capacitor, where the wiper can be used for the ground contact. The minimum-maximum capacitance ratio should be at least 5–6 in order to obtain a coverage for all phase angles and values of R. The initial maximum capacitance values are not critical and will only slightly affect the circuit Q and the values of L. Typical values for 30 MHz are 300–400 pF, for 100 MHz 40–50 pF, and for 200 MHz 10–15 pF. The L's are usually airwound inductors physically large enough to handle the RF currents at the power level in question. The L values can be calculated as:

$$L1 = \frac{1}{(2\pi f)^2 C(\text{min})}, \qquad L2 = \frac{1}{(2\pi f)^2 C(\text{max})}$$

Example: f = 100 MHz, C(min) = 7 pF, C(max) = 40 pF, then:

$$L1 = \frac{1}{628^2 \times 7} = 362 \text{ nH}, \qquad L2 = \frac{1}{628^2 \times 40} = 63 \text{ nH}$$

The same function can be accomplished with a single inductor and a differential capacitor, where one section is at its minimum capacitance while the other section is at its maximum. Their capacitance ratios are roughly the same as the inductance ratios in the network previously described. This circuit is in the familiar pi network configuration and is widely used for testing amplifier stability. The 30:1 mismatch provided by this circuit (Figure 9-7) can be reduced to any desired amount by inserting a power attenuator between the circuit (ATN in Figure 9-7) and the amplifier output through the VSWR sensor. The attenuator must, of course, be able to handle the power level in question. However, remember that an attenuator only dissipates part of the power fed into it. A 1 dB attenuator for example dissipates only 10% of the power and, therefore, one with a 100 W rating could be used at a power level of 1 kW, providing its resistor elements can handle the current. An attenuation in dB to produce a specific VSWR between a signal source and a 30:1 load mismatch can be figured as follows: First we must obtain a value for the magnitude of the voltage reflection coefficient (Γ) as:

$$|\Gamma| = \frac{\text{VSWR} - 1}{\text{VSWR} + 1}. \text{ Then RL} = 10 \log_{10} \frac{1}{|\Gamma|^2},$$

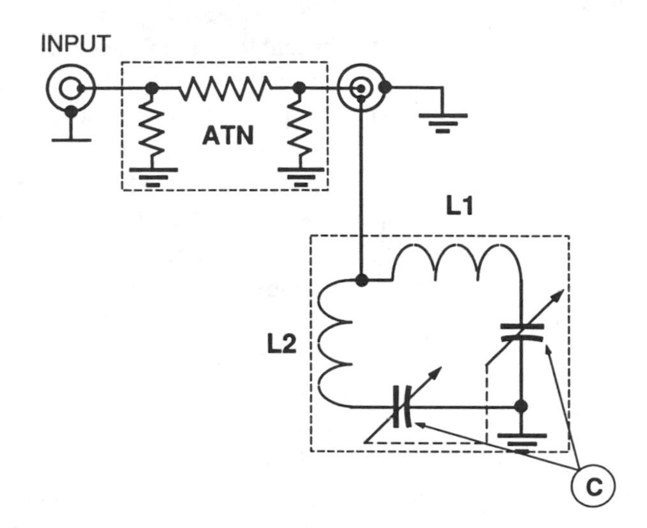

FIGURE 9-7
An LC circuit designed to simulate 30:1 load mismatch. L1 and L2 should not have mutual coupling for proper operation. The attenuator (ATN) has been added to provide mismatches at various levels of SWR.

where RL = return loss.

For the condition of a load return loss of 0 dB (load is open or shorted), then the value of attenuation in front of the open/shorted load needed to achieve a desired VSWR is equal to one-half the return loss that is created by the desired VSWR.

For example, if we wish to create a 5:1 "load" VSWR when the actual load is a short circuit, $|\Gamma| = 4/6 = 0.67$ and $RL = 10 \log_{10} [1/(.67)^2] = 3.5$ dB. Then the value of ATN is 3.5 dB/2 = 1.75 dB.

The VSWR simulator can be made to function over about 30% bandwidth. In a case of such broad band design, the value of L1 should be calculated for the high frequency limit and the value of L2 for the low frequency limit. Thus for testing multi-octave bandwidth amplifiers, a series of these networks would be required. An advantage of this type set-up to create load mismatches is that it can be adjusted to any phase angle, and different phase angles are needed to simulate loads that may exist in applications such as laser drivers, plasma generators, communications equipment, and certain medical instrumentation. For the VSWR sensor described, a PC layout, component placement diagram, and component values are available.[4]

OUTPUT FILTERING

Filtering for output harmonic reduction is required virtually with all solid state RF power amplifiers and especially those intended for radio communications. The specifications depend on the application, frequency of operation and the power level. Wideband amplifiers have much higher harmonic content than narrow band ones, the worst cases existing at the lowest frequencies. An amplifier with a bandwidth of 10–175 MHz, for example, would not only pass the second harmonic up to 85–90 MHz and the third harmonic up to about 60 MHz but these frequencies will be amplified along with the fundamental frequency. The harmonics generated at higher than these frequencies fall outside the amplifier's bandwidth and their amplitudes will gradually diminish at increasing frequencies. The rate of decrease depends on the amplifier's gain characteristics outside its intended passband. At 10 MHz, the same amplifier would pass and amplify signals all the way up to the seventh even and fourth odd order harmonic. Viewing the fundamental waveform on an oscilloscope, we would find it to resemble a squarewave more than a sinewave, which is normal considering the wide distribution of harmonics. In a single-ended amplifier, the second harmonic is the most troublesome but in a push-pull design the even order harmonics are suppressed (see Chapter 7). In a push-pull amplifier of 4–5 octave bandwidth, the amplitude of the second harmonic can be 30–50 dB below the fundamental depending on the circuit balance, but the third order harmonic may be attenuated only 10–12 dB.

Although bandpass filters can be used for RF power amplifier output harmonic filtering in narrow band systems, such filtering is almost exclusively done with low-pass filters. The purpose of this section is to familiarize the designer in using the so-called normalized tables for designing low-pass filters. These tables are available in a number of textbooks and other publications, and can be used for designing filters of high-pass, bandpass or low-pass varieties. Only a few simple calculations are necessary to derive the normalized values from the tables. The filters designed in this manner can be later computer analyzed to verify their accuracy to any required specifications. Although the examples given here are for the HF band, this technique of filter design is applicable up to frequencies where lumped constant elements cease to be practical.

However at high UHF and microwave frequencies, microstrip techniques can be employed to realize the components needed in the low-pass filter design. Chip capacitors can be used to frequencies as high as 800–1000 MHz, but lumped inductors become ineffective above 500 MHz. When calculating lengths of line in microstrip to obtain distributed L's and C's, one must remember that microstrip is not a true transverse electric magnetic (TEM) mode of propagation. The velocity of propagation is reduced by the medium in a manner similar to that encountered in true TEM mode transmission lines such as coaxial cable and stripline, however, to lesser extent. Velocity reduction was discussed briefly in Chapter 6 where it was stated that $\lambda_m = \lambda_o/\sqrt{\epsilon_r}$. An identical relationship exists for velocity: $v_m = v_o/\sqrt{\epsilon_r}$.

In the case of microstrip, one usually talks about an "effective" relative dielectric constant ϵ_{eff} which is related to ϵ_r by a formula involving the W/h ratio of the microstrip line.[18] The formula says

$$\epsilon_{eff} = \frac{\epsilon_r + 1}{2} + \frac{\epsilon_r - 1}{2}\left(1 + 12\frac{h}{w}\right)^{-1/2}$$

It is apparent from the formula that ϵ_{eff} approaches ϵ_r as the term h/W becomes small, which is the condition occurring as the characteristic impedance of the line approaches zero. It is common when designing microstrip circuits to use curves that show the effective relative dielectric constant as a function of W/h of the microstrip as shown in Figure 9-8.[19] By using the "effective relative

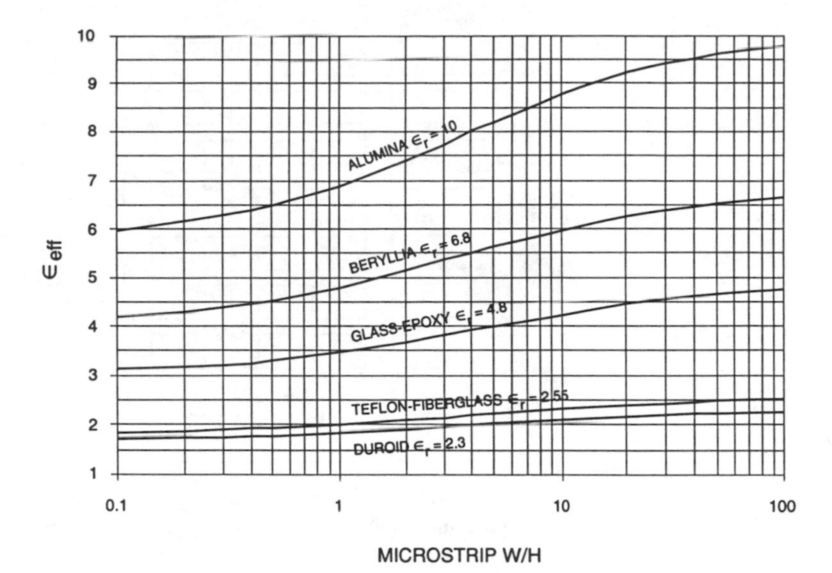

FIGURE 9-8

Effective dielectric constant versus the width-to-height ratio for some of the most popular microstrip substrate materials.

dielectric constant," one can still refer to a velocity v_{eff} in the medium that is given by the expression:

$$v_{eff} = v_o / \sqrt{\epsilon_{eff}},$$

where v_o = velocity of light in a vacuum (3×10^8 m/sec) and ϵ_{eff} = effective relative dielectric constant.

The length l of a microstrip line for a specific inductance is[17]:

$$l = L\, v_{eff}/Z_o$$

where L = inductance in nH, v_{eff} = effective velocity in the medium given in the expression above, and Z_o is the characteristic impedance of the line in ohms. If we select the desired value of L to be 2.5 nH and use microstrip line made with alumina ($\epsilon_r = 10$) having a $Z_o = 10$ ohms, then $\epsilon_{eff} = 8.9$ (see Figure 9-8) and $v_{eff} = 30/\sqrt{8.9} = 10.1$ cm/ns. From the formula stated above, $l = (2.5 \times 10.1)/10 = 2.53$ cm (0.99"). A plot of inductance for different line impedances is shown in Figure 9-9. For convenience, ϵ_{eff} normalized to ϵ_r is also plotted on the vertical axis. Figure 9-9 clearly shows how the value of ϵ_{eff} approaches ϵ_r as line impedance approaches zero.

Note that in the example given above, $Z_o = 10$ ohms and Z_{in} and Z_{out} are assumed to be 50 ohms. Figure 6-13 as well as other references[7, 8] gives W/h = 10 for 10 ohm line if the dielectric material has a relative dielectric constant of 10. In practice, if the line impedance is not more than 10% of Z_{in} and Z_{out}, i.e., less than 5 ohms, the value of ϵ_{eff} can be assumed equal to ϵ_r. Because the characteristic impedance of 50 Ω is an industry standard for RF power test systems and equipment, all subsequent material and data refer to this impedance unless otherwise noted.

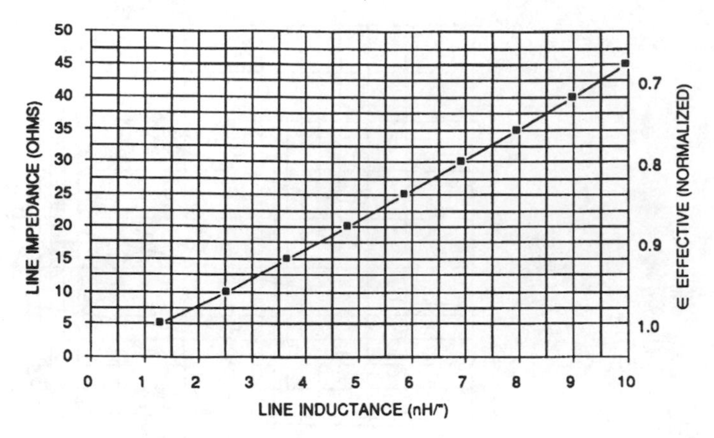

FIGURE 9-9

A graph showing a conversion of microstrip line impedance to inductance per unit length. The scale at the right shows the ϵ_{eff} relative to ϵ_r vs. line impedance.

TYPES OF LOW PASS FILTERS

There are two basic types of low pass filters commonly used to suppress harmonics of the desired frequency in the RF output of solid state amplifiers, the *Chebyshev* and *Butterworth* designs. There are several variations of each, which include the elliptic function filter (also called the Cauer-parameter filter), the constant-K filter, and the m-derived filter. Each has its own characteristics, performance, advantages and disadvantages. In addition these can be divided into inductive input and capacitive input categories. In RF power applications the most commonly used configurations are either a straight Chebyshev or one modified to provide the elliptic function characteristics. The elliptic function design has a sharper cut-off, but lower far band attenuation (Figures 9-10 and 11). It also provides deep and sharp notches in the out of band attenuation. Some of these notches can be fine-tuned to the specific harmonic frequencies to improve the out of band attenuation, but at a cost of decreased return loss. It is also only practical in fixed frequency or very narrow frequency range applications. With a plain Chebyshev or Butterworth filter, it would be difficult to obtain sufficient attenuation, especially for the second harmonic, except in the case of a well balanced push-pull circuit which provides even order harmonic suppression by itself due to the cancellation effect of the push-pull configuration, as mentioned earlier.

Low pass filters have—in general—the characteristics that unwanted harmonics are attenuated by reflecting them back to the source. This results in an alteration of the impedance for the harmonics, which originally was 50 Ω without the filter. If the low pass filter is a series L input type, it presents a high impedance for the harmonics. If it is a shunt C input type, the harmonics will see a low impedance load. In each case the linearity and efficiency of the signal source are affected depending on the original harmonic amplitude.

It has been stated in the literature[16] that a series L input low-pass filter, when used in conjunction with a solid state amplifier has superior efficiency characteristics. On the other hand, the shunt C input filter would result in better linearity and harmonic suppression, but it has been noted that the shunt C input results in abnormally high RF currents in the line connecting the signal source to the filter. Since in the shunt C input filter the harmonics are directed to ground, the amplifier will see a lower than optimum output load line, especially at the harmonic frequencies.

The effect of the shunt C input is reduced when inductance is added between the amplifier output and the filter input. Also, a transmission line will convert the shunt C input to a series L input type. In a series L input filter the harmonics are either dissipated in the series inductance of the filter or reflected back to the amplifier output, thus presenting a higher than normal load line to the amplifier. How severe the above characteristics are for each filter depends on the initial harmonic content of the amplifier and on the frequency of operation in the passband of the filter. The exact conditions under load mismatch in each case depend on the phase angle of the mismatched load in addition to the phase delay of the filter.

One solution to avoid the problems described in the preceding paragraph is to use a *diplexer* in the amplifier output.[11] A diplexer is nothing but a dual set of filters, one low pass and one high pass. Their cut-off frequencies are designed so that the low pass filter passes only the fundamental frequency to the primary load. The high pass filter presents a high impedance to the fundamental, but passes the harmonics and "dumps" them into a secondary load. This would be an ideal solution, since 50 Ω loads would be presented to both the fundamental frequency and the harmonics. The secondary load is only required to handle the power level of the harmonics, which is typically up to about 10% of the power level at the primary load. In spite of its advantages, the diplexer is practical only in single frequency or extremely narrow spectrum applications, except for certain military equipment where space and economy are of less importance.

THE DESIGN PROCEDURE

The design of low-pass filters using tables of normalized element values is relatively straightforward. All that is required is to set the specifications (refer to Figure 9-10):

1) The cut-off frequency f_p (in MHz).
2) The stop-band frequency f_s (in MHz).
3) The reference low-pass $(W_s)= f_s/f_p$.
4) The required stop-band attenuation A_s (in dB) at frequency f_s.
5) The allowable return loss RL (in dB) or voltage reflection coefficient $|\Gamma|$ (in %), which determines the band-pass ripple A_p (in dB).
6) The input and output impedances (Z).

Tables for normalized element values are available from several sources.[7, 10, 11, 12] From the tables one can select a filter type and determine its degree (how

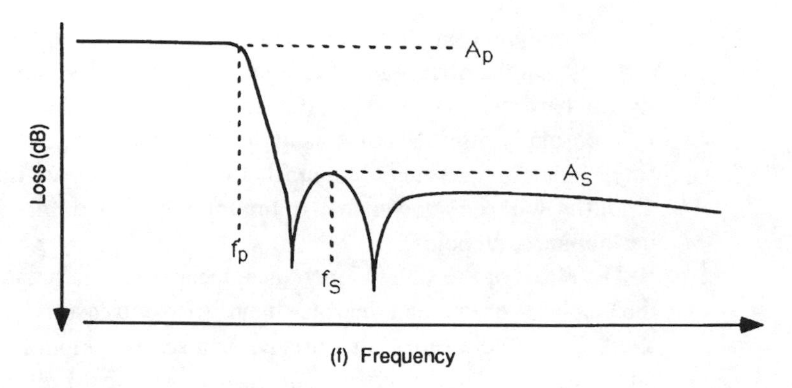

FIGURE 9-10

A typical response curve of a six element elliptic function low-pass filter. The A and f designations refer to the definitions given in text.

many elements are required to satisfy the conditions specified). Sometimes a compromise is necessary in order to reduce the number of elements and the complexity of the filter, which results in only a minor degradation in performance characteristics. Normally eight filters are required to cover the 1.6–30MHz HF frequency spectrum without gaps even if the sharper cut-off elliptic designs are used. However, sometimes frequency bands such as the ham radio spectrum up to 30 MHz (1.6, 3.5, 7.0, 14, 21, and 30 MHz) can be covered with only six filters of the same type because continuous coverage is not required.

Let us examine a low-pass filter design with $f_p = 32$ MHz and the lowest usable frequency of 24 MHz for example. Then f_s would be 2x24 or 48 MHz, resulting in $\Omega_s = f_s/f_p = 48/32 = 1.50$. A_s is set at –45 dB and the maximum allowable reflection coefficient $|\Gamma|$ at 10%. Due to the reasons discussed earlier, a series L input type filter was selected. There are two different configurations of such filters: one designed with series LC shunt elements and the other with shunt LC series elements. The latter is far more popular and is easier to implement in practice, since it uses a lower number of inductors. In order to make the filter a series L input type, it must be an even degree, meaning that it must have an even number of elements. For an even degree filter, the tables of normalized values for elliptic low-pass filters indicate that at least six elements are required for $A_s = -45$ dB, $|\Gamma| = 10\%$ and $\Omega_s = 1.50$. If the reflection coefficient $|\Gamma|$ is converted into a decimal: 5/100 = 0.050, the mismatch loss or passband ripple can be calculated as $-10\log_{10}(1-(0.05)^2) = 0.011$ dB. This (as well as the actual values) is theoretical, assuming an infinite Q for the elements. There will not be a large change if the loaded Q is 100, but a Q of 20 would result in a considerable loss in A_p and a 1–2 dB decrease in A_s. The largest influence of the reduced Q can be noticed in the depth of the notches, which are not of primary importance.

Next we find the reference L and C values:

$$L_{ref} = \frac{R}{2\pi f_p} = \frac{50}{201} = 249 \text{ nH and } C_{ref} = \frac{1/R}{2\pi f_p} = \frac{0.020}{201} = 99.5 \text{ pF}$$

These values multiplied by the numbers from the normalized tables will result in the actual L and C values required for the filter elements. This makes the values of the actual elements of the filter shown in Figure 9-11 to be as follows: L1 = 225 nH, C1 = 119.0 pF, L2 = 280.0 nH, C2 = 43.0 pF, L3 = 295 nH, C3 = 23.0 pF, C4 = 131.0 pF, and C5 = 70.0 pF. The component values for six filters covering the 1.6 to 30 MHz spectrum are given in Table 9-1. In a practical design these numbers can be rounded to the nearest standard values. The inductors are somewhat limited in their tolerances, since fractions of turns of course cannot be realized and the capacitor tolerances are 5% at best as standard stock items. Figure 9-14 is a computer plot of the expected response of the filter with the component values listed above.

THE COMPONENTS

Although paralleled multiple capacitors in filters are not recommended in some literature[2], the concept was used in this design. No abnormalities were noticed, when the responses were checked with a sweeper and a spectrum analyzer, although this may not be the case at higher frequencies. Paralleling of multiple capacitors provides a means to increase their current carrying capabilities and allows the use of inexpensive disc types. In addition, it is easier to compose values closer to the non-standards required in many instances. High voltage (3000 V) types were selected, since under certain load mismatch conditions they may be subjected to high RF voltages. Typically the RF voltage ratings of such capacitors are approximately 30–35% of their D.C. voltage ratings. The inductors should be located so that the adjacent ones are at 90° angles to each

Table 9-1

	30 MHz	21 MHz	14 MHz	7.0 MHz	3.5 MHz	1.6 MHz
f_p (MHz)	32.0	23.0	14.5	7.50	4.00	2.10
f_s (MHz)	48.0	34.0	20.0	10.0	5.80	3.20
$\Omega_s(f_s/f_p)$	1.50	1.48	1.38	1.34	1.45	1.52
A_p (Passband ripple)	0.044	0.044	0.044	0.044	0.044	0.044
A_s (dB)	−45	−45	−45	−45	−45	−45
C_{ref} (pF)	100.0	140.0	220.0	420.0	760.0	1430.0
L_{ref} (nH)	250.0	349.0	525.0	1050.0	1970.0	3650.0
Normalized values for C_N or L_N	**Actual Values = C_N Cref (pF) or L_NLref (mH)**					
$C_1 = 1.190$	119.0	161.0	262.0	500.0	905.0	1700.0
$C_2 = 0.426$	42.7	60.0	94.0	179.0	324.0	610.0
$C_3 = 1.312$	131.2	184.0	288.0	551.0	997.0	1875.0
$C_4 = 0.232$	23.2	32.5	51.0	97.5	177.0	332.0
$C_5 = 0.696$	69.6	97.5	153.0	292.0	529.0	995.0
$L_1 = 0.899$	225.0	314.0	472.0	944.0	1770.0	3280.0
$L_2 = 1.118$	279.5	390.0	587.0	1174.0	2200.0	4080.0
$L_3 = 1.181$	295.0	412.0	620.0	1240.0	2325.0	4310.0

The normalized values as well as the actual values of the low-pass filters are shown in the lower part of the table. The upper part shows the specified numbers from which the normalized and actual values are derived.

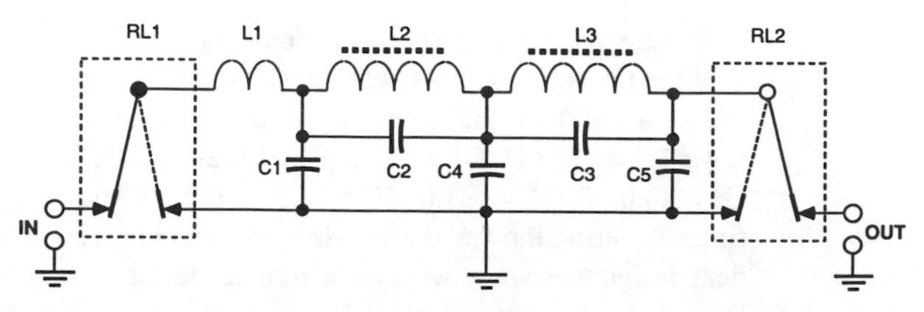

FIGURE 9-11
Schematic of a six element L-input elliptic function low-pass filter. The relays RL1 and RL2 were added to make possible switching of multiple filters.

other, which minimizes the possibility of mutual coupling. L1 here is an air-wound inductor and L2, L3 are wound on phenolic, non-magnetic toroids only for their more convenient shape factor. At lower frequencies (below 15 MHz) all inductors can be toroidal to reduce their physical sizes. The toroid material permeabilities (μ_r) should be 10 or less, limiting the selection to powdered iron since the minimum μ_r of ferrites is around 30.

Filter switching at HF can be done with relays, (Figures 9-11 and 9-12) but they should be of a low contact inductance type such as Omron G2R or equivalent manufactured by Magnecraft and several other companies. A photo of the low pass filter illustrated in Figure 9-12 is shown in Figure 9-13. At higher frequencies, where excessive parasitic inductances become increasingly critical, PIN diode switches are probably the only choice for a designer.

Calculation of the RF voltages and currents in each element would tell us the wire size required for the inductors and what the ratings for the capacitors should be. For certain elements such as L1, C1 and C5 (Figure 9-11), a manual

FIGURE 9-12
An example of the layout of a low-pass filter described in the text. Note the 90° positioning of the inductors and the paralleled capacitors in C1 through C5. Gray area represents a ground plane on top of the board. RL1 and RL2 are provided for switching of multiple filters.

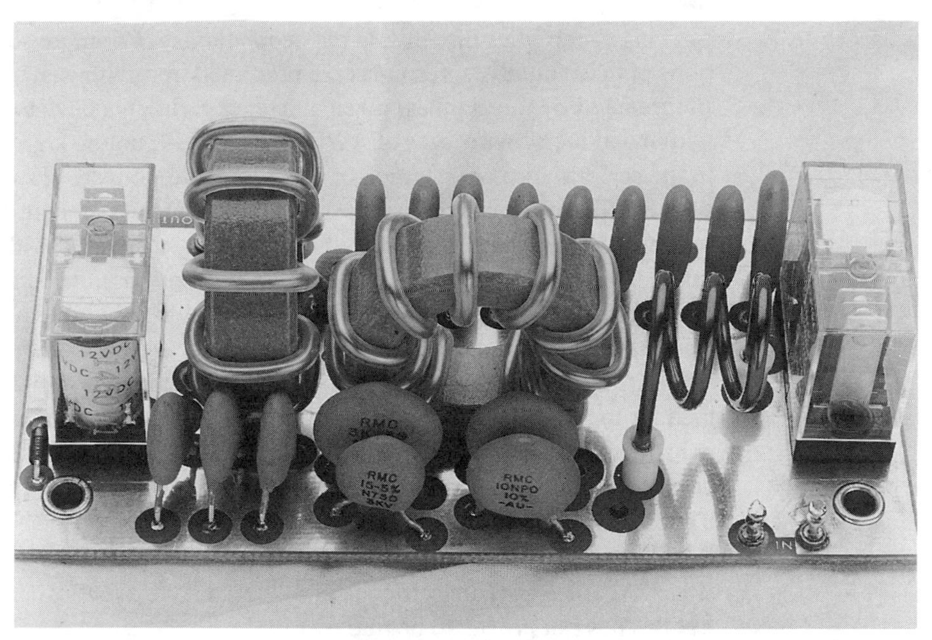

FIGURE 9-13
Photograph of switched filter shown in Figure 9-11.

calculation would be fairly simple and straightforward. However, for the remainder it would take several pages and will not be presented within the context of this book. There are some programs for the PC and Macintosh on the market,[13, 14] which optimize existing designs and plot the S11 and S21, etc. Other programs[15] designate a more comprehensive design and calculate the nodal currents and voltages as well.

In the design example L2 and C4 are exposed to the highest currents and C3 (Figure 9-11) is subjected to the highest RF voltage. At a power level of 2000 W for example, the numbers are 11.7, 11.4 A and 556 V respectively. Since the current carrying capability of copper wire under RF conditions is frequency dependent due to the so-called "skin effect," it would be advisable to employ as heavy gauge wire as practically possible to increase the conductor's surface area. (Skin effect is discussed in more detail in Chapter 6.) In addition to heating effects, the conductor size affects the inductor Q values. The skin depth (d) of a copper conductor is approximately 0.00035" at 100 MHz. Based on this reference, the skin depth versus frequency can then, for practical purposes, be figured as:

$$d = 0.00035 \times \sqrt{\frac{1}{f/100}}$$

where f = actual frequency in MHz and 100 = reference frequency in MHz. The final numbers for 1.6 and 30 MHz are obtained as 0.0028″ and 0.00064″ (0.071 mm and 0.016 mm) respectively. Usually five to six skin depths are considered adequate for good engineering practice, although the RF current carrying capabilities diminish in the deeper layers. For more scientific calculations of these numbers, formulas are presented in references 7, 8, and 9.

Prompted by the numbers given above, experiments have been conducted with the inductor wire sizes of AWG #16 and #14, but no significant differences in the performance or operation of the filter were noticed. However, regarding the high RF voltages in high power filter elements, it is recommended that the wire wound on ferrite or powdered iron toroids be covered with some type of high temperature sleeving such as TFE to prevent possible arcing to the core. High voltage, inexpensive disc capacitors suitable for high power filter applications up to 30–40 MHz are available from several manufacturers.

The filter in the example described in this chapter has been tested and operated for long periods of time at a power level of 2 kW. Common PC board construction techniques were used with a continuous ground plane on the component side, except for clearances provided for feed-throughs (Figure 9-12). The PC board material is G10 epoxy-fiberglass with a dielectric thickness of 0.062″ and 2 oz. copper on each side. In the early design stages, it was questioned whether G10 dielectric material would have excessive losses at 30 MHz, but this has not proven to be true.

At VHF (up to 200 MHz) power levels of 2 kW would be difficult to obtain without the use of high quality components and possibly other than printed circuit board construction techniques. Toroidal inductors as used in a HF design may be impractical and multilayer ceramic capacitors such as ATC type 100E or Tansitor type MPH are recommended. At UHF, the high power filter design requires a completely different mechanical concept, such as etching or deposit-

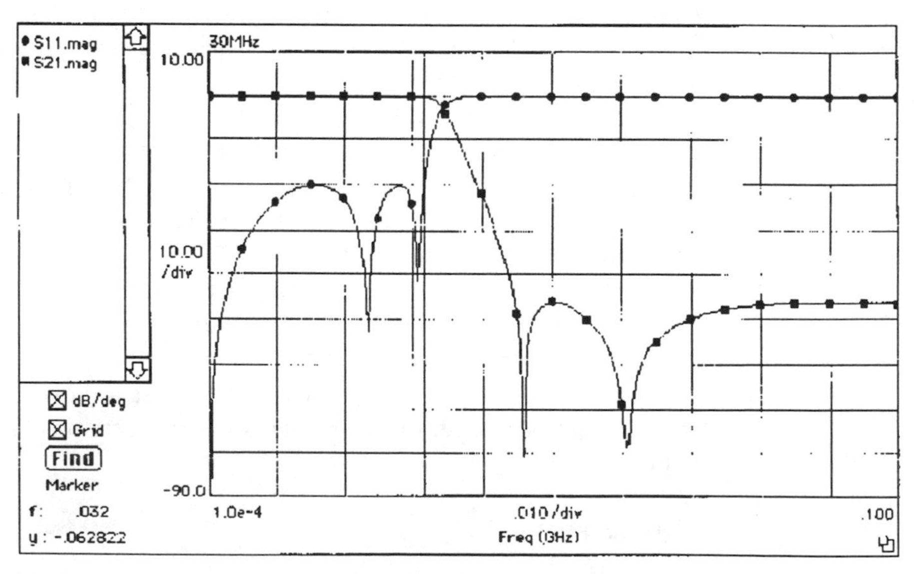

FIGURE 9-14

A computer analysis of the response of the 30 MHz filter shown in Figures 9-10 and 11 with the component values given in the text. The marker is set at 32 MHz and the response is shown up to 100 MHz. S21 is the actual filter response versus frequency and S11 is the return loss in dB.

ing the inductors and capacitors on a low loss substrate. Leadless chip capacitors would also be required to provide any degree of repeatability. A single unit can always be "tweaked in" to meet the specifications, but there may be a problem if the item has to be mass produced. This is especially true regarding the inductors, which must definitely be of an air wound type or stripline construction. As the frequency increases, the power handling capabilities of the components decrease exponentially due to higher losses and shallower skin depth of the conductors.

Filter switching with relays, which can be done at HF with certain types, is not possible at higher frequencies due to the high series inductances of the relay contacts. PIN diode switches are about the only way to switch signals in RF power applications at VHF and UHF, but at higher power levels this technique is costly due to the number of diodes required per filter.

References

[1] Kraus, Bostian, Raab, *Solid State Radio Engineering*, New York: John Wiley & Sons, Inc., 1980

[2] William E. Sabin and Edgar O.Schoenike, *Single-Sideband Systems and Circuits*, New York: McGraw-Hill Book Company, Inc., 1987.

[3] William I. Orr, *Radio Handbook*, 19th Edition, Indianapolis: Editors and Engineers, (Howard W. Sams & Co.), 1972.

[4] H. O. Granberg, "VSWR Protection of Solid State RF Power Amplifiers," *RF Design*, February, 1991.

[5] W. B. Bruene, "An Inside Picture of Directional Wattmeters," *QST*, April, 1959.

[6] Alan R. Carr, "A High Power Directional Coupler," *RF Design*, September, 1989.

[7] Edward C. Jordan, *Reference Data for Engineers: Radio, Electronics, Computers, and Communications*, Seventh Edition, Indianapolis: Howard W. Sams & Co., 1979.

[8] *Filters, A Handbook on Theory and Practice*, White Electromagnetics, Inc., 1963 Edition.

[9] David Festing, "Realizing the Theoretical Harmonic Attenuation of Transmitter Output Matching and Filter Circuits," *RF Design*, February, 1990.

[10] Rudolf Saal, *Handbook of Filter Design*, 1979 Edition, Telefunken Aktiengesellschaft, 715 Backnang (Wurtt.), Gerberstrasse 34, P.O. Box 129, Germany.

[11] *Wave Filters: Their Design and Specifications*, ADC Products, 6405 Cambridge St., Minneapolis, MN 55426.

[12] Geffe R. Philip, *Simplified Modern Filter Design*, 1963 Edition, New York: John F. Rider Publisher, Inc.

[13] Eagleware, 1750 Mountain Glen, Stone Mountain, GA 30087.

[14] Nedrud Data Systems, P.O. Box 27020 Las Vegas, NV 89126.

[15] PSpice by MicroSim Corporation, 20 Fairbanks, Irvine, CA 92718.

[16] J. Mulder, "Applications Laboratory Report ECO 7114," Philips Components, Discrete Semiconductor Group, 1971.

[17] "Step Recovery Diode Doubler," Application Note 989, Hewlett Packard, Palo Alto, CA,1982.

[18] Dr. I.J. Bahl and D.K. Trivedi, "A Designer's Guide To Microstrip Line," *MICROWAVES,* May, 1977.

10

Wideband Impedance Matching

INTRODUCTION TO WIDEBAND CIRCUITS

Multi-octave impedance matching is almost exclusively done with transformers, although bandwidths up to 1-2 octaves may be possible with complex LC networks in conjunction with negative feedback etc.[1, 2, 6, 10] Any type of wideband impedance matching results in compromises in an amplifier performance since device impedances vary with frequency. In addition low impedance RF transformer impedance ratios can only be realized with integers 1:1, 1:4, 1:9, etc. Other impedance ratios are possible, but the structures usually become very complex and some bandwidth will be lost due to increased leakage inductance as a result of the numerous interconnections required. In the input, compromises result in reduced power gain, increased return loss and VSWR, whereas the output shows reduced efficiency, lowered stability against load mismatches and poorer linearity. RLC networks are often inserted between the device input and the matching transformer[1, 2] to compensate for the impedance versus frequency slope as well as for the gain vs. frequency slope.

Using these corrective networks together with negative feedback and additional networks associated with it, it is possible to design amplifiers covering up to five or six octaves from low band to VHF or even UHF.[3, 4, 7] There is very little that can be done in the output to compensate for the output impedance/frequency slope due to excessive power loss. Fortunately, the output impedance variation with frequency with both the MOSFETs and BJTs is usually much smaller than that of the input. Sometimes a low value inductance or a microstrip between the device output and the matching transformer will improve considerably the efficiency at the high end of the frequency range by providing compensation for the device's output capacitance. Normally only "overcompensation" of the output transformer will do an adequate job. This means added capacitance across the transformer primary and in some cases also across the secondary.

A wideband RF transformer performs one or more of any combination of the following functions:

a) Impedance transformation.
b) Balanced to unbalanced transformation.
c) Phase inversion.

RF transformers are most often referred to their impedance ratios rather than primary-to-secondary turns ratios. The former are simply the turns ratios squared. In these applications, we are mostly interested in manipulating impedance rather than voltage or current with the transformers. Basically RF transformers can be compared to low frequency transformers, except that with increasing frequency a parameter called leakage inductance becomes an important factor. Some type of magnetic core is required to extend coverage at the low end of the frequency band. Either powdered iron or ferrite cores are acceptable depending on the frequency range.

Ferrites are the most common magnetic materials used for RF transformers today. There are two basic types of ferrites. *Nickel-manganese* compositions have high permeabilities (μ_r = relative permeability) and are used in low frequency applications, whereas *nickel-zinc* ferrites have lower high frequency losses, but their Curie points can be as low as 130°C and they can be manufactured only with μ's of less than approximately 1000. (Curie point is a temperature where magnetic material loses its magnetic properties). Low μ_r ferrites in general have higher volume resistivity than the high μ_r ones, which means lower eddy current losses.

Detailed information of the behavior of ferrites at RF is rarely available from ferrite manufacturers. Core eddy current losses and winding dielectric losses heat up the core and its temperature must be held well below the Curie point; otherwise, the magnetic properties of the material will be permanently altered. It must be noted that in order not to saturate the core, operational flux densities must be kept well within the linear portion of the B-H curve of the material. Saturation mainly occurs at low frequencies, where most of the coupling is through the core, which would lead to nonlinear operation, generation of heat and harmonics. The area inside the B-H curve normally represents the relative loss and, therefore, narrow curves would be preferred for low loss designs. This situation, however, is confusing since these curves are usually created under D.C. conditions and do not really give the required data for an RF designer.

High μ ferrites, although having higher saturation flux densities than the low μ ones, saturate easier under RF conditions. One reason for this is that high μ cores require a smaller number of turns than low μ ones to satisfy the minimum magnetizing inductance requirement. Thus, it is advisable to employ a ferrite core with relatively low μ and added number of turns in the windings at least to the extent that the added interwinding capacitance can be tolerated at the higher frequencies. As a general rule, the winding reactance should be at least twice the impedance across it. A general formula for calculating the maximum flux density of a ferrite core is:

$$B_{max} = \frac{V_{max}}{2\pi fAn}(10^2),$$

where

B_{max} = Maximum flux density (gauss),

V_{max} = Peak voltage across the winding,

f = Frequency in MHz,

A = Core cross sectional area in cm^2,

n = Number of turns.

With RF transformers either the primary or the secondary can be used for the B_{max} calculations but the 50 Ω side (if applicable) is commonly used for convenience and standardization. Then $V_{max} = \sqrt{2PR}$, where P = RF power level and R = resistance (50 Ω). Example: If V_{max} = 50V, f = 2.0 MHz, A = 1.0 cm^2 and n = 4, then $B_{max}(50/50.2)(10^2) = 99.6$ gauss.

Note that in certain types of transmission line transformers the RF voltage (V_{max}) used in the B_{max} calculations is lower than the value obtained from the V_{max} formula given above. The maximum voltage across the winding(s) must be divided by the number of line segments connected in series in the transformer configuration in question[1,2]. The division ratio for example would be 2 for the transformer in Fig. 10-9B and 4 for one in 10-9C. For the transformer in Fig. 10-9D the division ratio is 3 since two of the four line segments are connected in parallel. Conversely, the same result can be reached in the formula for B_{max} if the full voltage across the 50 Ω terminals is used for V_{max} as the numerator and n is multiplied by the number of line segments in series.

Since high permeability ferrites in general saturate easier than low permeability ones, it is good practice to limit their maximum flux densities as follows:

1) B_{max} of 40–60 gauss per cm^2 of cross sectional area for ferrites with μ's of 400–800.
2) B_{max} of 60–90 gauss per cm^2 of cross sectional area for ferrites with μ's of 100–400.
3) B_{max} of 90–120 gauss per cm2 of cross sectional area for ferrites with μ's of < 100.

Regarding the B_{max} vs. μ figures above, it is assumed that the magnetic path is solid e.g. without air gaps, such as found in toroids and balun cores.

At low frequencies, the leakage inductance is virtually unknown and most designers are unaware of such a term. However, it is the parameter that limits the high frequency response of an RF transformer. The performance of RF transformers becomes more critical at low impedance levels, where tight coupling between the windings is of utmost importance. The leakage inductance is a product of the coupling between the primary and secondary and any exposed area in either winding. It is also affected by interconnection lead lengths and mutual coupling. The leakage inductance (or reactance) is difficult to calculate, but it can be measured for each individual case with a vector impedance meter, a vector voltmeter, or a network analyzer. Ideally, when one winding is shorted with a low inductance path, measurement in the other winding should show essentially zero R and phase angle, but in practice this is never the case. A deviation from zero in the value of the resistive component and phase angle can be used to calculate the leakage inductance or rather the high frequency performance of the transformer. It is difficult to relate the leakage inductance directly to the RF performance of a transformer because it is impedance level dependent.

At VLF (50–500 KHz), where high μ cores are required, one may encounter a problem, which may appear unexplainable. It is called *magnetostriction*, which means a magnetic resonance of the ferrite core. The core can chatter and disintegrate at its resonant frequency. There are many resonant modes such as longitudinal and torsional, etc. The only cure to this is to select the physical core size and shape which has its resonances outside the critical frequency spectrum.

CONVENTIONAL TRANSFORMERS

The simplest type of RF transformer is the so-called *conventional* type. There are several kinds of conventional type RF transformers, some of which are more suitable for certain applications than others. In all the basic principle is roughly the same, namely that low frequency coupling between the primary and the secondary is provided through the flux of magnetic media (core) as in audio transformers. At high frequencies tight capacitive coupling between the windings is essential and the magnetic core has little effect except in the form of dielectric losses. It is obvious then, that the quality of the magnetic media employed is a very important factor in designing all types of RF transformers.[3, 4, 5, 8, 9]

Thus we come to another compromise: whether to use higher permeability material in the core and suffer the high frequency losses or design around the losses from the increased stray capacitances caused by additional turns in the windings required when using low permeability cores. A few tenths of a dB of unnecessary power loss in an output transformer can mean a significant increase in power consumption and device dissipation.

The conventional transformer is inferior in performance to a transmission line transformer. The difference is mainly in the power handling capability, loss factor, and bandwidth. The conventional RF transformer, however, can be constructed for a wider range of impedance ratios than the transmission line type. Some ratios will have wider bandwidths than others due to the number of turns required to achieve the desired turns ratio. There are no fractional turns, as in all transformers. If the wire passes through the core, one full turn is completed. Figure 10-1 shows a model of a conventional RF transformer where:

L_{LP} = leakage L, primary

L_{PP} = parallel L, primary

L_{LS} = leakage L, secondary

L_{PS} = parallel L, secondary

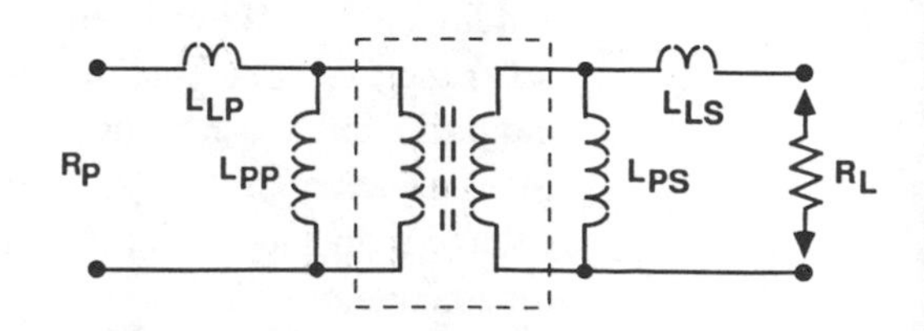

FIGURE 10-1
Equivalent circuit of a conventional, discrete winding RF transformer. See text for nomenclature.

Stray capacitances have been omitted since a relatively low impedance case is assumed and the capacitive reactance arising from applicable construction techniques rarely becomes appreciable in comparison with the low values of resistances involved. Figure 10-2 shows a conventional RF transformer that finds wide usage at high impedance levels (200 Ω and higher) in low power designs. It is a configuration in which one winding is simply wound on top of the other, which usually provides good enough coupling at these impedance levels up to UHF. The most convenient core shape is a two-hole balun, although toroids can be seen in some designs if a sufficient number of turns is provided on the periphery for the coupling required. As in all RF transformers, the wire size also has an effect on the coupling between the primary and secondary. The heavier the wire size, the tighter the coupling will be. But this increases the mutual winding capacitance, resulting again in a compromise. This capacitance can be lowered by using a high μ magnetic core, in which case the core losses would be higher. Since the mutual winding capacitance has a larger effect at higher impedance levels, the designer must determine which approach is most beneficial for a specific application.

Probably the most popular conventional type of RF transformer is shown in Figure 10-3.[12, 13] The one turn winding consists of metal tubes going through sleeves or stacks of toroids of suitable magnetic material. The tubes are electrically connected together in one end of the structure and separated in the opposite end, where the connections to the one turn winding are made. In practice these connections are usually made with pieces of single sided metal clad laminate with proper patterns etched in the metal. This kind of construction also results in a physically sturdy structure with all of its components intact. To make up a transformer, a required number of turns of wire is threaded through the two tubes to form a continuous multiturn winding. This results in a tight coupling between the two windings with relatively low mutual winding capacitance, thereby allowing its use at very low impedance levels.

The wire ends of the multiturn winding can be exited from either end of the transformer, whichever is physically most convenient. The most popular

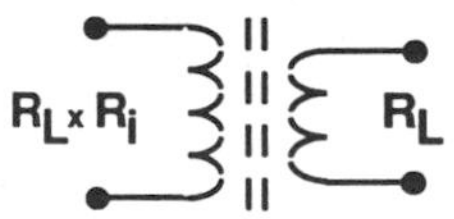

FIGURE 10-2
The simplest form of conventional transformer. The windings are usually randomly wound one on top of the other. It finds its uses at high impedance levels, 200 Ω and up, which dictates the frequency response of the unit.

FIGURE 10-3
The most common conventional type of RF transformer. One winding consists of metal tubes shorted in one end, thus forming only one turn. This limits the impedance ratios to integers 1:1, 4:1. 9:1, etc. It is fairly efficient at impedance levels down to 2–3 Ω if properly constructed, and may have a bandwidth up to 50 MHz.

arrangement is to have the primary and secondary terminals at opposite ends as shown in Figure 10-3. The transformer type shown in Figure 10-3 has the disadvantage that due to the one turn winding, only integer-squared impedance ratios such as 4:1, 9:1, 16:1, etc. are possible. It would be logical to think that fractional integers are possible by threading the winding wire through one tube one more time than the other, but this offsets the balance and the transformer will not function properly. The bandwidth is actually determined by the impedance ratio. A 9:1 impedance ratio transformer is usable up to 50–60 MHz, but higher impedance ratios reduce the bandwidth rapidly because of the increasing leakage inductance. A 25:1 transformer of this type performs poorly at 30 MHz, and a 36:1 unit is usable only to 15–20 MHz.

The form factor, i.e., the length to width ratio, is important. If the transformer structure is short, the coupling between the windings is lowered and the leakage inductance is increased. In the other extreme, if the unit is long, the mutual winding capacitance is increased and the physical length of the multiturn winding may result in resonances within the desired spectrum. Another disadvantage with these transformers is that when used in an amplifier output, the one turn winding makes the magnetic core saturate at a low flux density (see the B_{max} given earlier). Despite all these disadvantages, one turn transformers are widely used in both input and output matching in the 2–30 MHz frequency range and at power levels up to 100–150 W, and as input matching transformers to even higher frequencies. A clear advantage with this type of transformer is its simple construction, which makes it inexpensive and easily mass-produceable.

Other variations of the conventional transformer are shown in Figures 10-4 through 10-6. In these variations, impedance transformation is obtained by connecting a number of windings in parallel on one side and in series on the other side.[15] The transformer shown in Figure 10-3 has one turn in the low impedance winding, limiting the possible impedance ratios to full integers. The windings are made of segments of coaxial cable and the structure is formed into a shape of a "U" or a circle (see Figures 10-5 and 6). The leakage inductance is lower than with most other conventional type transformers, making it usable up to 200–300 MHz. The high frequency end is limited by the physical size of the structure because the length of the high impedance winding must as a general

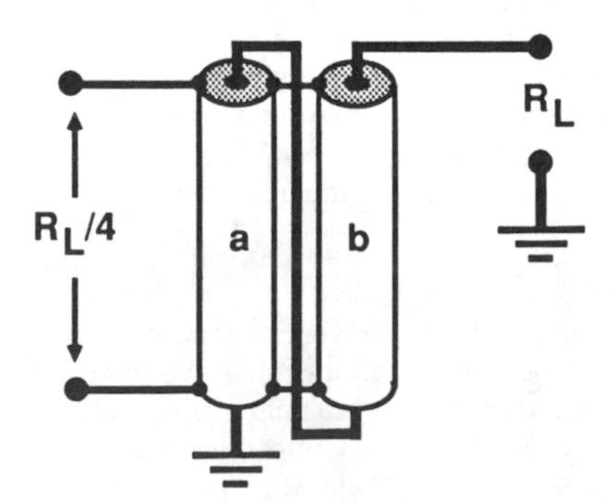

FIGURE 10-4

Another form of conventional type transformer. a and b are segments of coaxial cable, which in practice are "bent" in order to get the terminals of the low impedance winding close together (see Figure 10-5).

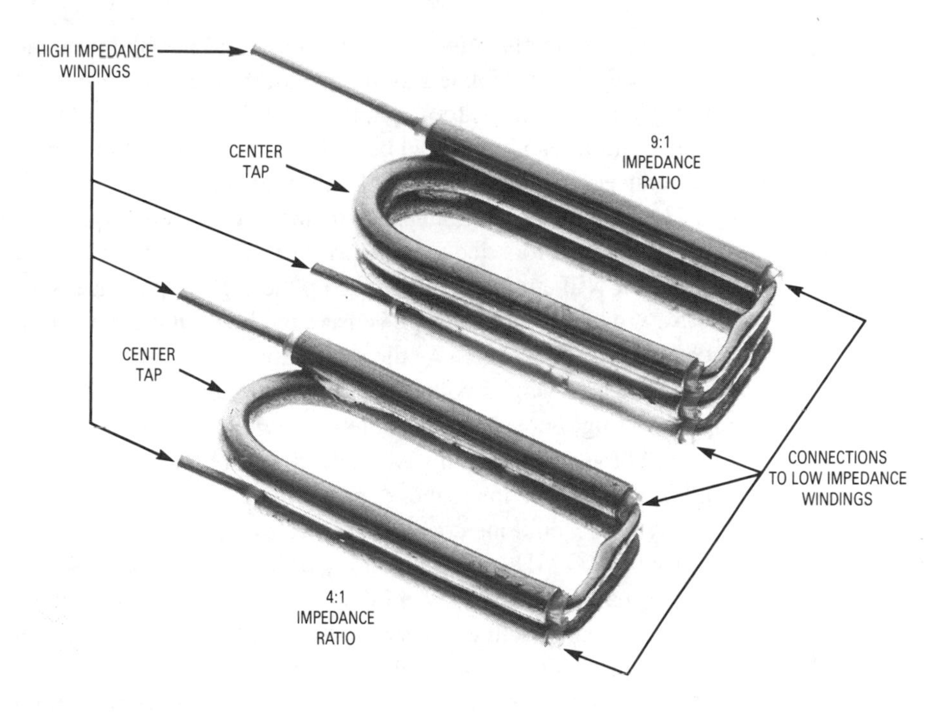

FIGURE 10-5
One possible physical realization of the transformer described. Note the height of the segment stacks with increasing impedance ratios. This produces a delay from the connection points of the low impedance winding to the uppermost segment.

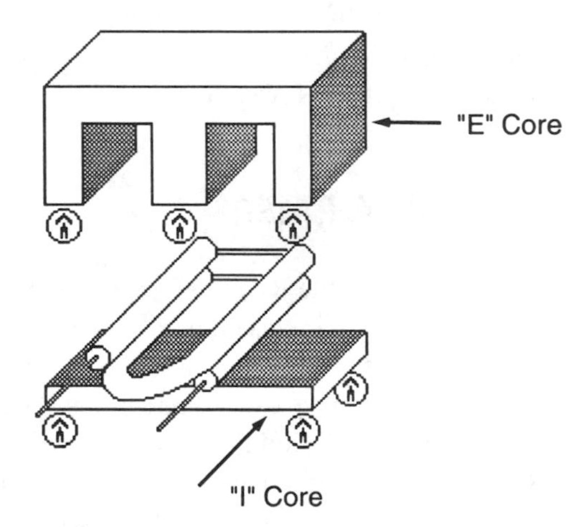

FIGURE 10-6
The transformer shown in Figures 10-4 and 10-5 provided with a magnetic core ("E" and "I") to broaden its low frequency response. The arrows indicate points where epoxy can be applied to make the unit a solid structure.

rule be kept below 1/8 wavelength at the highest frequency of operation, in order to avoid major resonances. Thus, the physical length of a "U"-shaped 4:1 unit is limited to about 3.5 cm and a 9:1unit to 2.5 cm for operation up to 200 MHz. The characteristic impedance of the coaxial cable determines the coupling coefficient between the windings and the optimum closely follows the line impedances calculated for transmission line transformers to be described. If the cable impedance is too high, the result is reduced bandwidth. If it is too low, the maximum bandwidth can be realized, but at a cost of capacitive reactance and reduced efficiency in case of output matching.

The transformer segments can be made from semi-rigid coaxial cable with all outer conductors tied together to form the low impedance side. The inner conductor will automatically make up the high impedance winding (Figure 10-5). If a "U"-shaped design is used, the bending radius should be as small as possible, but it is limited by the minimum recommended for the specific cable used. The best way to connect the inner conductor segments together would be spot welding, but soldering (preferably with high temperature solder) is adequate. Some commercially available units employ tiny PC boards at the front end of the cores to make these connections.

A typical 3 cm long coaxial cable transformer has a low frequency response of around 100 MHz without employing a magnetic core. With an "E" and "I" core of material (as shown in Figure 10-6) having a value of $\mu = 125$, for example, the response will be lowered to 3–10 MHz depending on the impedance ratio. There is only a physical limit to the highest practical impedance ratio. If too many line segments are stacked, the structure becomes high, which makes it difficult to make the electrical connections to all segments without introducing excessive phase delay to the uppermost ones. This of course depends on the cable diameter, but for a power level of 200–300 W a cable diameter of 0.090" or 2.3 mm (a standard with most manufacturers) can be considered a minimum. This would make the highest practical impedance ratio 9:1. If 16:1 or higher is required, a smaller diameter cable must be used, and the power handling capability is consequently lowered.

TWISTED WIRE TRANSFORMERS

A unique and versatile RF transformer can be realized with twisted wires.[14, 17] Enameled magnet wire is commonly employed since it has a thin but good temperature resistant insulation. It is also available with Teflon™ insulation for use in very high temperature applications. The characteristic impedance of a twisted wire transmission line is determined by the wire size, dielectric constant of the insulation and the number of twists per unit length. The latter has the least effect on the line impedance[14] (assuming the wires do not separate from each other in the winding process). A simple method of approximating line impedance is by measuring its capacitance per unit length and comparing it against a line of known impedance.

The most common twisted wire transmission line is a single pair of wires. If the wire size used is #28 AWG, the characteristic impedance will be approximately 50 Ω. Lower line impedances are possible by using heavier gauge wire or by replacing each single wire with a multiple of smaller gauge wires. In those cases where multiple numbers of smaller gauge wires are used to form a twisted wire transmission line, location of the wires with respect to each other should maintain a symmetry as shown in Figure 10-11A and B.

The twisted wire transformers discussed here do not have a defined line impedance except in case of Figure 10-7D. From Figures 10-7B and C, one can notice the versatility of these transformers. In addition, many more odd impedance ratios are possible.[17] Figure 10-7A is a normal 1:1 balun, but a magnetizing winding has been added (center). If the balun's load is balanced—as in the case of feeding FET gates in a push-pull amplifier, for example—the magnetization current flows through only one winding and only one half of the load. This causes undesirable phase and amplitude unbalance in the balun, restricting the bandwidth. The balance can be restored with a third or tertiary winding (shown in Figure 10-7A) to shunt the magnetization current around the load.

Figure 10-7D is a standard 4:1 transmission line transformer, in which the required line impedance is $R_L \times 2$ or 25 Ω for $R_L = 12.5\ \Omega$. This can be best achieved with two twisted pairs of #28 AWG magnet wire, each pair connected in parallel by shorting at both ends. The two pairs are twisted together to form the low impedance transmission line. It is customary to locate the pairs with respect to each other as shown in Figure 10-11. In the twisted wire transformers shown in Figures 10-7B, C, and Figures 10-8A through D (as well as others not shown), there are no defined line impedances, as stated earlier. Since there are no data available, a designer should experiment and make measurements with various interconnection combinations of the twisted wires. Although not shown in the figure, all terminals are referenced to ground.

Another example of the versatility of twisted wire transformers is that they

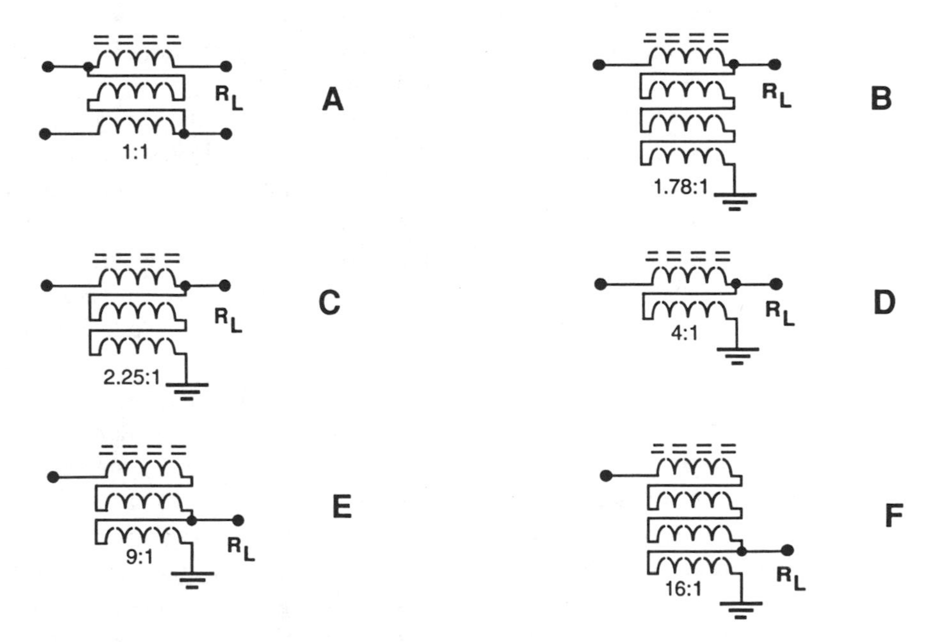

FIGURE 10-7

Conventional RF transformers using multiple twisted wires. A wide variety of impedance ratios are possible depending on the number of wires used and connection configurations. Although not shown, the terminals are referenced to ground.

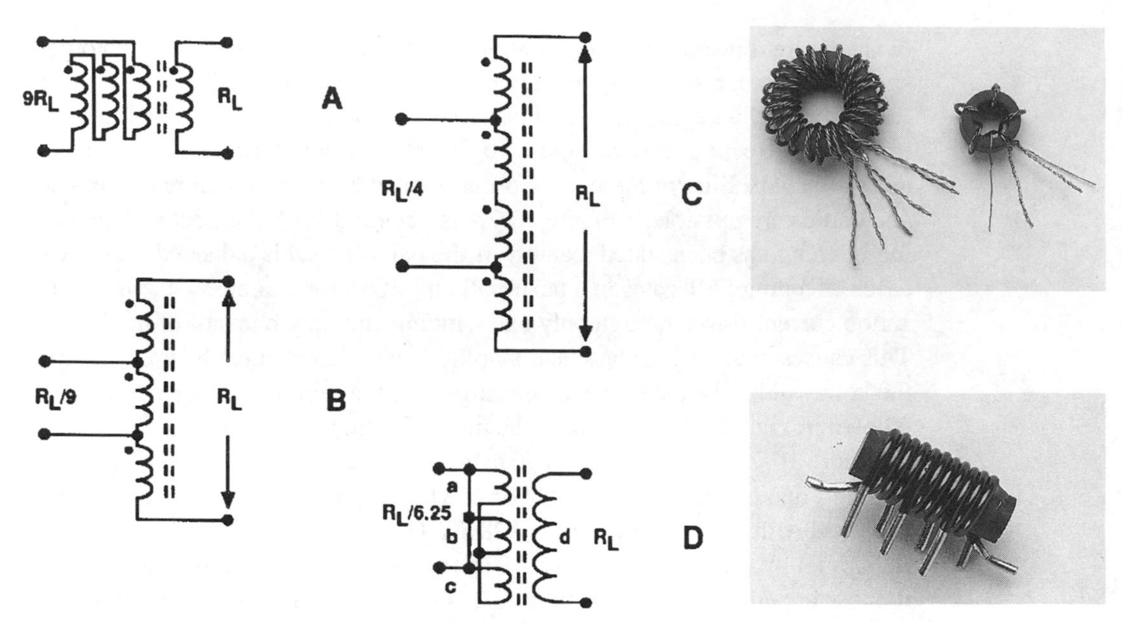

FIGURE 10-8

Conventional transformers that provide balanced-to-balanced function and isolated primaries and secondaries. A to C are twisted wire types, while D is a unique single wire transformer with its paralleled low impedance windings interlaced between the turns of the high impedance winding. The schematic for the 2.25:1 balanced transformer at left in the photo accompanying C is not shown.

can also be connected in balanced-to-balanced, and even in isolated primary and secondary configurations, to provide a number of impedance ratios. Some typical designs are shown in Figure 10-8A to C. Many other fractional integer impedance ratios are also possible with additional numbers of wires. These units can make compact interstage matching elements in push-pull circuits and are ideal especially when D.C. isolation between the stages is required. Figure 10-8C can be considered to represent a transmission line transformer if the line impedance is correct (25 Ω in this case), which is also the case with the ones shown in Figures 10-7A and D. The twisted wire transformers have bandwidths higher than most other conventional transformers. Up to 7 octaves have been measured at 50 Ω and lower impedance levels, and at least 1 octave higher when the impedance levels are higher and the transform ratios low. Advantages of these transformers are their versatility to odd impedance ratios. Disadvantages are limited power handling capability and, in some cases, difficulty of construction with all the multiple interconnections.

Although the transformer shown in Figure 10-8D is not one of a twisted wire type, its description fits better here than with other conventional transformers since it uses capacitive coupling to a larger degree than magnetic coupling. It represents a unique concept, where several 2–3 turn low impedance windings are connected in parallel and interlaced between the turns of the high impedance winding. Heavy gauge enameled wire (#18–16 AWG) is usually employed, which increases the capacitive coupling between the windings and makes the unit a self supporting structure. The windings are wound on some type of cylindrical core such as a length of ferrite rod (see photo) and all the winding connections are made when the transformer is mounted to a PC board. Multiple impedance ratios are possible depending on the number of turns in the low impedance windings. The number of turns in winding "d" (Figure 10-8D) should equal a+b+c–1 in order not to have extra uncoupled turns to windings a, b and c. These transformers have been in commercial use in equipment operating up to 175 MHz and at power levels of 100–120 W. Variations such as flat ribbon wound units have been experimented with, but their fabrication is more difficult and no significant improvement in performance has been found. Obvious advantages of the single wire transformer are its extremely compact size versus power handling capability and the D.C. isolation between the primary and secondary.

TRANSMISSION LINE TRANSFORMERS

Transmission line transformers[1, 2, 6, 10] are quite different from the conventional ones in many ways:

1) The line impedance must be correct for the type of transformer in question in order to take advantage of its optimum performance.
2) At high frequencies, the series reactance combines with the interwinding capacitance and the circuit behaves as a transmission line, greatly extending the high frequency response.
3) The power transferred from the input to the output is not coupled through the magnetic core (except at very low frequencies) but rather through the dielectric medium separating the line conductors. This is an important point regarding the transmission line transformer principle.
4) From 3), it follows that a relatively small cross sectional magnetic core can operate unsaturated at very high power levels.

In practice, transmission line transformers can be realized with twisted enameled wires, coaxial cables, paralleled flat ribbons (separated by a dielectric medium), or a microstrip on a two sided substrate. The practicality and convenience in each case depends on the exact application and frequency spectrum. The simplest transmission line transformer is a quarter wavelength line whose characteristic impedance (Z_o) is chosen to give the correct impedance transformation. This transformer is a relatively narrow band device and valid only at frequencies for which the line is an odd multiple of a quarter wavelength. This transformer is pictured with power splitters and combiners in section 4. A 1:1 balun is shown in Figure 10-9A, in which the line impedance (Zo) = R_L. The low frequency performance is limited by the amount of impedance offered to common mode currents. This impedance should be at least twice the load impedance and can be increased with a core of suitable magnetic material. The inductance of a conductor is in direct proportion to its relative permeability. As the line length limits the high frequency response of transmission line trans-

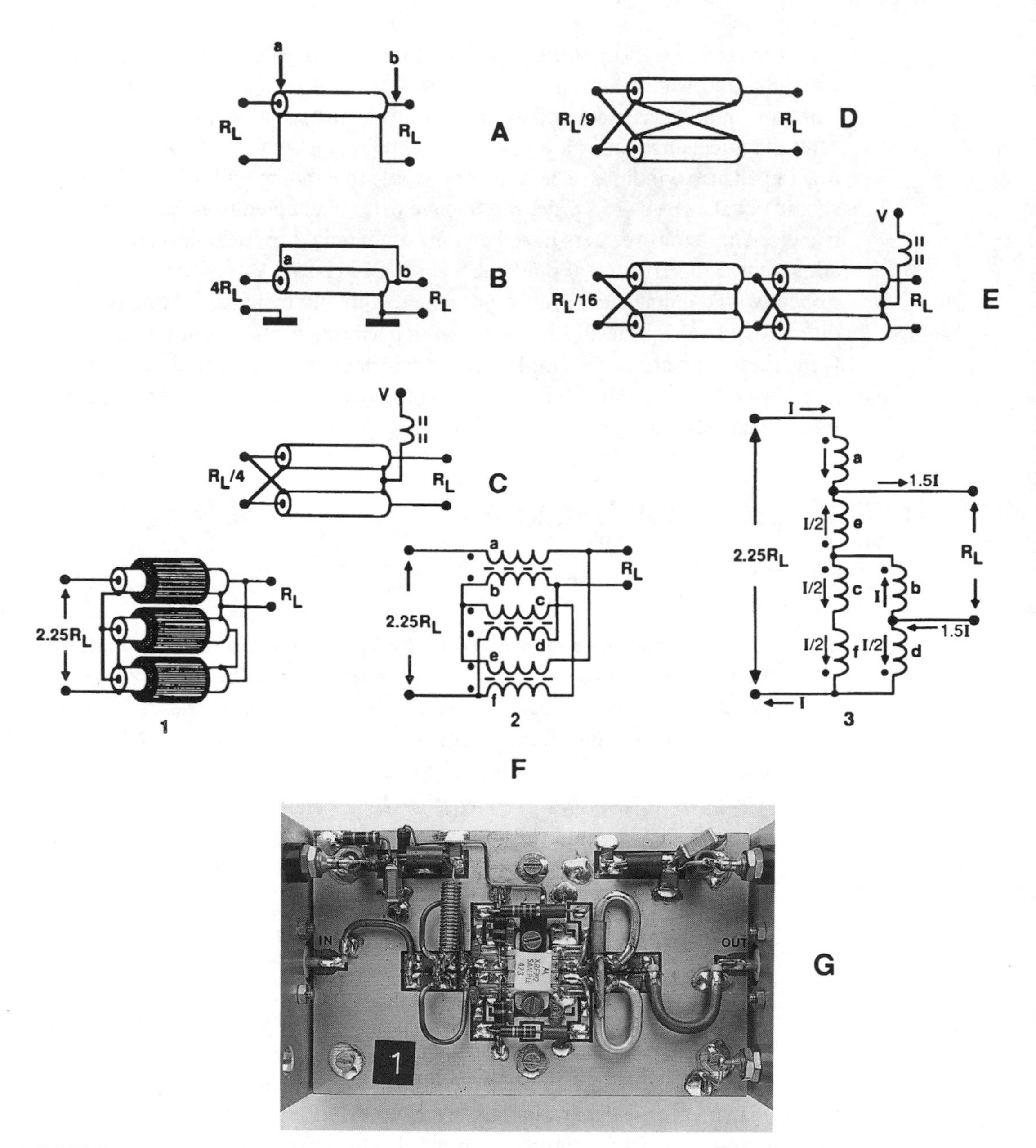

FIGURE 10-9

Some examples of transmission line transformers. For simplicity, most are shown without magnetic cores and can be used as such in many VHF and UHF applications. All the lines must be formed into a physical shape which minimizes the lengths of the interconnections for reduced leakage inductance. The photo at G shows a UHF push-pull amplifier using transmission line transformers in its input and output matching. The input at the left uses a 4:1 as shown in C and the output at the right uses a 1:9 transformer as shown at D.

formers, these two seem to be in direct conflict and we must remember the 1/8 wavelength rule discussed earlier, which applies to all RF transformers.

The most commonly used material for transmission lines in these transformers is coaxial cable with Teflon™ dielectric. It can be either semirigid type or flexible of which both have equal velocity propagations, at least in theory.

The velocity factor must be known for calculating the maximum line length allowable. The multiplier for the velocity factor is obtained as:

$$v_r = 1/\sqrt{\epsilon_r}$$

where ϵ_r = relative dielectric constant of the insulating medium. Then for Teflon™ cable with its $\epsilon_r = 2.5$, the velocity factor multiplier is 0.633. Unlike a microstrip line, where two dielectric materials (air and the main substrate) form the medium and the width-height ratio is a variable, coaxial cable has a constant velocity factor as a function of characteristic impedance.

If points a and b are connected in Figure 10-9A as shown in Figure 10-9B, we arrive at an unbalanced 4:1 design. For the minimal leakage inductance, it is important that this connection be kept short, which can be done by bending the line to get the connection points close together. In this case the line Z_0 should be the geometric mean of the input and output impedances or $\sqrt{50 \times 12.5}$ or $25\ \Omega$. The same is true with other impedance ratio transformers. Derivations of this transformer are shown in Figures 10-9C and E. They are of a balanced-to-balanced configuration, where two or four lines are employed. A common magnetic core can be used for both if the coupling between the two can be kept minimal, but separate cores are usually recommended. Since this transformer has a 4:1 impedance ratio, the optimum line impedance is again $25\ \Omega$. D.C. can be fed through the "center tap" if it is left "floating," e.g. not by-passed to ground, and a balun normally seen to provide a balanced-to-unbalanced function can be omitted. Otherwise, another D.C. feed method must be chosen. This also applies to the 16:1 ratio transformer shown in Figure 10-9E, which employs two 4:1 transformers in series, where the same rules are in effect.

The line impedance of the high impedance 4:1 segment is 25 Ω, which was previously determined to be the required value. The line impedance of the second section would be $\sqrt{12.5 \times 3.12} = 6.25\ \Omega$, thus making the design of the second section somewhat impractical. Line impedances of such a low value as 6.25 Ω are difficult to achieve, although it would be possible to parallel two 12.5 Ω coaxial cables, for example, which is standard practice. Coaxial cables with impedances of 12.5, 16.7, and 25 Ω are beginning to be standard items with cable manufacturers today. For many applications however, the line impedance is not critical if some bandwidth degradation is acceptable.

Figure 10-9D shows a 9:1 balanced-to-balanced transformer which has good performance if the interconnections can be kept short. Keeping interconnections short is more difficult than with the 4:1 transformer since there are more interconnections and the impedance levels are lower. Here the optimum line impedance is $\sqrt{50 \times 5.55}$ or $16.6\ \Omega$. Unlike the 4:1 unit, the balanced 9:1 transformer always requires a balun in the end that is to be terminated with an unbalanced source or load, and it does not have a balanced point to allow D.C. feeding through the lines.

As mentioned earlier, a limitation of squared integer transformation ratios is the biggest disadvantage of the transmission line transformer. There are ways to get around this, but the designs get complex and bulky requiring additional lines and connections between them, resulting in greatly reduced bandwidths in some cases. One such design for three different configurations[1, 16] is shown in Figure

10-9F. It is shown in a simplified form in Figure 10-9F (#3), which makes analyzing its operation easier than using configuration #2 for example. An analysis of the current distribution between each winding was performed and it revealed a ratio of 1.5:1 between the primary and the secondary, which is equal to the turns ratio and results in an impedance ratio of 1.5^2 or 2.25:1. Assuming $R_L = 50\ \Omega$ (in which case the source would be 112.5 Ohms), the optimum line impedance is 50/1.5 or 33 Ω. This transformer has a balanced-to-balanced circuit configuration, requiring a balun if interfaced with an unbalanced source or load in either a step-up or step-down mode. An excellent and detailed analysis of this transmission line transformer is presented in the references.[16]

EQUAL DELAY TRANSMISSION LINE TRANSFORMERS

In normal 4:1 transmission line transformers, the high frequency response is limited by phase errors introduced between the interconnection points such as a – b in Figures 10-9 and 10 A, B. If the connection from a to b were made with a transmission line of equal impedance and length as the main line, the phase difference between the input and the output would be eliminated.[6, 7, 9] The transformer topology would remain the same, except the a – b connection would have the same phase delay as the main transformer line. This transformer can be viewed as two coaxial lines with their input terminals in series and output terminals connected in parallel. It is also the case with equal delay transformers of any other impedance ratio, where one line is always used only to provide a delay of a controlled amount. For this reason, this subclass of transmission line transformers is called *equal delay transmission line transformers*.

These equal delay networks can serve applications from 1 MHz to at least 500 MHz depending on the impedance levels involved. The transformer input and output connections can be physically separated, which is advantageous in many cases. Figure 10-10B is a pictorial and schematic representation of a 4:1 equal delay transformer. If a third line is added to the 4:1 design (Figure 10-10C), a 9:1 impedance transformer results. Likewise, four lines would produce a 16:1 transformer (Figure 10-10D) and so on. For wide band purposes, most of the transmission lines must be surrounded by magnetic material, generally in the form of toroids or sleeves. The amount of magnetic material required in each line depends on the level of the impedance transformation.

Although the line impedances are equal, the highest impedance transform line requires one unit of magnetic material, the next one two and the following one three and so on. By unit we mean a measure of cross sectional area of similar magnetic material. Note that these designs are all unbalanced-to-unbalanced transformers although baluns (c and d in Figure 10-10F) can be added to obtain a balanced interface. Suppose we add a magnetic core to the bottom line of a 4:1 transformer. Now we can disconnect the grounds of the parallel connected lines (still keeping the shields connected) and connect a balanced, floating load between the center conductors and the shields to form a 4:1 balun.

The stray capacitances to ground can be balanced by connecting the center conductor of one coax to the shield of the other one and a transformer as in

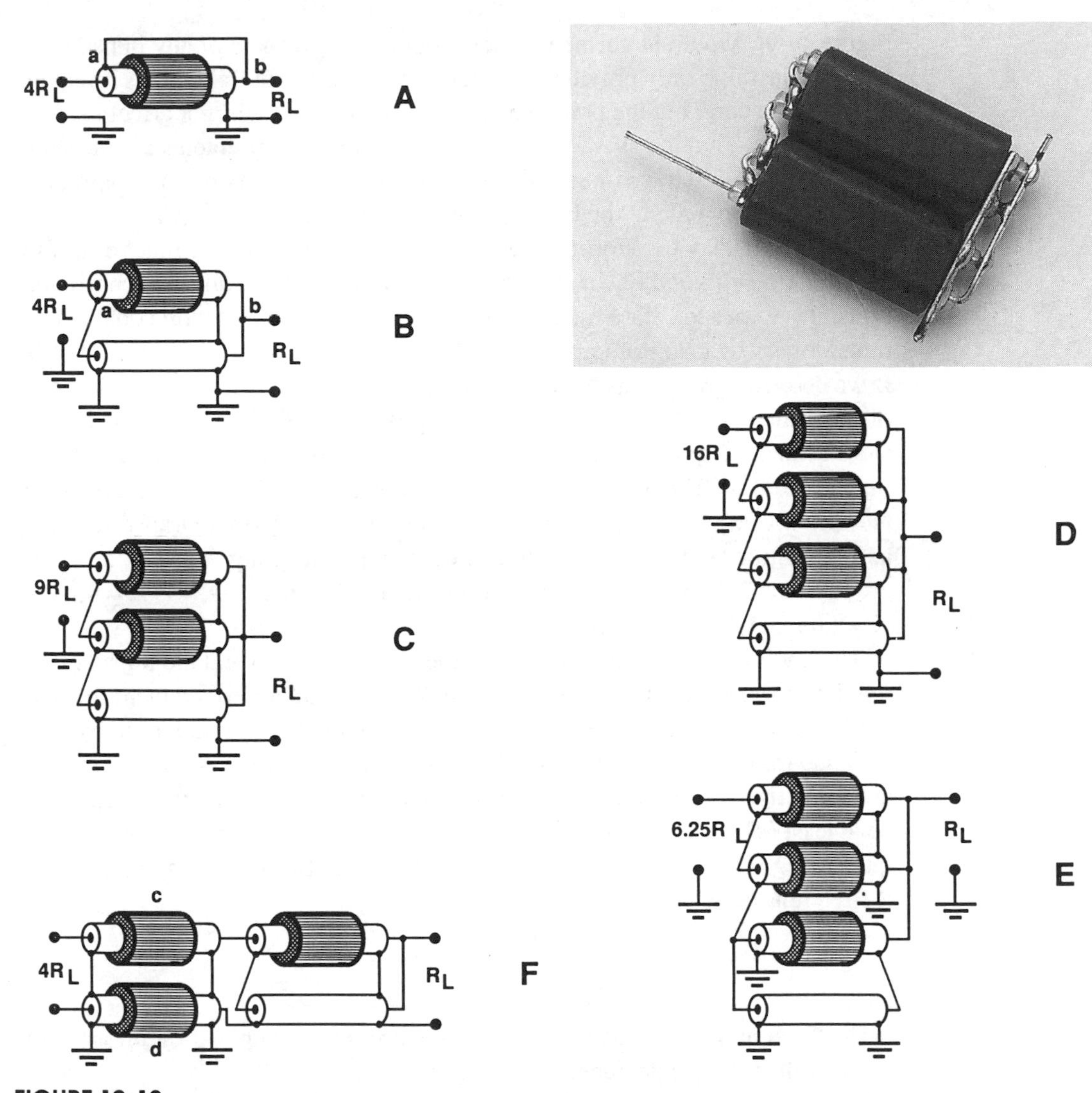

FIGURE 10-10

Examples of equal delay transformers. A is a basic standard transmission line type shown as a comparison against B. B, C, and D are basic configurations, whereas E uses a sub-group of lines to provide a fractional integer impedance ratio. F is a 4:1 unit with baluns (c and d) added for balanced interface.

FIGURE 10-11

Cross-section of a correctly arranged twisted wire lines with two pairs of wires (*a*) and four pairs of wires (*b*). **O** represents one conductor of the line and **X** the other.

Figure 10-9C would be formed. In equal delay transformers of any impedance ratio, the last line only provides delay and has no external fields; thus it requires no magnetic core, but the presence of magnetic material will not affect its performance. The line characteristic impedance (Z_0) requirements are the same as for the standard transmission line transformers, e.g., Z_0 equals the ratio of voltage to current along the line or simply Z_{in}/N, where $N = \sqrt{Z_{in}/Z_{out}}$.

The equal delay transformers basically have the full integer limitations of the standard transmission line networks. However, due to their physical configurations, it is easier to create fractional integer impedance ratios with equal delay transformers by using subgroups of additional lines as shown in Figure 10-10E. If we describe group A as the main transformer, which provides the full integers of impedance transformation, adding group B lines with their low impedance sides connected to the high impedance side of group A results in fractional impedance transform ratios. The resulting impedance ratios can be calculated as $N = (n_A + 1/n_B)$, where n_A = impedance ratio of group A (main transformer) and n_B = impedance ratio of group B. For example, if group A has 1 line and group B has 2 lines, the transform ratio is 2.25:1. Further, if A = 2, B = 4, N = 5.0625:1, and A = 2, B = 2, N = 6.25:1.

The line impedances are dictated by the transform ratio and the impedances required for the main transformer (group A). How much improvement in bandwidth the equal delay transformer gives compared with the standard transmission line transformer depends largely on mechanical factors. Also even if both are correctly compensated, the insertion loss of the equal delay transformer can be at least 0.1 dB less than the standard transmission line transformer in the frequency region up to 500 MHz. A detailed analysis of this topology is given in the references.[7]

References

[1] "Design of HF Wideband Power Transformers," Application Information, Philips/Amperex, June, 1970.

[2] "On the Design of HF Wideband Transformers," Parts I and II, Electronic Application Reports # ECO6907 and ECO7213, Philips Components, Discrete Semiconductor Group, 1969-72.

[3] D. Maurice and R. H. Minns, "Very-wide Band Radio Frequency Transformers," Parts I and II, *Wireless Engineer*, June-July, 1947.

[4] H. L. Krauss and C. W. Allen, "Designing Toroidal Transformers to Optimize Wideband Performance," *Electronics*, August, 1973.

[5] "Use of Ferrites for Wide Band Transformers," Application Note, Fair-Rite Products Corporation.

[6] R. K. Blocksome, "Practical Wideband RF Power Transformers, Combiners and Splitters," *Proceedings of RF Expo West*, January, 1986.

[7] Daniel Myer, "Equal Delay Networks Match Impedances over Wide Bandwidths," *Microwaves & RF*, April, 1990.

[8] M. Grossman, "Focus on Ferrite Materials: They Star as HF Magnetic Cores," *Electronic Design*, April, 1981.

[9] W. A. Lewis, "Low-Impedance Broadband Transformer Techniques in the HF and VHF Range," Working Paper, Collins Radio Co., June, 1965.

[10] D. N. Haupt, "Broadband-Impedance Matching Transformers as Applied to High-Frequency Power Amplifiers," *Proceedings of RF Expo West*, March, 1990.

[11] J. N. Nagle, "Use Wideband Autotransformers in RF Systems," *Electronic Design*, February, 1976.

[12] H. O. Granberg, "Combine Power without Compromising Performance," *Electronic Design*, July, 1980.

[13] H. O. Granberg, "Broadband Transformers and Power Combining Techniques for RF," Application Note AN-749, Motorola Semiconductor Sector, Phoenix, AZ, 1975.

[14] P. Lefferson, "Twisted Wire Transmission Line," *IEEE Transactions on Parts, Hybrids and Packaging*, Vol. PHP-7, #4, December, 1971.

[15] H. O. Granberg, "Building Push-Pull Multioctave VHF Power Amplifiers," *Microwaves & RF*, November, 1987.

[16] Udo Barabas, "On an Ultrabroad-Band Hybrid Tee," *IEEE Transactions on Microwave Theory and Techniques*, Vol. MTT-27, #1, January, 1979.

[17] G. Soundra Pandian, "Broadband RF Transformers and Components Constructed with Twisted Multiwire Transmission Lines," Instrument Design Development Centre, Indian Institute of Technology, Delhi, India, December, 1983.

11

Power Splitting and Combining

INTRODUCTION

When the required power output level exceeds the capabilities of a single power amplifier stage, multiple amplifier stages or "modules" can be combined to produce the required output. Combiners are closely related to wideband transformers in design and construction. The main difference is in the manner the lines or windings are connected. A power splitter is simply a lower powered version of a combiner and used in reverse. The splitter divides the input signal into multiple equal amplitude outputs to be applied to the inputs of each module. The power combiner then recombines the module outputs to be fed into a single load. Since the power splitter has the same configuration as the combiner, the discussion here will concentrate on combiners only.[1, 2, 3, 4]

Because of the multiple port configuration, the performance of splitters and combiners is difficult to test. One way to test combiners, for example, is to terminate all outputs except one and make the measurements between the one unterminated output and the single input port. Each output port would be sequentially tested in this manner, which tells us the amount of insertion loss, return loss, and the phase angle over the desired frequency spectrum. To test the isolation characteristics between the output ports, the input and all output ports but two should be terminated. Isolation can then be measured between all output ports by sequentially switching the terminations and the active ports. A quick test would be to connect all the multiple output ports together (back-to-back), which leaves two open ports compatible with a standard 50 Ω system. However, this would not give us information on the port-to-port isolation characteristics of the output.

The test methods outlined in the previous paragraph are applicable to basically any type of splitter or combiner to be described. Testing can be done with a network analyzer or an RF test station consisting of a signal source, load, and appropriate means for measuring forward and reflected power. The RF test station is probably the more accurate of the two methods, since the subject devices can be driven with a more realistic level of RF. In either case this allows us to examine the amount of return loss, insertion loss and in the case of the network analyzer, also the phase relationship. The phase error can be as much as 25° and this would create only a 0.22 dB loss. We can see that phase errors of around

10°, which are typically maximum are negligible. The effects of the source amplitudes are also exaggerated. Considering a two port system, a 50% difference in the source's power outputs results only in approximately 0.2 dB loss in the combined output. These figures may vary depending on the type of combiner. From the discussion above we can conclude that in practice the effect of amplitude unbalance between the input ports of a combiner (especially in a case of a failed source) is a far more dominant factor than the phase relationship. The trigonometric formulae to calculate the power loss due to phase errors are not presented here, but are available in the references.[4]

BASIC TYPES OF POWER COMBINERS

A wideband power combiner must perform the following basic functions:

a) Provide low insertion loss over the required bandwidth.
b) Provide isolation (minimum coupling) between the input ports.
c) Provide a low return loss at the input ports over the required bandwidth.

The operating bandwidth of combiners must be as wide or wider than that of the amplifier modules in order not to restrict the overall bandwidth of the combined amplifiers. Emphasis must be placed on the importance of employing an input power splitter of the same type as the output combiner. An in phase splitter and a 180° combiner combination for example would result in zero combined output power, since the outputs from each amplifier module would cancel.

There are several different types of power combiners, each one having its advantages regarding the frequency spectrum, bandwidth and other preferred features:[1, 7, 9, 10, 11]

1) A zero degree device, which means that ideally there is no phase shift between the input ports in relation to the combined output. This type combiner can be designed for even or odd number inputs up to a practical limit. The practical frequency range is up to about 500 MHz. These combiners are the most common ones used at lower frequencies because of their versatility and straight forward design.
2) A 180° device, which means that the two input ports or sets of input ports are 180° out of phase. It is only applicable to even numbers of inputs, e.g., 2, 4, etc. The out of phase situation must also be taken into account when designing the input power splitter. The practical frequency range is up to about 100 MHz.
3) 90° hybrid, which is basically a two port unit and by nature has a narrower bandwidth capability than the configurations in 1) and 2). They are applicable from low frequencies to microwaves in proper design configurations.
4) A so-called "Wilkinson combiner," which has relatively narrow bandwidth characteristics, but is simple and inexpensive. The practical frequency range can be up to microwaves (1 to 2 GHz).

IN PHASE AND 180 DEGREE COMBINERS

Transmission line techniques are commonly used in the in phase and 180° power combiners for lowest losses and widest bandwidths. One primary function of an RF power combiner is to provide port-to-port isolation, of which 30 dB is typical and acceptable. By this the output of one of the amplifier modules will be sufficiently isolated from the others; if a failure in one occurs, the remaining amplifiers will not be affected and will still be operating into the original load impedance. Amplitude unbalance is usually created by a completely disabled module (the "source' for the combiner) or, sometimes, more than one disabled source. The power output with various numbers of disabled sources can be calculated as:

$$P_{out} = \left(\frac{P}{N}\right)N_1$$

where P = total power of operative sources, N = total number of initial sources, N_1 = number of operative sources. For example, if we have a four port system designed to deliver 1kW with 250 W modules with one module disabled, the power output would be (750/4)3 = 562 W. The difference power or 188 W would be dissipated in the balancing resistors, which divides according to the type combiner in question. In a straight zero degree combiner (Figures 11-1A, B, C) all resistors are of equal value and the power dissipated in each would be (188/3) = 62.5 W in a four port system.

In a two port combiner system having a maximum output power P_{max}, if one amplifier fails the output power will decrease to a value 6 dB below P_{max}. Half of the power loss is due to lack of power from the disabled module and an additional 3 dB is lost because the power from the remaining module now divides equally between the balancing resistor and the output load. In this case the balancing resistor must dissipate 1/4 of the original P_{max}. Values of the balancing resistors depend on the number of combiner ports and how many ports are assumed disabled at one time, as can be seen in the expression for P_{out} given above. Sometimes these resistors, which must be of the non-inductive type, are referred as "dump loads" since power due to phase or amplitude unbalance is directed to them. In most cases even if one module fails, the system is forced into a shut-down mode. Then the balancing resistors do not have to dissipate significant amounts of power. The comments just made about power loss and power dissipated in "dump loads" applies to all types of combiners described in subsequent paragraphs of this chapter.

A failure detection method using pick-up coils is illustrated in Figure 11-1C. The signal pick-up coils (pc) can be small toroids wound with multiple turns of wire, which form the secondaries for RF voltage step-up transformers, whose primaries are the leads of the balancing resistors (e.g., two watt carbon type) threaded through the toroids. The RF voltages in the secondaries generated by the unbalance due to a module failure can then be rectified and processed to operate the shut-down circuitry. If the a, b and c outputs (Figure 11-1C) are kept separated, each one can be made to operate an indicator to inform which

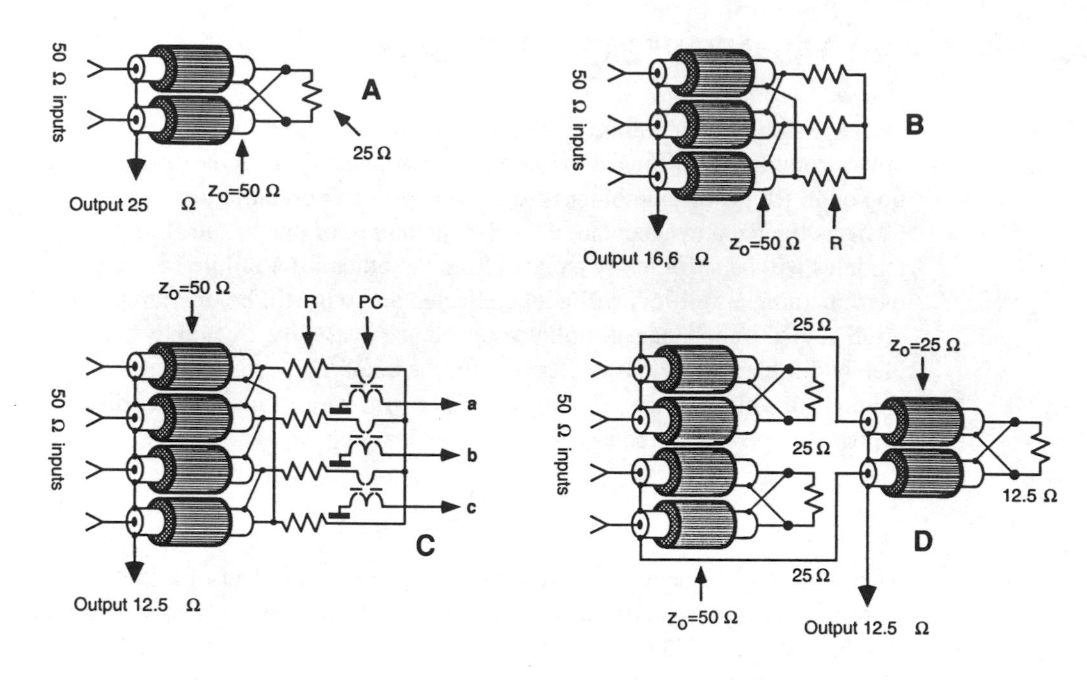

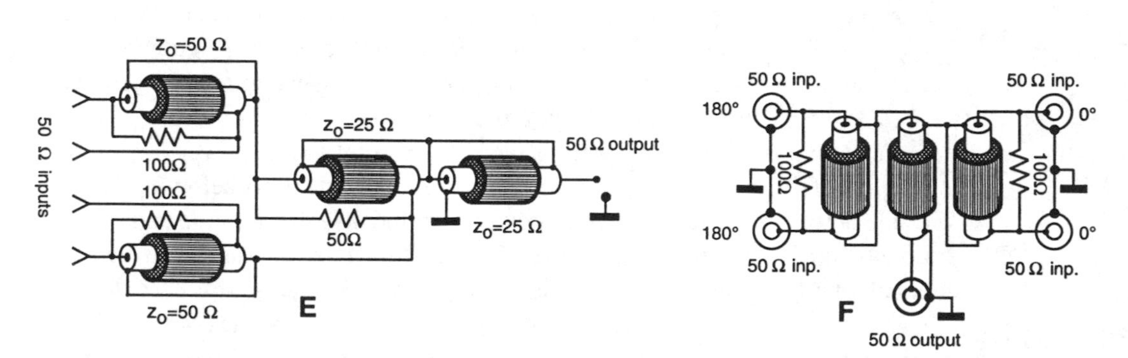

FIGURE 11-1

(A), (B), and (C) represent so-called straight in-phase combiners. Even and odd numbers of input ports are possible with them. (D) and (E) are staggered or "totem pole" structures, which are adaptable only to even numbers of ports. (F) is a 180° combiner that requires no step-up transformer into 50 Ω. Note that step-up transformers are not shown with (A), (B), (C), and (D).

module has failed. The resistor (R) values are not critical, since the power will be shut-off within a millisecond or so making the load mismatch unimportant. If on the other hand, operation under reduced power conditions is desired, the balancing resistors will have to handle continuous power levels as described earlier. This requires the use of high power resistors that can be heat sunk, and which must be of the so called "floating type," meaning that the resistor element must be electrically isolated from its mounting structure. Here, again, a system like that shown in Figure 11-1C can also be applied.

In the event of a failure of one module (which is the most common case), the balancing resistor values can be determined from the formula:

$$R = \frac{Z_{in}}{N}$$

where N = number of input ports. This makes the values 25, 16.6, and 12.5 Ohms for a two, three and four port combiners respectively, which applies to Figures 11-1A, B and C. Examples of 180° power combiners are shown in Figures 11-1E and F. Figure 11-1E represents a so called "totem pole" structure and Figure 11-1F is a system consisting of a pair of two port hybrids and a balun. A comparison of the 180 degree and in phase combiners has shown that the 180° type can be superior to the zero degree combiners (Figure 11-1A-D) in input VSWR, but the latter has better port-to-port isolation characteristics. In the cascaded or "totem pole" structures shown in Figures 11-1D and E, the balancing resistor values follow the values of the two port hybrids and transmission line transformers.

Figure 11-1F shows a unique system, which does not require a step-up transformer. Its simplicity compared to other four port combiners is attractive. It uses half the number of transmission lines used in Figure 11-1D, for example, and due to the lack of the 4:1 step-up transformer, its bandwidth characteristics are enhanced. It should be noted that this combiner is covered by a recent U.S. patent.[6] Anyone wishing to use the idea for commercial purposes should consult with the holder of the rights to the patent. Descriptions of step-up transformers for these power combiners can be found in Chapter 10.

Impedance ratios such as 2:1 and 3:1 can be implemented with a 4:1 transformer wound with coaxial cable where the corresponding taps are made to the coax braid.[2, 3] Improved performance over wider bandwidths is possible with fractional integer equal delay transformers, designs for which are given in the references.[7] Figure 11-2 is a photograph of a commercially available[14] four port splitter-combiner intended for use at 1.6–30 MHz and up to power levels of 1 kW. It is of the type shown in Figure 11-1C, but the combiner lacks the power pick-up coils. The 50 Ω lines are realized by metal tubes with a proper size insulated conductor threaded through the tube, which makes the construction inexpensive.

Port-to-port isolation figures of 27 to 40 dB have been measured on both the in phase and the 180° combiners over a bandwidth of several octaves. Note the compactness of the units, where the step-up transformer is right next to the combiner structure. There are a number of manufacturers of high power combiners for use at frequencies up to UHF, but most of the units are housed in enclosures with connectors and do not suit many designer's needs because of their physical outlines or cost. In addition to the combiner configurations shown, many derivatives of the two-port hybrid have been designed. Some of the various types are shown in the references cited.[5, 7, 8, 9, 10, 11]

90 DEGREE HYBRIDS

One class of hybrid that is a variation of the two-port is the quadrature hybrid. The 90° hybrid combiner or quadrature coupler can be realized in several different forms. A quadrature combiner usually refers to a passive device with one input port, two output ports (or vice versa) 90° apart in phase and an isolated port. A few such networks that can be considered as quadrature couplers are:

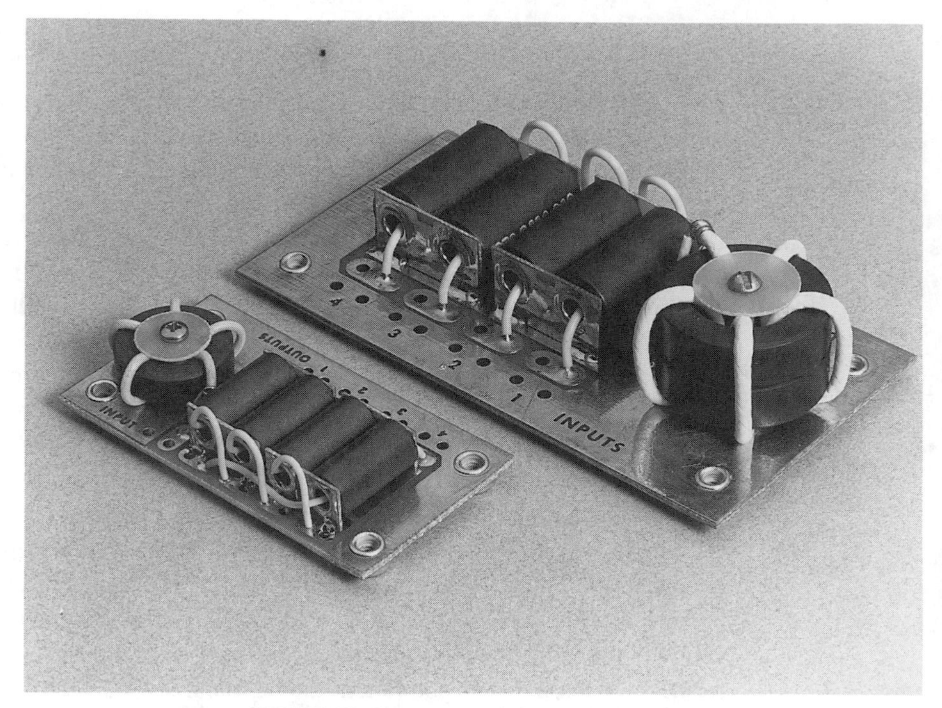

FIGURE 11-2
Commercial four port splitter-combiner for use at 1.6-30 MHz and power levels up to one kilowatt.

1) A line hybrid based on the principle shown in Figure 11-3.
2) The 3/2 λ ring coupler (commonly known as "Rat-Race") (Figure 11-4).
3) A branch coupler, widely used in small signal mixer circuitry, but is also applicable to use as a power combiner (Figure 11-5).
4) A "Wilkinson" combiner, which consists of a series of λ/4 transmission lines as shown in Figure 11-6. It is not a quadrature coupler like the ones described in items 1, 2 and 3, but it is included with these because all four types are based on the principle of a delay generated by a quarter wavelength transmission line.

LINE HYBRIDS

The line hybrid shown in Figure 11-3A is one of the most common combiners used in the VHF and UHF frequency ranges.[6, 10] It consists of two transmission lines (microstrip) of λ/4 in length separated by a dielectric medium. In addition this structure is sandwiched between two ground planes, again separated by a dielectric medium. The mutual impedance between the two lines is designated as Z_{even}, whereas the impedance from the lines to ground is designated as Z_{odd}. Z_{even} is the impedance that controls the coupling coefficient between the lines and is typically $Z_{in}/2$. Then Z_{odd} must be calculated for a value that gives

$$\sqrt{Z_{even} \times Z_{odd}} = Z_{in}(50\Omega)$$

If $Z_{even} = 25\ \Omega$ as in this case, then Z_{odd} must be 100 Ω for $Z_{in} = \sqrt{2.5 \times 10^3} = 50\ \Omega$. The Z_{even} and Z_{odd} values can be modified for greater isolation or extended bandwidth, but the $\sqrt{(Z_{even} \times Z_{odd})}$ relationship is always valid. Typical bandwidths of these couplers are about 15% with 1:1.5 VSWR and port-to-port isolations range from 20 to 30 dB.

In the past the dielectric materials used were Teflon™-fiberglass or epoxy-fiberglass with their dielectric constants around 2.5 and 5 respectively. These low values of dielectric constants limited the couplers to a lowest practical frequency of approximately 175 MHz. Even at these frequencies the couplers were bulky. Today with advances in the development of dielectric materials, materials are available with dielectric constants of 10 or higher, which makes the lower frequency couplers more practical. In order to realize any kind of a practical size or shape factor, the lines are usually folded several times as seen in Figure 11-3B, and the total electrical length calculated for l/4 according to the dielectric constant of the medium. Along with an opened unit, Figure 11-3B also shows a complete UHF coupler having dimensions of approximately 1.5" square and 1/8" thick (38 × 38 × 3.5 mm).

A variation of the line combiner can be designed with lumped constant elements for use at low frequencies. It behaves much like its counterpart, the stripline quadrature hybrid. In a true representation of the stripline design, the capacitors C (Figure 11-3C) should be split and their center taps grounded, which would simulate Z_{odd}. Z_{even} would be determined by the mutual line impedance and the electrical length of the line, which should also be $\lambda/4$.

In practice, the transmission line can be made of twisted enameled wires or

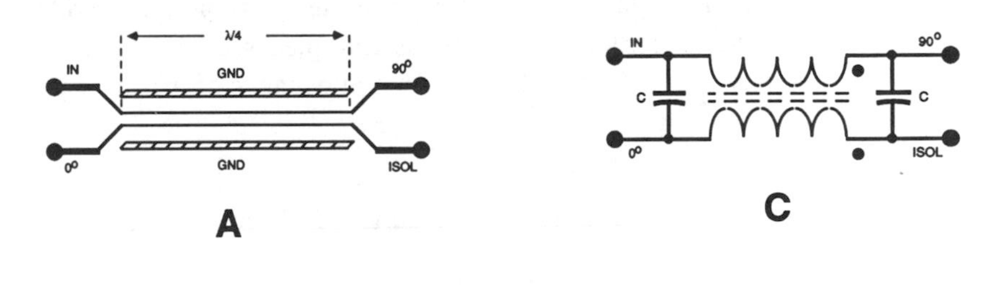

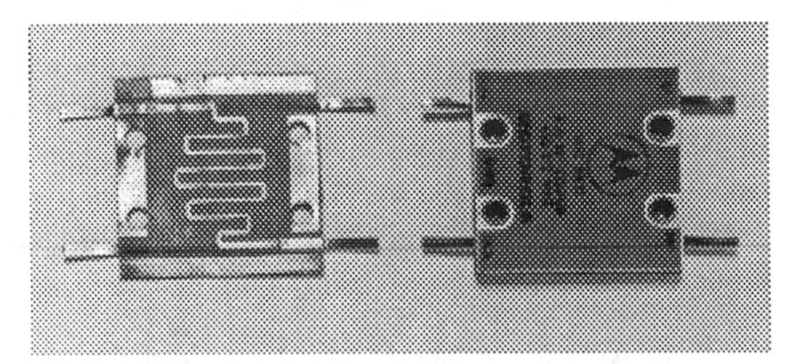

FIGURE 11-3
The line hybrid (A and B) is widely used at VHF and UHF since it can be made compact by folding the lines. Its lumped constant equivalent is shown in C.

two lengths of low impedance coaxial cable with their braids connected together and floating, leaving the center conductors to form two symmetrical lines. Another variation of this type quadrature coupler, especially suited for low frequency use, is the so-called Fisher's hybrid,[11] which resembles more closely the hybrid shown in Figure 11-3C. In it a very tight coupling between the lines is required. The physical lengths of the lines do not need to be $\lambda/4$ since they are electrically lengthened by the presence of the magnetic medium. The values of C depend on the line coupling coefficient and they may not be necessary at all in some cases. The phase relationship is the same as in the stripline hybrid and the port isolations are comparable as well. The bandwidth achievable depends on the line inductance, and thus the properties of the magnetic core, as well as the line mutual capacitance (coupling). Graphs of insertion loss and port unbalance versus line coupling coefficient are shown in the references.[13] Using the technique described, it may be possible to develop quadrature hybrids for wider bandwidths than the 10–15% normally attributed to this type of coupler.

RING HYBRIDS

The hybrid ring, 3/2 λ hybrid or "rat race" (Figure 11-4) as it is commonly called,[9] is a directional coupler that can be used to sample RF power traveling in different directions, and is thus adaptable also for mixer, power splitter, and combiner applications. The ring hybrid is usually constructed with microstrip transmission lines, a technique which limits its use to about 1000 MHz and above. Although coaxial cable designs can be realized that would extend the frequency range down to 200–300 MHz, such structures would become bulky. A simple hybrid ring consists of a transmission line in which input, output and isolation ports are connected in four places. If the impedances (Z_{in}) of these ports are 50 Ω, the characteristic impedance of the transmission line ring is

$$Z_o = \sqrt{(N)(Z_{in}^2)}$$

where N = number of active ports = 2. Then

$$Z_o = \sqrt{5 \times 10^3} = 70.7\Omega$$

All four ports are separated from each other by quarter wavelength sections making the remaining part of the ring three-quarter wavelengths long. The total circumference, then, is 1.5 wavelengths.

Among its other uses, the hybrid ring is commonly applied as a power splitter or combiner. If a signal is applied to port 1 (Figure 11-4A), the power will be equally divided between ports 2 and 4 and their phase relationship will be 180°. On the other hand, power incident at port 2 will also be equally divided between ports 1 and 3, but the two output signals will be of similar phase. When port 2 is used as the input in a combiner, any power reflected at output port 3 due to a mismatch arrives at the other output port 1 by two paths. One signal travels a

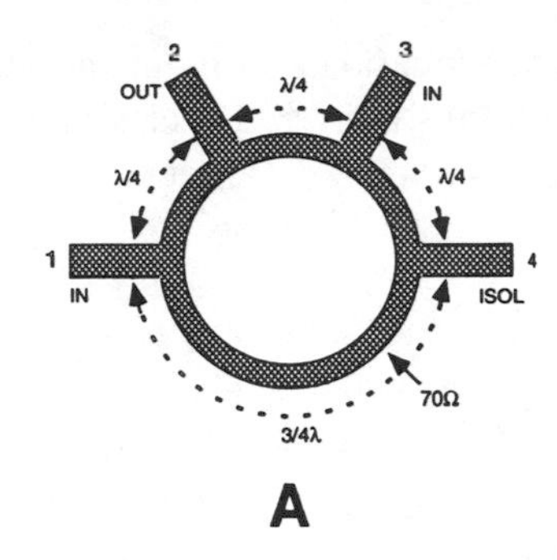

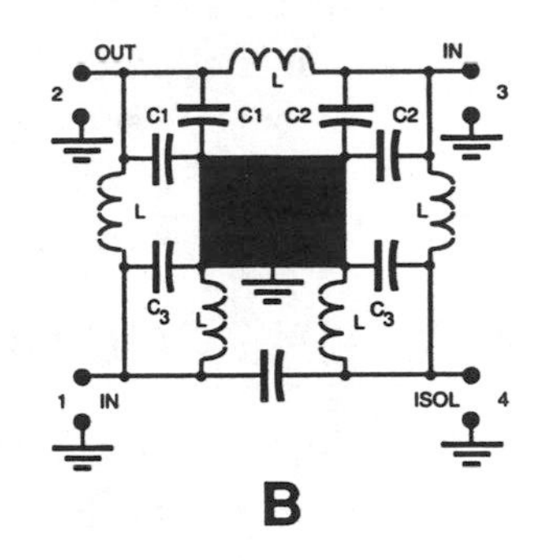

FIGURE 11-4

Ring hybrid (A) and its lumped equivalent (B). In (B), C_1 and C_2 can be paralleled into single units twice the value.

half wavelength in a counter clockwise rotation from port 3. The clockwise signal appears at port 1 delayed by a full wavelength.

The half wave difference in arrival and equal path loss results in cancellation of the two signals at port 4, with total cancellation resulting in highest port-to-port isolation. The reflected signal from any mismatch at port 3 arrives at port 4 in phase from both circular paths, where it is dissipated. This port is designated as the isolation port, where a termination absorbs any power due to the unbalance between output ports. The input signal from port 2 cancels at port 4 because the two-way paths differ by a half wavelength. Advantages of a ring hybrid include simplicity of construction and a reasonable tolerance for variations in line impedance. In addition, the power at the output ports can be adjusted by varying the impedances of the interconnecting lines. A simple hybrid ring can provide a good match and excellent isolation over a 8–10% bandwidth.

The ring hybrid can also be designed with lumped constant elements, which would extend its low frequency response to audio frequencies. The technique of deriving a lumped-circuit equivalent is simple. It entails replacing each length of transmission line by its π equivalent as shown in Figure 11-4B. We can see three π networks representing the $\lambda/4$ sections: C3-L-C1, C1-L-C2 and C2-L-C3. The lowest branch L-C4-L forms the 3/2 λ section. Since all capacitors will be of equal value, C1 and C2 can be paralleled into a single unit, which makes them 2xC1 and 2xC2 respectively. All branches must have a Z_o of $\sqrt{2}\,(Z_{in})$ or 70.7 Ω as the line sections in the microstrip design. All inductors in the circuit are also of equal value, which can be calculated as:

$$L = \frac{\sqrt{2}(Z_{in})}{2\pi f_o} \text{ and } C = \frac{1}{2\pi f_o \sqrt{2}(Z_{in})}.$$

Example: Assume we wish to design a lumped constant ring combiner for a frequency of 100 MHz. Then

$$L = \frac{70.7}{628} = 0.11\mu\text{H and } C\frac{1}{(628)(70.7)} = \frac{1}{444 \times 10^3} = 22.5 \text{ pF}.$$

C1's and C2's combined would then be 45 pF each. The isolation and insertion loss characteristics are greatly dependent on the component tolerances, loss factors (Q), and the symmetry of the total structure. Typical numbers for 100 MHz are 20–25 dB and 0.4–0.45 dB respectively.

BRANCH LINE COUPLERS

The easiest coupler to construct is the branch line type, most often realized in the two branch, 3 dB unit of Figure 11-5A. In this coupler, the main transmission line is coupled to a branch line by two quarter wave lines, each spaced a quarter wave apart. The branch-line coupler is easily made in a microstrip configuration for coupling values from 3 to 9 dB. It is a quadrature coupler with its output signals 90 degrees out of phase, when used as a power combiner. The overall bandwidth of branch-line couplers is rather narrow (comparable to the hybrid ring), but it can be increased by adding more branches. This is not often done, however, because additional branches have higher impedances, and in addition to increased losses, the microstrip tolerances become critical.

The use of transformer sections in the main lines results in characteristic impedances of the branch lines equal to those of the input and output arms. The impedance of these transformer sections ("a" in Figure 11-5A), which are also called main lines, can be figured as:

$$Z_o = \sqrt{Z_{in} \times \frac{Z_{out}}{2}}.$$

Then for a 50 Ω system,

$$Z_o = \sqrt{50 \times \frac{50}{2}} = 35.4\,\Omega$$

The impedance of the branch lines ("b" in Figure 11-5A) will be $Z_o = Z_{in,}\ Z_{out}$ or 50 Ω.

Both the hybrid ring and the branch line coupler are very versatile devices. With the ring coupler, impedance transformation is possible by varying the characteristic impedances of the λ/4 line sections. Similarly the transformer or the main line impedance can be varied in a branch-line coupler to result in impedance transformation between the input and output ports. This function may only be necessary when these devices are used for non-power-splitting-combining applications such as mixers.

Although the branch-line coupler is physically smaller than the hybrid ring since its loop periphery measures only one full wavelength, its practical low frequency limit is in the low microwave region. If the lines are constructed of coiled coaxial lines for example, this would bring the lowest practical frequency into the 150–200 MHz region. Lower frequency versions can be realized with lumped element designs as shown in Figure 11-5B. The principle is the same as with the hybrid ring, namely that each λ/4 transmission line section is replaced with a π LC network. In the hybrid ring equivalent, there are three such networks, whereas the branch-line equivalent requires four, two for 35.4 Ω

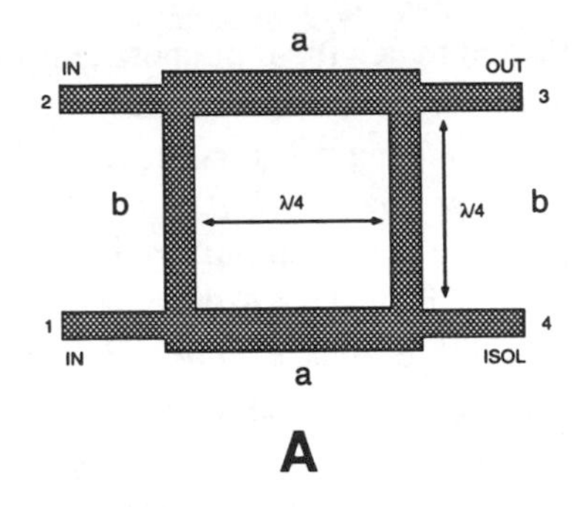

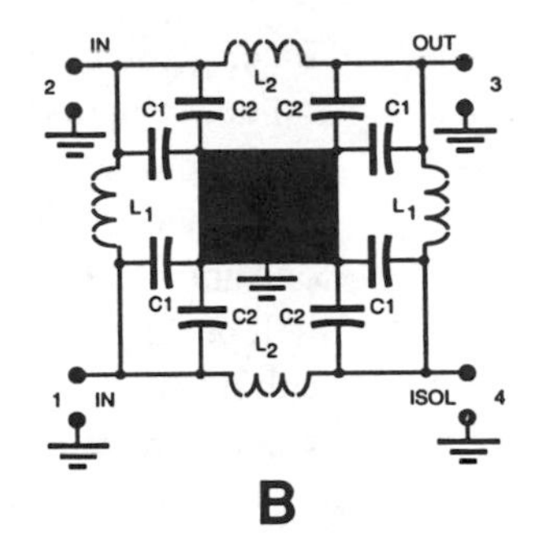

FIGURE 11-5

The branch line hybrid (A) and its lumped equivalent (B). As in the ring hybrid equivalent, the paralleled capacitors (C_1 and C_2) can be combined into single units (B).

impedance and two for 50 Ω. The formula to calculate the component values has been presented earlier. Thus, for 100 MHz the 35.4 Ω branches (a) will be:

$$L = \frac{35.4}{628} = 56 \text{ nH and } C = \frac{1}{(628) \times (35.4)} = 45 \text{ pF.}$$

Similarly the 50 Ω branches (b) are L = 80 nH and C = 32 pF. The combined capacitances ($C_1 + C_2$) will be 45 + 32 pF = 78 pF. As pointed out in the references,[12] the component tolerances are critical. Chip capacitors and air-wound inductors were used in the design at 140 MHz as described in the references.[13]

WILKINSON COUPLERS

Another power combiner that uses quarter wave transmission lines is the Wilkinson hybrid.[10, 12] However, it really cannot be called a hybrid since the word "hybrid" refers to a device with two input or output ports. The Wilkinson coupler is a reciprocal network and sums N coherent sources to a common port, all in one step. It can be designed for any number of ports (Figure 11-6A) and is, thus, commonly called a N-way coupler. In a manner similar to other types of combiners discussed in this chapter, this device can be used for power splitting as well as combining. Unlike quadrature hybrids, all inputs are in equal phase *relative to the output,* but create a delay of 90°. Since the line length between any two ports is λ/2, the power arriving at each port is 180° out of phase with the power from any other port and, therefore, cancels. This provides the isolation between inputs.

The port-to-port isolation and VSWR are theoretically perfect at mid-band (where the lines provide a 90° phase shift) and degrade at frequencies away from band center. The amount of isolation and the VSWR in practice are primarily dependent on the phase relationship (line impedance and length) of the transmission lines. Typical numbers for the isolation are 20–25 dB and for the VSWR 1.2:1. The balancing resistors (R) have an important role in this combiner. They serve to help isolate and match the input ports, although normally no power is dissipated in them, but any unbalance between input ports would result in power dissipation in the balancing resistors. In case of a two port combiner, if the power to one input port is completely lost due to a failure of one amplifier module, then half of the power output from the remaining module will be dissipated in the balancing resistor. This is the case with all two port combiners described. (Refer to straight in-phase combiners and Figure 11-1C, where a system shut-off under the condition of an amplifier failure is described.)

The Wilkinson coupler has the advantages of perfect output port amplitude balance (due to its symmetry) and equal phases at all of its input ports. Its chief advantage for high power applications is the series of combined balancing resistors. Perfect isolation requires completely noninductive resistors, which must be heat sunk. The Wilkinson N-way combiner can be constructed with coaxial, stripline, or microstrip transmission lines. It offers a relatively low cost method of combining a number of signals with a fair amount of isolation and low VSWR. However, it performs at a relatively narrow bandwidth, comparable to the 90° hybrids. If wider bandwidths than 15–20% are required, multiple lines of appropriate lengths can be cascaded with stepped characteristic impedances for an optimum design with Chebyshev response. But this will add to the line IR losses, and may only be practical up to 3–4 sections. For N number of input ports, the line impedances are calculated as: $Z_o = \sqrt{(N)(Z_{in}^2)}$, which makes the Z_o for a two port system 70.7 Ω, for a three port system 86.6 Ω, for a four port system 100 Ω, and so on. The balancing resistor values (which are equal) can be obtained as:

$$R = \frac{Z_o^2}{NZ_{in}}$$

A lumped equivalent of a two port Wilkinson power combiner is shown in Figure 11-6B. Its operation is based on the same principle as the ring and branch-line coupler equivalents, i.e., that the λ/4 sections are simulated with π

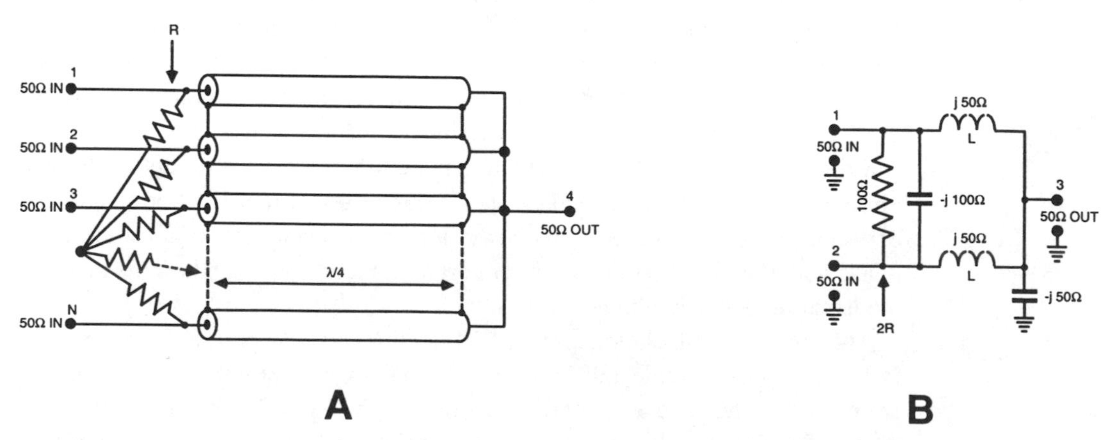

FIGURE 11-6

The Wilkinson combiner (A) and its lumped equivalent (B). In practice, (B) may not be feasible for more than two input ports.

networks. The two 50 Ω balancing resistors have been combined into a single 100 Ω unit because the center tap is "floating." Similarly, the C's between the input ports have been combined to a –j100 Ω reactance, since it is not necessary for their center tap to be grounded. The reactances shown can be converted to the required values of capacitance and inductance for the center of the operating frequency band. The lumped constant version of the Wilkinson combiner works well with comparable isolation and VSWR characteristics to its transmission line counterpart if the component tolerances can be kept within 1–2%.

References

[1] R. K. Blocksome, "Practical Wideband RF Power Transformers, Combiners and Splitters," *Proceedings of RF Expo West*, January, 1986.
[2] H. O. Granberg, "Combine Power without Compromising Performance," *Electronic Design*, July, 1980.
[3] H. O. Granberg, "Broadband Transformers and Power Combining Techniques for RF," Application Note AN-749, Motorola Semiconductor Sector, Phoenix, AZ, 1975.
[4] William E. Sabin and Edgar O. Schoenike, *Single-Sideband Systems and Circuits*, New York: McGraw-Hill Book Company, Inc., 1987.
[5] Jim Benjamin, "RF Power Combination Using Hybrid Junctions," Working Paper, ITT Semiconductors, 1967.
[6] US Patent # 4,647,868, "Push-Pull Radio-Frequency Power Splitter/Combiner Apparatus," Assigned to General Electric Co., March, 1987.
[7] Samuel Y. Liao, *Microwave Devices and Circuits*, Englewood Cliffs, NJ: Prentice Hall, Inc., 1980.
[8] G. Matthaei, L. Young, E. M. T. Jones, *Microwave Filters, Impedance-Matching Networks and Coupling Structures*, Norwood, MA: Artech House Books, 1980.
[9] Ernie Franke, "The Hybrid Ring," *Ham Radio*, August, 1983.
[10] Ernest J. Wilkinson, "An N-Way Hybrid Power Combiner," *PGMTT Transactions*, January, 1960, pp. 116-118.
[11] R. E. Fisher, "Broadband Twisted Wire Quadrature Hybrids," *IEEE Transactions on Microwave Theory and Techniques*, Vol. MTT-21, May, 1973.
[12] Alfred A. Morse, "Wilkinson, Proximity and Branch-Line Couplers," *Microwaves*, January, 1978, pp. 70-79.
[13] R. Chattopadhyai, et al, "140 MHz Lumped Element Hybrid," Indian Telephone Industries, Bangalore, India.
[14] RF Power Systems, Phoenix, Arizona, FAX (602) 971-9295.

12

Frequency Compensation and Negative Feedback

FREQUENCY COMPENSATION

The purpose of frequency compensation in amplifiers is to equalize the input impedance of a transistor so that the matching element can look into a relatively constant R and Z over a given bandwidth. Frequency compensation in narrow band designs, using L & C matching elements, is not often used or required since the bandwidth is limited to 5% or 10% by the matching element. All wideband designs with bandwidths greater than 10%, in general, are combinations of L & C or microstrip and wideband transformers or wideband transformers alone. Although transistor impedance matching over bandwidths of half an octave or more are possible with complex L & C or microstrip designs, they are not considered feasible or good design practice today. Since the input impedance of a transistor (BJT or FET) varies with frequency much more than does the output impedance, it is usually necessary to compensate only the input.[1, 2, 3, 4, 5] At power levels higher than a few watts, where the output impedance level has a low value, output compensation would be impractical due to losses in the compensation networks. However, output compensation is sometimes done with only series inductance, for example, in the case of a capacitive output or with shunt capacitance with an inductive output. We cannot use both L's and C's with wideband transformers because shunt capacitance is used to compensate for leakage inductance.

Losses must be tolerated in certain interstage matching situations . If the power amplifier (PA) operates at a power level of 150–200 watts and has a power gain of 6–7 dB, then the driver power output would be 30–50 watts and would have (for a 12 volt design) an output impedance of around 1.5 Ω. Assuming the PA input has a frequency compensation network, part of the drive power is dissipated in this network in addition to what is dissipated in the matching network itself, which in a case such as the one just stated would result in considerable power loss. Such loss would lower the overall efficiency of the system and possibly result in a requirement for an additional amplifying stage in the chain.

Wideband amplifiers generally use push-pull designs because it is much easier to achieve low emitter-emitter or source-source inductances than low emitter/source-to-ground inductances (important in a single-ended design). In addition, the input/output impedances are higher which makes it easier to achieve wideband impedance matching networks.

Transistor input impedance is high at low frequencies and low and more reactive at high frequencies. The change is around 40–80% per octave depending on the frequency spectrum and device type. This is true for both BJTs and FETs, although the input impedance of a FET for a given electrical size is higher, particularly at lower frequencies. If the device input crosses over from capacitive to inductive within the desired frequency band, the task of designing compensation networks becomes even more difficult.

In case of a capacitive input, a shunt LR combination (Figure 12-1), R3/L3 and R4/L4 can be used as an initial compensating network. Ideally, reactances of the inductances L3 and L4 are very large at the high frequency end of the band and the shunt circuit has negligible effect. At the low frequency end of the band, the reactances of L3 and L4 become low, leaving only R3 and R4 effective. Since the reactances of the series inductors L1 and L2 are also at their minimum values, the series combination of R3/R4 will in fact be in parallel with the output of T1, presenting an artificial load to it. At high frequencies the reactances of L1 and L2 are adjusted to a value, which in series with C1/R1, and C2/R2, results in a load to T1 that is comparable with the low input impedances of Q1 and Q2. C1/R1 and C2/R2 are actually used for gain leveling more than for frequency compensation. The idea is that the reactances of C1 and C2 are low at high frequencies, where the power gain is lowest.

At low frequencies the higher device power gain is lowered for a more even response by the increased reactances of C1 and C2, leaving R1 and R2 as the main power carriers to the bases. Thus R1/R3 and R2/R4 form π attenuators with the input impedances of Q1 and Q2 serving as the second shunt leg. The values of all R's can then be calculated when the transistor input impedance and the desired power gain slope are known. Typical component values for the networks of Figure 12-1 applied to a 2–30 MHz, 200 watt amplifier design are:

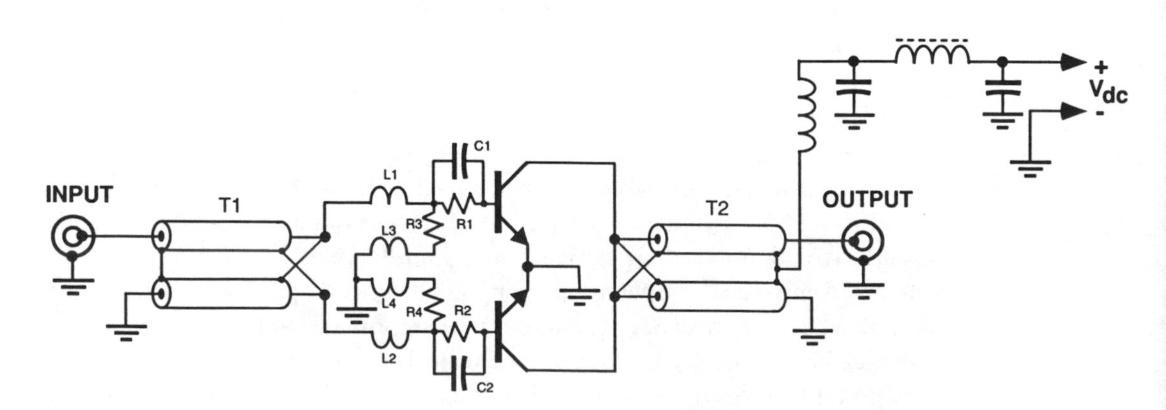

FIGURE 12-1

Schematic of a push-pull amplifier showing networks for input impedance compensation as well as for leveling of the power gain in wide band applications.

L1, L2 - 27-33 nH,
L3, L4 - 35-40 nH,
C1, C2 - 2000-2800 pF,
R1, R2 - 10-15 Ω,
R3, R4 - 8.2-12 Ω.

Detailed analysis of compensation networks such as the one shown in Figure 12-1 can be found in the references.[2, 3, 4] Although all three references describe circuits intended for low frequency applications, the same criteria holds at higher frequencies up to the point where the device input impedance turns inductive. At this point the component values must be re-adjusted. As previously mentioned, designing a network to match a "load" that changes from capacitive to inductive as a function of frequency is a difficult task. With internally matched transistors (see Chapter 2) the situation is different. For their specified frequency ranges, power gains and input impedances are much more constant than those of non-internally matched transistors. In many cases frequency compensation is not required at all since the maximum bandwidths that can generally be realized with internal matching are less than three octaves.

Very low Q, broad band circuits used to "match" the input or output of a transistor can be realized at low frequencies (below 100–200 MHz) using low Q matching networks, for example, broad band transformers. Generally it is possible when designing such circuits to express the input/output impedance of the transistor as the magnitude of Z_{in}/Z_{out} without regard for phase angle. The larger the value of $R_{in/Rout}$ relative to X_{in}/X_{out}, the more accurate will be this approximation. However, for narrow band circuits (which generally include all circuits above 200 MHz) one must design matching networks that utilize both the real and imaginary parts of the load impedance (in this instance, for example, the input/output of the amplifying transistor). Input/output impedances in complex form are usually given in RF device data sheets. Use of such data and the relatively narrow band matching networks required to match transistors at radio frequencies are discussed in detail in Chapter 7. In many cases computer software[9, 10, 11] is used to generate and optimize the elements of the matching network.

NEGATIVE FEEDBACK

Another impedance compensation and gain leveling method with advantages and disadvantages over LCR networks is known as *negative feedback*. Negative feedback means that part of the output power is fed back to the input out of phase, in which case part of the input voltage and the voltage fed back cancel. The advantages include simplicity and a stabilizing effect on the amplifier. The only disadvantage is that power is being dissipated in the feedback network, which lowers the overall efficiency of the system. The amount of power loss depends on the amount of gain reduction desired at low frequencies, i.e., the amount of feedback.

The out of phase feedback voltage is set to a certain amplitude with respect to the input voltage which holds at any power level providing the input impedance remains constant. Then the input voltage must exceed the voltage fed back in amplitude in order to produce output power. In addition to gain reduction, negative feedback lowers the effective input impedance of the device(s). Although the device input impedance itself remains unchanged, the out of phase voltage fed back to the input lowers the load impedance to the input matching element.

In a wide band amplifier, the amount of feedback voltage ideally should be inversely proportional to the frequency and of an amplitude such that the gain would be reduced the correct amount at all frequencies below the high end of the band. This is not possible with simple networks consisting only of R and L where the feedback voltage source is the collector (or drain) of the output transistor and the voltage is fed back to the base (or gate) directly.[7] Exceptions are low power designs, where impedance levels are relatively high. A collector-to-base feedback circuit is illustrated in Figure 12-2A. Note that the input impedance for the feedback voltage is set by T1. The same kind of feedback with a lower impedance source can be accomplished by adding a third winding in T2 (Figure 12-2B). Again feedback voltage is fed to the input through the primary of T1, which has a higher impedance level than the base. In Figure 12-2C a third winding is again added in T2. It has a very low impedance since the feedback voltage goes directly to the base, which has a low impedance itself. This is

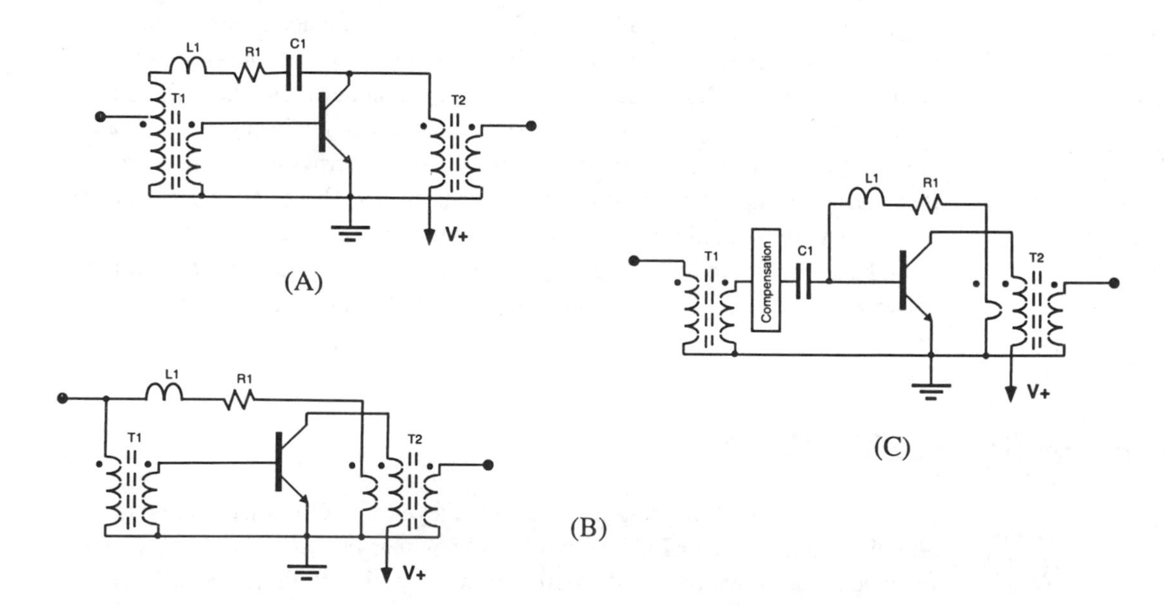

FIGURE 12-2

In circuit (A), negative feedback is derived directly from the collector. Adjustment of the feedback voltage source with respect to the base is done in T1 by providing a high impedance input point. In circuit (B), a lower impedance point than the collector is created by adding a third winding in T2. This allows the feedback voltage to be fed directly to the input of T1. In (C), the feedback voltage is fed to a low impedance point (the base), necessitating a low impedance voltage source. This is also done by adding a third winding in T2.

the most commonly used negative feedback arrangement with BJTs since the base impedance is well defined leaving only one variable, the third winding in T2. Instead of T2, the third winding for deriving the feedback voltage can be located in the collector/drain D.C. feed choke,[5, 6] which is sometimes more convenient because of its proximity to the input. The voltage swing across the choke is equal to that across the output transformer. However, its use as the source for feedback voltage increases flexibility of the circuit design since its impedance ratio to the feedback winding is easily adjustable without affecting output matching of the transistor.

The negative feedback loops for FETs are easier to determine because the FET is a voltage controlled device. For BJTs the voltages must be converted to current since the base voltage variations are small and it would be difficult to achieve sufficient accuracy with calculations. Thus the model for a negative feedback loop shown in Figure 12-3,[8] which is meant primarily for FET amplifiers, can also be used with BJTs in a modified format.

This model refers to the push-pull amplifier design given in Figure 12-4. At 10 MHz, which is the low frequency end of this example, magnetic cores are necessary in the input and output transformers, but are not shown in the schematic for simplicity. In the model, the feedback voltage is derived directly from the FET drains, which as explained earlier, will limit the optimization of the system in this respect. A peak in power gain of about 2.5 dB will remain around the middle of the 10–175 MHz spectrum. If a flatter gain response is required, methods shown in Figure 12-2A and C are recommended. Although there is a considerable phase deviation from 180 degrees at 175 MHz as a result of the series L, there will also be about 1 dB gain reduction at 175 MHz due to the finite reactances of the inductances. At low frequencies, where the amount of feedback is at its maximum, the phase error is negligible and the model of Figure 12-3 will result in fairly accurate values. In the model the series inductance, which is used to further shape the gain slope, has been omitted. This L can be treated as an additional variable and its value for the spectrum in question would probably be lower than the minimum achievable with the physical size of the circuitry. Ideally the reactance of the series L should be infinite at the high end of the spectrum and zero at the low end. C_1 and C_2 in Figure 12-4 are D.C. blocking capacitors and their values are not critical, but must be large enough to present a low reactance at the lowest frequency of operation.

It is assumed that Q1 and Q2 are each MRF151[6] devices. In the frequency region of 10 to 175 MHz, these could be replaced with a single push-pull device, namely the MRF151G. (The MRF151G is equivalent to two MRF151's

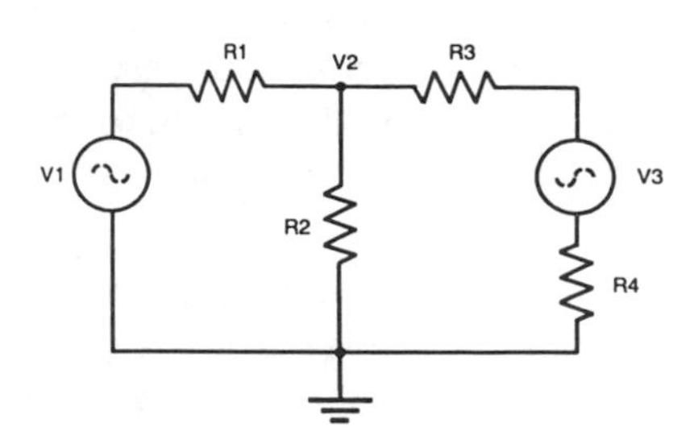

FIGURE 12-3
A simplified model of a negative feedback network which can be used to determine the loop parameters with sufficient accuracy. The design of this model is based on a series RLC loop.

in a single package, but tested to 175 MHz specifications, whereas the MRF151 is tested at 30 MHz although it is usable to at least 175 MHz).

From the data sheet and by simple calculations, we can establish that the nearest full integer impedance ratios of 9:1 and 1:4 are the closest practical at 175 MHz (with 50 Ω interface) for the input and output transformers respectively. Referring to Figure 12-3, from the data sheet we can also deduct the following parameters:

G_{PS} at 10 MHz ≈ 26 dB,

G_{PS} at 175 MHz ≈ 16 dB (lowered to 15 dB with feedback),

P_{in} 1 (f = 10 MHz, P_{out} = 300 W) = 0.75 W, V_{in} (RMS) = 2.03 V (V_2),

P_{in} 2 (f = 175 MHz, P_{out} = 300 W) = 9.50 W, V_{in} (RMS) = 7.23 V (V_1),

$V_3 = V_{out}$ (RMS) (drain to drain) = 61.25 V,

R_1, R_2 (transformer source and gate-to-gate impedances) = 5.5 Ω,

R_3 = feedback resistor,

R_4 = (output load) = 12.5 Ω.

The value of the feedback resistor is given by:

$$R_3 = \frac{(V_2 + V_3)}{\left(\frac{V_1 - V_2}{R_1}\right) - \left(\frac{V_2}{R_2}\right)} - R_4$$

$$= R_3 = \frac{(2.03 + 61.25)}{\left(\frac{7.23 - 2.03}{5.5}\right) - \left(\frac{2.03}{5.5}\right)} - 12.5 = 96.6\Omega / 2$$

or 48.3 Ω each resistor. The total power dissipated in the feedback resistors at the low frequency end of the spectrum of operation, which represents the worst case is:

$$(V_2 + V_3) \times \left[\left(\frac{V_1 - V_2}{R_1}\right) - \left(\frac{V_2}{R_2}\right)\right]$$

or 63.28 × 0.58 = 36.7 W, or 18.35 W per resistor. But this assumes that the series L has zero reactance. (There are no simple formulas available to calculate the values of the R and series L versus the frequency response, but some computer programs are able to plot the amplifier's response characteristics for given values of these elements).[9, 10] Any series reactance would be treated as added series R at a given frequency, and deducted from its original value. Since there will be a voltage drop across the reactance, the voltage across R will be lower resulting in a reduced dissipation.

At the low frequency end, it is customary to select the series L with its reactance approximately equal to the input impedance of the device, which in this case has a value of 80 nH (5 Ω) at 10 MHz and, for this value, the phase delay is negligible. At 175 MHz, the same inductance represents a reactance of

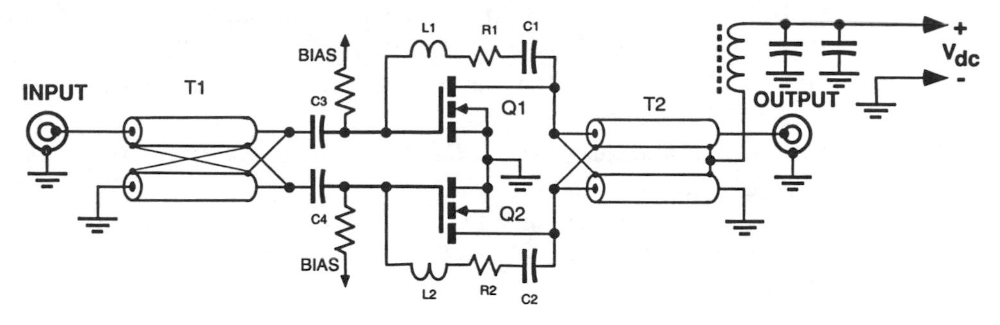

FIGURE 12-4
A FET RF push-pull amplifier with negative feedback. The component values for the RLC feedback networks can be established with the model of Figure 12-3.

90 W resulting in a phase delay of about 15°. This phase delay is normal and would become dangerous only if 180° is approached, resulting in the feedback turning to positive in phase which is likely to create instabilities. Such a phase shift could occur only if the initial value of L is unnecessarily high or if the amplifier bandwidth is 7-8 octaves or more, which may be the case in certain low power designs. The Q value of the series L, which is already reduced by the series R as X/R can be further controlled with a parallel R (R/X). In practice typical Q values for the series L are less than 10 in most cases.

From these examples we can see that the power loss at low frequencies is considerable and in this case amounts to 6–7% of the overall efficiency. The feedback resistor values can be rounded to 50 Ω, and the reactance of the series L is 5 Ω, but the dissipation factor of R is reduced only by 10%. Recalculating using the formula above, we get dissipation figures of: $(63.28 - 6.33) \times 0.58 =$ 32.5 W or 16.75 W for R_1 and R_2 of the push-pull amplifier in Figure 12-4. It can be noticed that at the low frequency end of the amplifier's frequency band the change in efficiency is minimal by adding the series L, but at the high end (175 MHz) the effective value of the feedback resistor is increased from 50 W to 140 W, which including the phase delay results only in an approximate 1 dB gain loss. If the loss of efficiency with negative feedback is not acceptable in an application, a combination of the RLC compensation technique and negative feedback (Figures 12-1 and 2) usually yields excellent results.[1, 5]

References

[1] William E. Sabin and Edgar O. Schoenike, *Single-Sideband Systems and Circuits*, New York: McGraw-Hill Book Company, Inc., 1987.
[2] A. Boekhoudt, "Applications Laboratory Report ECO 7501," Philips Components, Discrete Semiconductor Group, 1975.
[3] J. Mulder, "Applications Laboratory Report ECO 7114," Philips Components, Discrete Semiconductor Group, 1971.
[4] M. J. Koppen, "Applications Laboratory Report ECO 7308," Philips Components, Discrete Semiconductor Group, 1974.

[5] H. O. Granberg, "A Two Stage 1 kW Linear Amplifier," Application Note AN 758, Motorola Semiconductor Sector, Phoenix, AZ, 1974.

[6] *RF Device Data*, DL110, Rev 4, Volume II, Motorola Semiconductor Sector, Phoenix, AZ.

[7] Kraus, Bostian, and Raab, *Solid State Radio Engineering*, New York: John Wiley & Sons, Inc., 1980

[8] H. O. Granberg, "Building Push-Pull Multioctave VHF Power Amplifiers," *Microwaves & RF*, November, 1987.

[9] Eagleware, 1750 Mountain Glen, Stone Mountain, GA 30087, (404) 923-9999.

[10] Nedrud Data Systems, P.O. Box 27020, Las Vegas, NV 89126, (702) 255-8080.

[11] PSpice by MicroSim Corporation, 20 Fairbanks, Irvine, CA 92718, (714) 770-3022.

13

Small Signal Amplifier Design

This chapter will describe a simple, straight-forward approach to the design of low power RF amplifiers. The three basic ingredients of a design are the selection of a bias point and then the use of scattering parameters and noise parameters to complete a specific circuit. Selection of a transistor is assumed based on the required gain at frequency, noise at frequency, package type, and operating voltage available. A brief discussion of scattering parameters will be given, followed by noise parameters. Actual design examples will illustrate the step-by-step procedure.

SCATTERING PARAMETERS

The introduction of scattering parameters (commonly referred to as *S-parameters*) in the late 1960s resulted in a simple, systematic, and accurate theoretical method of designing low power RF amplifiers. Other two-port parameters, such as "y-parameters," have been used for many years but without widespread acceptance. Probably the most significant reason for the popularity of S-parameters has been their ready availability brought about by the creation of specialized test equipment used to perform S-parameter measurements and the large number of articles that have been written describing their use in a variety of practical design situations.[1, 2, 3, 6] Today, all low power RF transistors introduced to the marketplace are characterized with S-parameters, generally at several bias conditions and over a wide range of frequencies. Because S-parameters can be measured quickly and accurately with automated test equipment,[2] it is relatively easy for semiconductor manufacturers to supply to a customer any specialized data that may be required for specific bias conditions and frequencies not specified on the device's data sheet.

Scattering parameters tell you "everything you need to know" about small signal amplifier design with one exception—noise. Impedance matching, gain, input and output VSWR, and stability can be expressed by mathematical equations involving S-parameters. S-parameters are basically a means for characterizing n-port networks using the concept of traveling waves. A traveling wave created by a generator (source) and launched on a transmission line toward a load is referred to as an "incident" wave. Any mismatches encountered by the incident wave will result in a "reflected" wave which travels back down the transmission line toward the generator. For a two-port network such as a transistor, if the "network" is embedded in a 50 Ω measuring system, the "S-parameters" become simply the coefficients of the incident and reflected voltage waves, as described in Figure 13-1.

S_{11} and S_{22} in a 50 Ω system are the input and output voltage reflection coefficients which can be related to input and output VSWR by the formula

$$\mathrm{VSWR} = 1 + |\Gamma| / 1 - |\Gamma|, \tag{13-1}$$

where $|\Gamma|$ is the magnitude of the voltage reflection coefficient. The quantity $|S_{21}{}^2|$ is the power gain of the transistor at the specified bias conditions and frequency—and, of course, with 50 Ω source and load terminations.

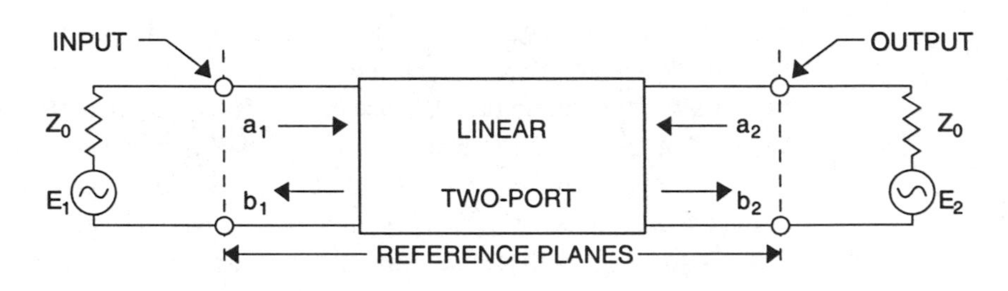

S_{11} = INPUT REFLECTION COEFFICIENT = $\left.\frac{b_1}{a_1}\right|_{a_2 = 0}$

S_{22} = OUTPUT REFLECTION COEFFICIENT = $\left.\frac{b_2}{a_2}\right|_{a_1 = 0}$

S_{21} = FORWARD TRANSMISSION COEFFICIENT = $\left.\frac{b_2}{a_1}\right|_{a_2 = 0}$

S_{12} = REVERSE TRANSMISSION COEFFICIENT = $\left.\frac{b_1}{a_2}\right|_{a_1 = 0}$

FIGURE 13-1
Two-port S-parameter definitions.

NOISE PARAMETERS

There are three basic noise parameters that completely describe the noise characteristics of a low power transistor. These are the minimum possible noise figure that can be obtained from the transistor—called NF_{min}, the equivalent noise resistance of the transistor—called R_n and the optimum source reflection coefficient—called Γ_{opt}. Sometimes people refer to four basic noise parameters, and this is because the quantity Γ_{opt} is a complex number and is often referred to by stating its magnitude and angle. Also the quantity R_n is sometimes normalized to a specific characteristic line impedance by dividing the quantity by Z_o. When this is done, the normalized noise resistance is always specified using the lower case letter "r," i.e.,

$$r_n = R_n / Z_o. \quad (13\text{-}2)$$

A given value of noise figure, NF, can be determined from the equation:[6]

$$NF = NF_{min} + 4r_n \{|\Gamma_s - \Gamma_{opt}|^2 / (1 - |\Gamma_s|^2)(|1 + \Gamma_{opt}|^2). \quad (13\text{-}3)$$

Once the three noise parameters are known, it can be seen from the above equation that the noise figure of a transistor amplifier for a specific bias condition and frequency is entirely dependent on the source impedance seen by the transistor, i.e., Γ_s. If you specify the value of NF, it can be shown[2] that the locus of points representing possible values of Γ_s are circles on the Smith Chart. The radius of a noise circle will increase with increasing values of NF, with the circle having zero radius being located at the point of Γ_{opt}. It can also be shown[2] that the centers of all the NF circles will lie along the Γ_{opt} vector which originates at the center of the Smith Chart and terminates at the location of Γ_{opt}.

Finally, it can be shown[2] that the centers of the noise figure circles are located at the points determined by the following equation:

$$C_{Fi} = \frac{\Gamma_{opt}}{1 + Ni} \quad (13\text{-}4)$$

where Ni is a noise figure parameter defined by the equation

$$Ni = \frac{NF_i - NF_{min}}{4r_n} \bullet |1 + \Gamma_{opt}|^2 \quad (13\text{-}5)$$

and NF_i is the value of the desired noise figure circle. Likewise, the radii of the circles are given by the expression

$$R_{Fi} = \frac{1}{1 + Ni}\left[N_i^2 + N_i\left(1 - |\Gamma_{opt}|^2\right)\right]^{1/2}. \quad (13\text{-}6)$$

The optimum source reflection coefficient (Γ_{opt}), the noise resistance (r_n), and minimum noise figure (NF_{min}) remain the same as previously described. Plotting noise figure circles is a tedious operation best done by computers with programs that work in conjunction with Smith Chart displays. However, if the circles are not given by the device manufacturer for the conditions you desire, you are left with few alternatives if you wish to perform a systematic design of a low noise amplifier. Most RF low power transistor manufacturers have automated equipment and computer programs for generating noise (and gain) circles and will provide users with the required information as part of the job of selling their transistors.

BIASING CONSIDERATIONS

Now that we've talked about S-parameters and noise parameters for an RF low power transistor, let's discuss bias conditions. Choosing the bias point is less difficult than designing a suitable bias network. First, the manufacturer supplies a curve showing f_τ versus collector current for a bipolar transistor. For good gain characteristics, it is necessary to bias the transistor at a collector current that results in maximum or near-maximum f_τ. On the other hand, for best noise characteristics a low current is generally most desirable. Finally, one must consider the maximum signal level expected at the input of the transistor. The bias point must be at a sufficiently high current (and voltage)level to prevent the input signal from swinging the collector current out of the "linear" region of operation. It is assumed that a transistor has been chosen having a sufficient operating current level to prevent the input signal from driving the transistor into the so-called saturated region of operation, which would also be an operating condition that would prevent Class A (or linear) operation.

If the amplifier is to work over a range of temperature, the circuit designer must attempt to design a bias network that maintains the D.C. bias point as the operating temperature changes. Two basic internal transistor characteristics are known to have a significant effect on the D.C. bias point. These are ΔV_{BE} and $\Delta\beta$. The base-emitter voltage of a bipolar transistor decreases with increasing temperature at the rate of about 2.5 mV/°C. Emitter voltage V_E tends to minimize the effect because as base current increases (as V_{BE} decreases) collector current increases and this causes V_E to increase also. However as V_E increases, collector current tends to decrease. A mathematical expression for this behavior is:

$$\Delta I_C = \Delta V_{BE}\, I_C/V_E \tag{13-7}$$

where ΔI_C = change in I_C,

I_C = quiescent collector current,

ΔV_{BE} = change in base-to-emitter voltage, and

V_E = quiescent emitter voltage.

Likewise, the transistor's D.C. current gain β typically increases with increasing temperature at the rate of about 0.5 % per °C. Further bias circuit complications arise from the fact that most semiconductor manufacturers control β the least of any major dc specification. It is not uncommon to have a bipolar transistor with a range of β that exceeds 5 or 6 to 1. That is to say, the ratio of guaranteed maximum β to minimum β is 5 or 6 to 1. A more normal range is 4:1 and only by special selection can the manufacturer achieve a guaranteed range of 2:1.

It can be shown[1] that a change in collector current for a corresponding change in β can be approximated by the equation:

$$\Delta I_c = I_{c1}\left(\frac{\Delta\beta}{\beta_1\beta_2}\right)\left(1+\frac{R_B}{R_E}\right) \tag{13-8}$$

where

I_{C1} = collector current at $\beta = \beta_1$,

β_1 = the lowest value of β,

β_2 = the highest value of β,

$\Delta\beta = \beta_2 - \beta_1$,

R_B = the parallel combination of the resistors R1 and R2 in a base bias network, and

R_E = the emitter resistor.

This equation indicates why one desires a minimum spread in D.C. current gain. The smaller the value of $\Delta\beta$ (both with temperature and from transistor to transistor), the lower will be the resulting change in collector current. However, once a transistor is specified, the only control left for the designer is the resistance ratio R_B/R_E, unless a more complicated bias network is chosen, such as a constant current source. Obviously, the smaller this ratio, the less the collector current will vary. However, the lower the value of R_B/R_E the lower is the current gain of the amplifier. A practical rule of thumb is to keep the ratio less than but close to the value of 10.

A typical bias circuit is shown in Figure 13-2. Bowick[1] walks the reader through the necessary calculations for this circuit as follows:

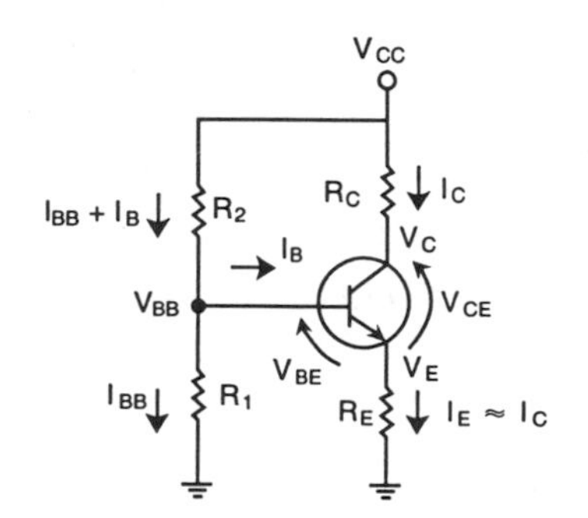

1. Choose the operating point for the transistor.

$$I_C = 10 \text{ mA}, V_C = 10 \text{ V}, V_{CC} = 20 \text{ V}, \beta = 50$$

2. Assume a value for V_E that considers bias stability:

$$V_E = 2.5 \text{ volts}$$

3. Assume $I_E \sim I_C$ for high-beta transistors.
4. Knowing I_E and V_E, calculate R_E.

$$R_E = \frac{V_E}{I_E} = \frac{2.5}{10 \times 10^{-3}} = 250 \text{ ohms}$$

5. Knowing V_{CC}, V_C, and I_C, calculate R_C.

$$R_C = \frac{V_{CC} - V_C}{I_C} = \frac{20 - 10}{10 \times 10^{-3}} = 1000 \text{ ohms}$$

6. Knowing I_C and β, calculate I_B.

$$I_B = \frac{I_C}{\beta} = 0.2 \text{ mA}$$

7. Knowing V_E and V_{BE}, calculate V_{BB}.

$$V_{BB} = V_E + V_{BE} = 2.5 + 0.7 = 3.2 \text{ volts}$$

8. Assume a value for I_{BB}, the larger the better (see text):

$$I_{BB} = 1.5 \text{ mA}$$

9. Knowing I_{BB} and V_{BB}, calculate R_1.

$$R_1 = \frac{V_{BB}}{I_{BB}} = \frac{3.2}{1.5 \times 10^{-3}} \text{ ohms} = 2133 \text{ ohms}$$

10. Knowing V_{CC}, V_{BB}, I_{BB}, and I_B, calculate R_2.

$$R_2 = \frac{V_{CC} - V_{BB}}{I_{BB} + I_B} = \frac{20 - 3.2}{1.7 \times 10^{-3}} = 9882 \text{ ohms}$$

FIGURE 13-2
Typical bias network.

POWER GAIN

Transducer power gain, G_t, is defined as the power delivered to the load divided by the power available from the source. It can be shown[1] that transducer (or amplifier) gain is given by:

$$G_t = \frac{|S_{21}|^2\left(1-|\Gamma_s|^2\right)\left(1-|\Gamma_L|\right)^2}{|(1-S_{11}\Gamma_s)(1-S_{22}\Gamma_L)-S_{12}S_{21}\Gamma_S\Gamma_L|^2} \tag{13-9}$$

This expression can be stated in different ways[6] (through mathematical manipulations) as follows:

$$G_t = \frac{1-|\Gamma_S|^2}{|1-\Gamma_{IN}\Gamma_S|^2}\bullet|S_{21}|^2\bullet\frac{1-|\Gamma_L|^2}{|1-S_{22}\Gamma_L|^2} \tag{13-10}$$

where

$$\Gamma_{IN} = S_{11} + \frac{S_{12}S_{21}\Gamma_L}{1-S_{22}\Gamma_L}, \tag{13-11}$$

or

$$G_t = \frac{1-|\Gamma_S|^2}{|1-S_{11}\Gamma_S|^2}\bullet|S_{21}|^2\bullet\frac{1-|\Gamma_L|^2}{|1-\Gamma_{OUT}\Gamma_L|} \tag{13-12}$$

where

$$\Gamma_{OUT} = S_{22} + \frac{S_{12}S_{21}\Gamma_S}{1-S_{11}\Gamma_s} \tag{13-13}$$

Equation 13-10 relates G_t to an input term ($\{1 - |\Gamma_s|^2\}/|1 - \Gamma_{IN}\Gamma_s|^2$), a device term ($|S_{21}|^2$) and an output term ($\{1 - |\Gamma_L|^2\}/|1 - S_{22}\Gamma_L|^2$) where the input term is dependent on output quantities. Equation 13-12 shows a similar expression except in this case the output term depends on input quantities. Likewise, it can be shown[6] that if the source reflection coefficient, Γ_s, is made equal to the conjugate of the transistor input reflection coefficient, Γ_{IN}, (that is, the transistor input is conjugately matched) we obtain an expression G_P called the operating power gain. The importance of G_P is that it is "independent" of the source impedance because we forced Γ_s to be equal to S_{11}^*. The equation for G_P is:

$$G_p = \frac{1}{1-|\Gamma_{IN}|^2}\bullet|S_{21}|^2\bullet\frac{1-|\Gamma_L|^2}{|1-S_{22}\Gamma_L|^2} \tag{13-14}$$

Equations 13-9, 13-10, or 13-12 can be solved for known values of load and source reflection coefficients. The problem, however, is that the load reflection coefficient depends on the source reflection coefficient and vice versa. The equation can be solved but only through an iterative process. At the root of our problem is the term S_{12}, which is the cause of the interaction of input and output. In some cases, S_{12} is sufficiently small to be considered equal to zero. Such a network is called a unilateral network. In some cases, S_{12} cannot be neglected. If we wish to find an exact solution ($S_{12} \neq 0$), we could turn to Equation 13-14 and develop a process for determining the load reflection coefficient. For the time being, we will assume our network is unilateral, work with the simpler equations and develop a technique for determining source and load impedances to obtain desired amplifier performance. Then, after we have digested this large bite, we can return to the situation where we cannot assume $S_{12} = 0$.

While the condition of $S_{12} = 0$ is generally never true in real life, it is often a good approximation. One way to verify if a network can be considered unilateral is to calculate a term called the "unilateral figure of merit." This quantity, called U,[6] is defined by the following formula:

$$U = \frac{|S_{11}|\,|S_{21}|\,|S_{12}|\,|S_{22}|}{\left(1-|S_{11}|^2\right)\left(1-|S_{22}|^2\right)} \tag{13-15}$$

If we define G_{tu} as the transistor power gain with $S_{12} = 0$ and G_t as the actual transistor power gain, the maximum error introduced by using G_{tu} instead of G_t is given by the expression[6]

$$\frac{1}{(1+U)^2} < \frac{G_t}{G_{tu}} < \frac{1}{(1-U)^2} \tag{13-16}$$

To illustrate the use of Equation 13-16, let's take the MRF571 at 1 GHz and a bias condition of 6 volts and 50 mA. From the data sheet,[9]

$|S_{11}| = 0.60$

$|S_{12}| = 0.09$

$|S_{21}| = 4.4$

$|S_{22}| = 0.11$

Equation 13-15 determines the value of U as

$$U = \frac{(.60)(.09)(4.4)(.11)}{[1-(.60)^2][1-(.11)^2]} = \frac{.0261}{(.64)(.988)} = 0.0413$$

and from Equation 13-16 we can calculate the minimum and maximum errors as

–0.35 dB and + 0.37 dB.

Frequently, the errors are less than 0.25 dB and, as such, are sufficiently small to justify using G_{tu}.

Now let's return to the expression for G_t (Equation 13-9) and assume $S_{12} = 0$. Our expression becomes:

$$G_{tu} = \frac{1-|\Gamma_S|^2}{|1-S_{11}\Gamma_S|^2} \cdot |S_{21}|^2 \cdot \frac{1-|\Gamma_L|^2}{|1-S_{22}\Gamma_L|^2} \cdot \tag{13-17}$$

This equation can be broken into three sources of gain, namely

$$G_o = |S_{21}|^2 \qquad (13\text{-}18)$$

which is the contribution of the transistor itself;

$$G_S = (1 - |\Gamma_s|^2) / |1 - S_{11}\Gamma_s|^2 \qquad (13\text{-}19)$$

which is the "gain" achieved by the input circuit; and

$$G_L = (1 - |\Gamma_L|^2) / |1 - S_{22}\Gamma_L|^2 \qquad (13\text{-}20)$$

which is the "gain" achieved by the output circuit. This is illustrated by the three gain blocks shown in Figure 13-3.

If the circuit design is narrow band and you desire maximum gain, all that is required is to set $\Gamma_s = S_{11}^*$ and $\Gamma_L = S_{22}^*$. If the circuit is broad band and one desires a certain amount of gain across a band of frequencies, then what is required is to use circuits that compensate for the variations in gain with frequency of the device itself. This is usually done in one of two ways: with feedback or with selective mismatching. In selective mismatching, the input and output "gains" are varied (by matching) to compensate for the gain variations with frequency of the transistor, which is represented by $|S_{21}|^2$.

As we vary Γ_s to other values, thereby causing G_s, the "gain" created by the input matching network, to vary between 0 and $G_{s,\,max}$, we find that for a given value of gain G_s, the locus of points representing values of Γ_s is a circle. And in a manner similar to noise circles, the center of the circle having zero radius is located at the point S_{11}^*. The radius of a gain circle will increase with increasing values of G_s, and again the centers of all the gain circles will lie along the S_{11}^* vector which originates at the center of the Smith chart and terminates at the location of S_{11}^*. An identical situation occurs for the output matching network. Another set of "gain" circles (G_L) can be drawn whose centers lie along the S_{22}^* vector which originates at the center of the Smith chart and terminates at the location of S_{22}^*. Typical gain circles are shown in Figure 13-4.

In a manner similar to noise circles, the gain circles for either the input net-

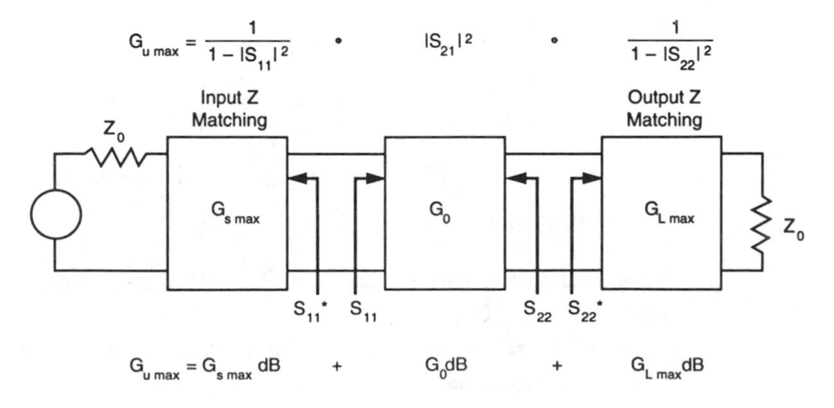

FIGURE 13-3

Amplifier representation for $S_{12} = 0$.

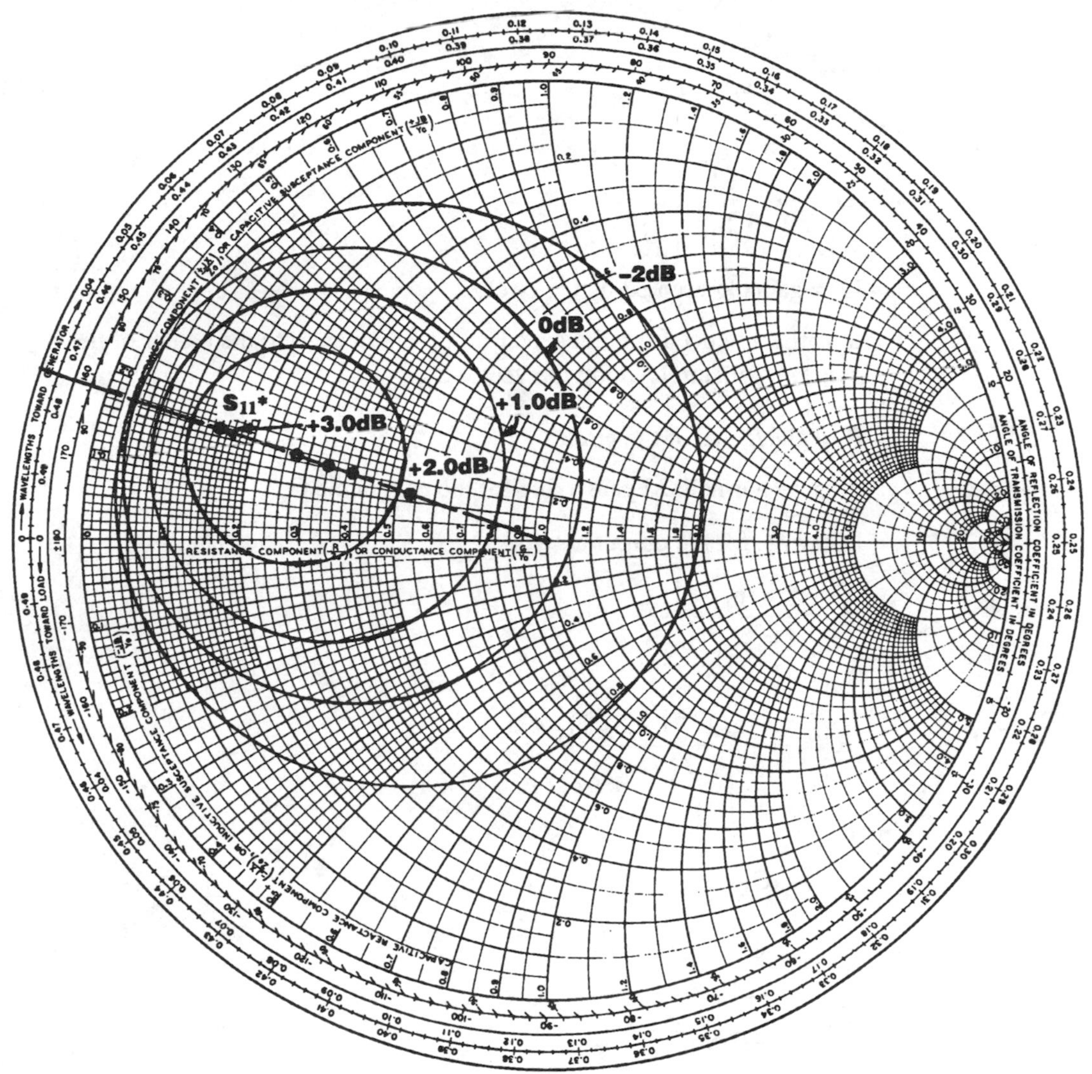

FIGURE 13-4

Typical input gain circles.

work or the output network can be drawn on a Smith Chart using the following formulas:[2]

(For the input network)

$$d_s = \frac{g_s|S_{11}|}{1-|S_{11}|^2(1-g_s)} \tag{13-21}$$

$$R_S = \frac{(1-g_s)^{1/2}(1-|S_{11}|^2)}{1-|S_{11}|^2(1-g_s)} \tag{13-22}$$

where

$$g_s = \frac{G_s}{G_{s,MAX}}, \tag{13-23}$$

and

$$G_s = \frac{1-|\Gamma_s|^2}{|1-\Gamma_s S_{11}|^2}, \tag{13-24}$$

G_s = gain represented by the circle,

d_s is the distance from the center of the Smith chart to the center of the constant gain circle along the vector S_{11}^*

R_s is the radius of the circle

g_s is the normalized gain value for the gain circle G_s.

Likewise, for the output network:

$$d_L = \frac{g_L|S_{22}|}{1-|S_{22}|^2(1-g_L)} \tag{13-25}$$

$$R_L = \frac{(1-g_L)^{1/2}(1-|S_{22}|^2)}{1-|S_{22}|^2(1-g_L)} \tag{13-26}$$

and

$$g_L = \frac{G_L}{G_{L,MAX}} \tag{13-27}$$

$$G_L = \frac{1-|\Gamma_L|^2}{|1-\Gamma_L S_{22}|^2} \tag{13-28}$$

where

G_L = gain represented by the circle,

d_L is the distance from the center of the Smith Chart to the center of the constant gain circle along the vector S_{22}^*,

R_L is the radius of the circle,

g_L is the normalized gain value for the gain circle G_L.

These circles represent different values of G_s, the "gain" created by the input matching network or G_L, the "gain" created by the output matching network. Note that negative "gain" circles can be drawn for both cases. We can use input and output "gain" circles in two kinds of amplifier designs: 1) designing an amplifier with a specified amount of gain; and 2) designing a broad band amplifier having a specified gain over a band of frequencies.

In either case, "gain" or "loss" from the input and/or output matching networks can be allocated in whatever manner desired provided the gains (or losses) are actually realizable. The maximum available gain can be determined at any frequency by conjugate matching and it is obvious one cannot achieve more gain than this value at a specified frequency. It is common to assign half the loss (or gain) to both input and output circuits although this is not essential.

STABILITY

Before launching into practical examples of design, let's return to the assumption of unilateral gain. It helped us to analyze the overall gain of a transistor stage by considering contributions from three parts. However, assuming that S_{12} has a value of zero ignores the problem of amplifier stability. It also leads to the erroneous conclusion that output matching has no effect on input matching. Amplifier design calculations which do not include device (and circuit) feedback are only an approximation which can lead to inaccurate solutions and possibly circuit oscillations when the design is realized.

So how does one achieve acceptable gain, acceptable noise figure and stability in the real world of modern high performance transistors that have values of S_{12} other than zero? The answer is straightforward. S-parameters come to the rescue again by allowing one to calculate device stability by determining a term called the Rollett Stability Factor K.[1, 6] To make the equation simple, you should first calculate an intermediate quantity referred to as D_s {called Δ in the references[6]}:

$$D_s = S_{11} S_{22} - S_{12} S_{21} \quad (13\text{-}29)$$

The stability factor K is then calculated as

$$K = (1 + |D_s|^2 - |S_{11}|^2 - |S_{22}|^2)/2\,|S_{21}|\,|S_{12}| \quad (13\text{-}30)$$

If K is greater than unity, then the device will be unconditionally stable for any combination of source and load impedance. If, on the other hand, K calculates to be less than 1, the device is potentially unstable and will most likely oscillate with certain combinations of source and load impedance.

S-parameters go one step further. They permit the calculation of "stability circles" which can be plotted on the Smith chart and which separate regions of stability and instability. Generally only a portion of the circle will be visible on the Smith chart. Then, when choosing source and load impedances, you must be careful to avoid values that lie within the regions of instability. Manufacturers who supply gain and noise circle data with their transistors also plot regions of instability which are typically indicated by dashed lines. Obviously these circles (or portions thereof) will not exist within the Smith chart boundaries for transistors with a value of $K>1$.

Calculating and plotting "instability" circles are straightforward operations involving S-parameters. However, the operations are tedious and, again, best performed using a computer program. This can be easily understood by referring to Figure 13-5 and to the equations for the center locations of the input instability circle and the output instability circle along with equations for their radii r_{s1} listed below:[1]

FIGURE 13-5
Stability circles.

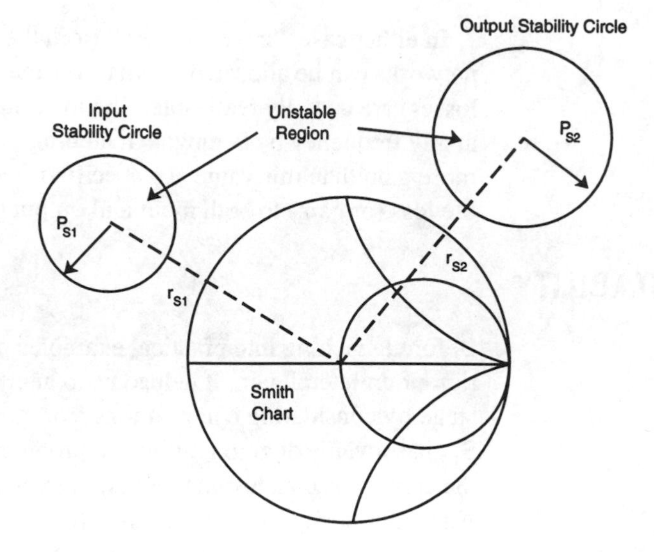

$$r_{s1} = \frac{C_1^*}{|S_{11}|^2 - |D_s|^2}, \tag{13-31}$$

where $D_s = S_{11}\, S_{22} - S_{12}\, S_{21}$

$$C_1 = S_{11} - D_s S_{22}^* \tag{13-32}$$

and r_{s1} = center location of the input stability circle.

Also

$$p_{s1} = \left| \frac{S_{12}\, S_{21}}{|S_{11}|^2 - |D_s|^2} \right|, \tag{13-33}$$

where p_{s1} = radius of the input stability circle.

Likewise,

$$r_{s2} = \frac{C_2^*}{|S_{22}|^2 - |D_s|^2}, \tag{13-34}$$

where

$$C_2 = S_{22} - D_s S_{11}^* \tag{13-35}$$

and r_{s2} = center location of the output stability circle.

And

$$p_{s2} = \frac{S_{12}\, S_{21}}{|S_{22}|^2 - |D_s|^2}, \tag{13-36}$$

where p_{s2} = radius of the output stability circle.

However, determining the proper source and load impedances is simplified to a large extent when the transistor can be treated as a unilateral network. And if we've satisfied ourselves about the stability of our circuit, then we will find it beneficial at least as a first approximation in our circuit design to treat the circuit in this manner whenever possible.

If it is not possible to assume $S_{12} = 0$, then Equation 13-14 can be used to develop a mathematical procedure for determining values of Γ_L and $\Gamma_s = \Gamma_{IN}^*$. First, take Equation 13-14 and manipulate it in a manner that allows recognition of constant operating power gain circles[6] having radii R_P of

$$R_p = \frac{\left[1 - 2K|S_{12}S_{21}|g_p + |S_{12}S_{21}|^2 g_p^2\right]^{1/2}}{\left|1 + g_p\left(|S_{22}|^2 - |D_s|^2\right)\right|} \tag{13-37}$$

where K is the previously identified Rollett Stability Factor (see Equation 13-30), $D_s = S_{11}\,S_{22} - S_{12}\,S_{21}$ (previously specified in Equation 13-29) and

$$g_P = G_P/|S_{21}|^2. \tag{13-38}$$

The locations of the centers C_P of the circles are

$$C_p = \frac{g_p C_2^*}{1 + g_p\left(|S_{22}|^2 - |D_s|^2\right)}, \tag{13-39}$$

where C_2 was previously given in Equation 13-35.

Note that the maximum operating power gain occurs when $R_P = 0$, and for this condition and for the case where $K > 1$ (the circuit is unconditionally stable) it can be shown[6] that

$$G_{p,MAX} = \frac{|S_{21}|}{|S_{12}|}\left(K - \sqrt{K^2 - 1}\right). \tag{13-40}$$

Remember we have already assumed that $\Gamma_s = \Gamma_{IN}^*$ and, under these conditions,

$$G_{P,max} = G_{t,max}$$

A step-by-step procedure for plotting a specific power gain circle would be:

a) Select the desired value of G_P.
b) Calculate g_P from Equation 13-38.
c) Calculate K from Equation 13-30.
d) Calculate D_s from Equation 13-29.
e) Determine R_P from Equation 13-37.
f) Determine C_P from Equation 13-39.

Once we select a value of Γ_L from a point on the gain circle, we can then determine $\Gamma_s = \Gamma_{IN}^*$ using Equation 13-11.

Because power gain circles involve load reflection coefficients, it is more common to plot constant available power gain circles which involve source reflection coefficients. The process is similar; the equations are slightly different and expressed below:[6]

Let

$$g_a = G_A / |S_{21}|^2 \tag{13-41}$$

and

$$C_1 = S_{11} - D_s S_{22}^* \tag{13-42}$$

Then

$$R_a = \frac{\left[1 - 2K|S_{21}S_{12}|g_a + |S_{21}S_{12}|^2 g_a^2\right]^{1/2}}{\left|1 + g_a\left(|S_{11}|^2 - |D_s|^2\right)\right|}, \tag{13-43}$$

and

$$C_a = \frac{g_a C_1^*}{1 + g_a\left(|S_{11}|^2 - |D_s|^2\right)}, \tag{13-44}$$

where $D_s = S_{11} S_{22} - S_{12} S_{21}$ (previously identified in Equation 13-29), R_a = radius of gain circle, and C_a = center of gain circle.

Constant available power gain circles involve Γ_s and, as seen earlier, constant noise figure circles also involve Γ_s. Thus, both sets of circles can be plotted together on a Smith Chart to provide trade-off information between gain and noise figure. These are the curves presented by device manufacturers in their low noise transistor data sheets.

A comparison of gain circles with noise circles, shown together in Figure 13-6, makes clear another fundamental point about low noise amplifier design. One cannot generally achieve minimum noise figure while at the same time achieving maximum gain. Designing a low noise amplifier, then, becomes a tradeoff of gain and noise figure to achieve an acceptable value of each. Again, it should be pointed out that drawing gain/noise circles is a tedious process best accomplished by use of computers with appropriate computer programs. Readers are referred to references 7 and 8.

SUMMARY OF GAIN/NOISE FIGURE DESIGN PROCEDURES

A step-by-step procedure for applying the theory discussed in the previous sections can be set forth as follows:

1. Once a transistor and its bias conditions have been selected, the S-parameters should be analyzed to determine if the simpler design procedures involving the assumption that $S_{12} = 0$ can be used. Equations 13-15 and 13-16 will place limits on the maximum error introduced by this assumption.
2. Next, use Rollett's Stability Factor—Equations 13-29 and 13-30—to identify the possibility of instabilities depending on source and load matching.

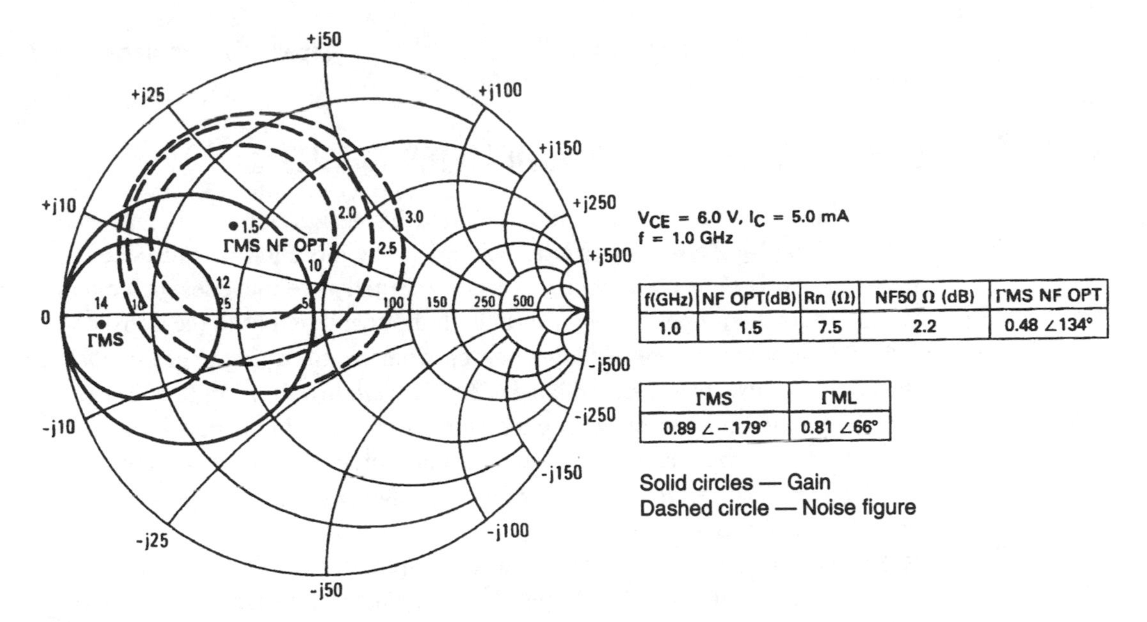

f(GHz)	NF OPT(dB)	Rn (Ω)	NF50 Ω (dB)	ΓMS NF OPT
1.0	1.5	7.5	2.2	0.48 ∠134°

ΓMS	ΓML
0.89 ∠ −179°	0.81 ∠66°

FIGURE 13-6
Gain and NF circles.

3. Subsequent steps depend on the desired results:
   a) If the application is narrow band and one desires maximum gain, then conjugate match input and output.
   b) If a specific gain is required at a single frequency, then use the gain circles provided by the device manufacturer (or draw the appropriate available gain circle using Equations 13-41 through 13-44). After the gain circle is drawn, then select a value for Γ_s and calculate $\Gamma_L = \Gamma_{OUT}^*$ using Equation 13-11. If it can be assumed that $S_{12} = 0$, you could divide the "gain" or "loss" between the input /output matching networks using Equations 13-21 through 13-28 and determine appropriate values for source and load terminations. It should be remembered in this case that the input and output of the amplifier will not be matched to Z_0. Therefore, if a low VSWR is a requirement for the design, this approach should not be taken.
   c) If a noise figure and gain at a frequency are needed, use both gain and noise figure circles provided by the device manufacturer and select an appropriate value of Γ_s and again calculate Γ_L as previously stated in 3b.
   d) If broad band performance is required, examine the $|S_{21}|^2$ performance of the transistor over the frequency range of interest and determine the amount of gain or loss that must be provided by the matching networks to keep the overall gain the same at the band edges. Plot these gain circles on a Smith Chart using Equations 13-41 through 13-44. By trial and error (or the use of a computer optimization program) determine a matching network that will satisfy both "gain/loss" circles simultaneously.

In the next section, we will look at examples intended to further clarify these procedures by working through specific problems.

ACTUAL STEPS IN LOW POWER AMPLIFIER DESIGN

Now that we have the tools behind us to actually design our circuit, let's step back and look at the processes left to complete the job. First we must choose a transistor. Next we must choose a bias point. Finally, we must design a circuit which offers the proper impedances at both input and output to the transistor. All of these steps depend in a large part on what we are trying to accomplish. Is the circuit narrow band? Broad band? Must it have low noise figure? These factors determine how we impedance match the transistor. In the matching process, there are two distinct parts: first, we must determine what the desired source and load impedances are; second, we must design networks that present these impedances to the transistor.

Remember that maximum gain at a single frequency is achieved by conjugate matching both input and output of the transistor. Lowest noise figure is achieved by creating the proper source impedance to the transistor that results in minimum noise. Tradeoffs between gain and noise figure are achieved by selecting source and load impedances that result in the desired gain and noise at the same time. Suitable performance over a band of frequencies is accomplished by presenting to the transistor source and load impedances that will achieve proper overall circuit results as frequency varies.

Whether or not you ignore S_{12} obviously depends on its magnitude and the accuracy you wish to accomplish in your design. Wherever possible it is recommended that you assume $S_{12} = 0$ at least in the initial design effort. And it is also suggested that if S_{12} cannot be ignored, one should use computer programs such as TOUCHSTONE or MMICAD (see Chapter 8), particularly if the design requires controlled noise and gain performance over a band of frequencies.

The best way to understand the procedures described in the preceding paragraphs is to work through specific examples. In the following section, examples will be given to illustrate determining source and load impedances for maximum gain at a single frequency, both with and without consideration of S_{12}. The third example will illustrate how to achieve a specified amount of gain, again for the same conditions as in Examples 1 and 2. A fourth example goes through the procedures for a broadband design. Finally, an example is given of a design achieving low noise while maintaining adequate gain.

DETERMINING DESIRED VALUES OF SOURCE/LOAD IMPEDANCES

Example #1: Narrow Band—Match For Optimum Gain, $S_{12} = 0$

First, let's take an example where we assume $S_{12} = 0$. Let's also assume the source and load impedances are 50 Ω, and we are interested only in maximum gain at a single frequency. The frequency of interest is 1 GHz. We select as our transistor the MRF571. Our bias will be 6 volts and 50 mA because the manufacturer's data sheet[9] shows that f_τ is near its peak at 50 mA and values of scattering parameters are given for this particular bias point.

From the data sheet we find

$S_{11} = 0.6$ at an angle of 156°,

$S_{22} = 0.11$ at an angle of -164°,

$S_{12} = 0.09$ at an angle of 70°,

$S_{21} = 4.4$ at an angle of 75°.

Thus $\Gamma_s = S_{11}^* = 0.6$ at an angle of –156° and $\Gamma_L = S_{22}^* = 0.11$ at an angle of +164°. These are plotted in Figure 13-7. Gs, the "gain" contributed by the input

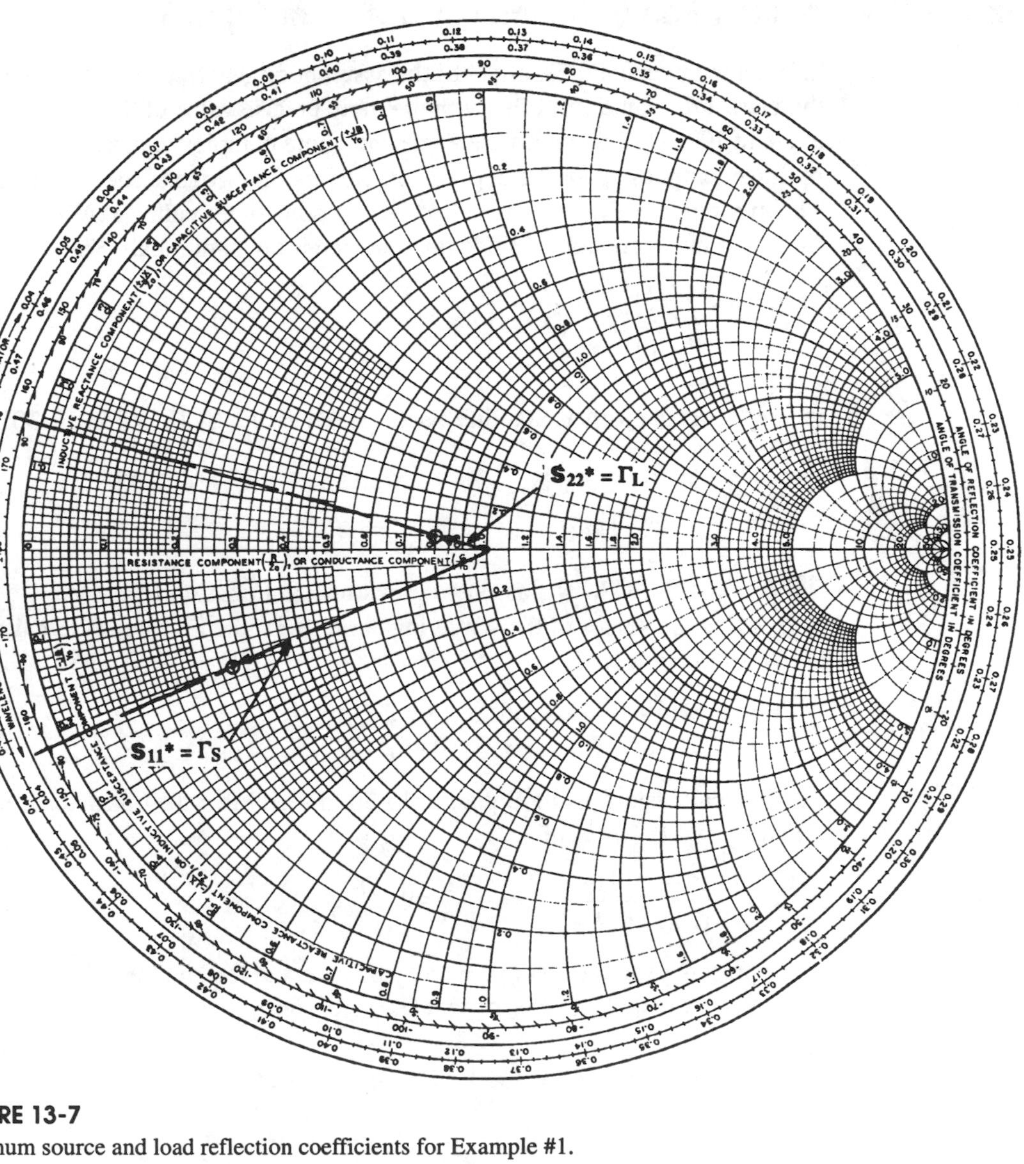

IMPEDANCE OR ADMITTANCE COORDINATES

FIGURE 13-7
Optimum source and load reflection coefficients for Example #1.

circuit is calculated from Equation 13-19 to be 1.56 or 10 $\log_{10}1.56$ which is 1.93 dB. Likewise for the output circuit where the magnitude of S_{22} is 0.11, the "gain" (G_L) contributed by the output circuit match is calculated from Equation 13-20 to be 1.01 or 0.05 dB. It is obvious in this example that the transistor is essentially matched in the output and little is gained from further matching.

Looking at $|S_{21}|^2$ we can determine the gain contributed by the device itself by using Equation 13-18. It is

$$G = (4.4)^2 \text{ or } 10 \log (4.4)^2 = 12.9 \text{ dB}$$

Thus the total gain expected from conjugate matching, then, is

$$G_{TU} = G_s + G_o + G_L = 1.94 + 12.9 + 0.05 = 14.9 \text{ dB}.$$

Example #2: Narrow Band—Match For Optimum Gain, $S_{12} \neq 0$

In order to understand the effect of the assumption that $S_{12} = 0$, let's calculate the gain of the same amplifier with the more precise value of $S_{12} = 0.09$ at an angle of 70°.[9] The formulas are somewhat more complex and, in addition, the optimum output impedance depends on the input impedance and vice versa. The maximum available gain, called $G_{P,max}$ (sometimes called MAG), is given by the formula (Equation 13-38) if the device is unconditionally stable:

$$G_{p,MAX} = \frac{|S_{21}|}{|S_{12}|}\left(K - \sqrt{K^2 - 1}\right),$$

where "K" is the Rollett Stability Factor given earlier by Equation 13-30:

$$K = \frac{1 + |D_s|^2 - |S_{11}|^2 - |S_{22}|^2}{2|S_{21}|\,|S_{12}|}$$

and $D_s = S_{11} S_{22} - S_{12} S_{21}$ as stated previously in Equation 13-29.

By using the values of the scattering parameters, the first step is to calculate D_s from Equation 13-29:

$$\begin{aligned} D_s &= (.6\angle 156°)(.11\angle{-164°}) - (.09\angle 70°)(4.4\angle 75°) \\ &= .066\angle{-8°} + .396\angle{-35°} = 0.46\angle{-23°}. \end{aligned}$$

Then we calculate K using Equation 13-30 to verify that the transistor is unconditionally stable (value of K>1):

$$K = \frac{1 - .36 - .0121 + .212}{(2)(.09)(4.4)} = \frac{.840}{.792} \cong 1.07.$$

In the present example, using the "S" parameters given and the subsequent calculated value of "K," we see that $G_{P,max}$ becomes

$$G_{p,MAX} = \frac{4.4}{.09}\left(1.07 - \sqrt{.145}\right) = 33.7 \text{ or } 15.3\text{dB}.$$

It is easy to see from the above calculations how much more complex the situation becomes when S_{12} cannot be assumed equal to zero. In the instance given, the difference in gain is approximately 0.4 dB and it appears marginal to have assumed $S_{12} = 0$. We could have determined the magnitude of the error at the outset by using Equation 13-16 as follows. First, calculate "U" from Equation 13-15:

$$U = (0.6)(4.4)(0.09)(.11) / [1-(.6)^2][1-(.11)^2] = 0.041$$

Then from Equation 13-14, we can determine the limits of possible error for assuming $S_{12} = 0$. The lower limit is $1/(1+U)^2 = 0.919$, while the upper limit is $1/(1-U)^2 = 1.108$. Expressed in decibels, these limits become + and –0.36 dB.

We can also see the effect that S_{12} will have on the optimum source impedance by letting thc load reflection coefficient remain the conjugate of S_{22} but then calculate Γ_s from Equation 13-11. When we do, we will find that Γ_s becomes 0.55 at an angle of –159°, only a slight change from the previous value for Γ_s when $S_{12} = 0$, which was 0.6 at an angle of –156°.

Example #3: Narrow Band, Specified Gain < Optimum Gain, $S_{12} = 0$

Before going to broad band circuit design, let's assume for the same transistor and same frequency as Examples #1 and #2 we want a specified value of gain less than maximum, say 12 dB. We already know from example #1 that the optimum match for the load results in only 0.05 dB gain increase. We also know that the transistor gain is 12.9 dB, which tells us that we must create a match on the input such that the gain contribution from the input network is approximately –1.0 dB.

We can use Equations 13-21 through 13-24 to create the –1 dB gain circle for the input network. We know (from Example #1) $G_{s,max}$ is 1.94 dB or, stated as a number, it is 1.56. Thus g_s becomes $g_s = G_s/G_{s,max} = 0.79/1.56 = 0.506$, $d_s = 0.37$ (from Equation 13-21) and $R_s = 0.545$ (from Equation 13-22). These values are plotted in Figure 13-8. Remember in plotting the –1dB gain circle, d_s is expressed in terms of the magnitude of S_{11}. Any point on the –1 dB gain circle will provide the desired value of Γ_s but for reasons of convenience in matching, the point selected is point "A." At this point, $\Gamma_s = 0.44$ at an angle of 100°. We can now compute Γ_L from Equation 13-13, which for the case of $S_{12} = 0$ becomes $\Gamma_{OUT} = S_{22}$ and we know that Γ_L is the conjugate of Γ_{OUT}. Thus $\Gamma_L = 0.11$ at an angle of +164°. We recognize Γ_L as having the same value for example #3 as for Example #1.

Example #4: Broadband Design, $S_{12} = 0$

Our goal will be to design the MRF571 into a broadband circuit having 14 dB of gain operating from 500 to 1000 MHz. The amplifier is to be driven from a 50 Ω source and is to drive a 50 Ω load. We will again assume the bias to be 6 volts and 50 mA. Remember, this is less than the maximum available gain as seen in example #1. From the table of S-parameters, we find $|S_{21}|^2$ results in

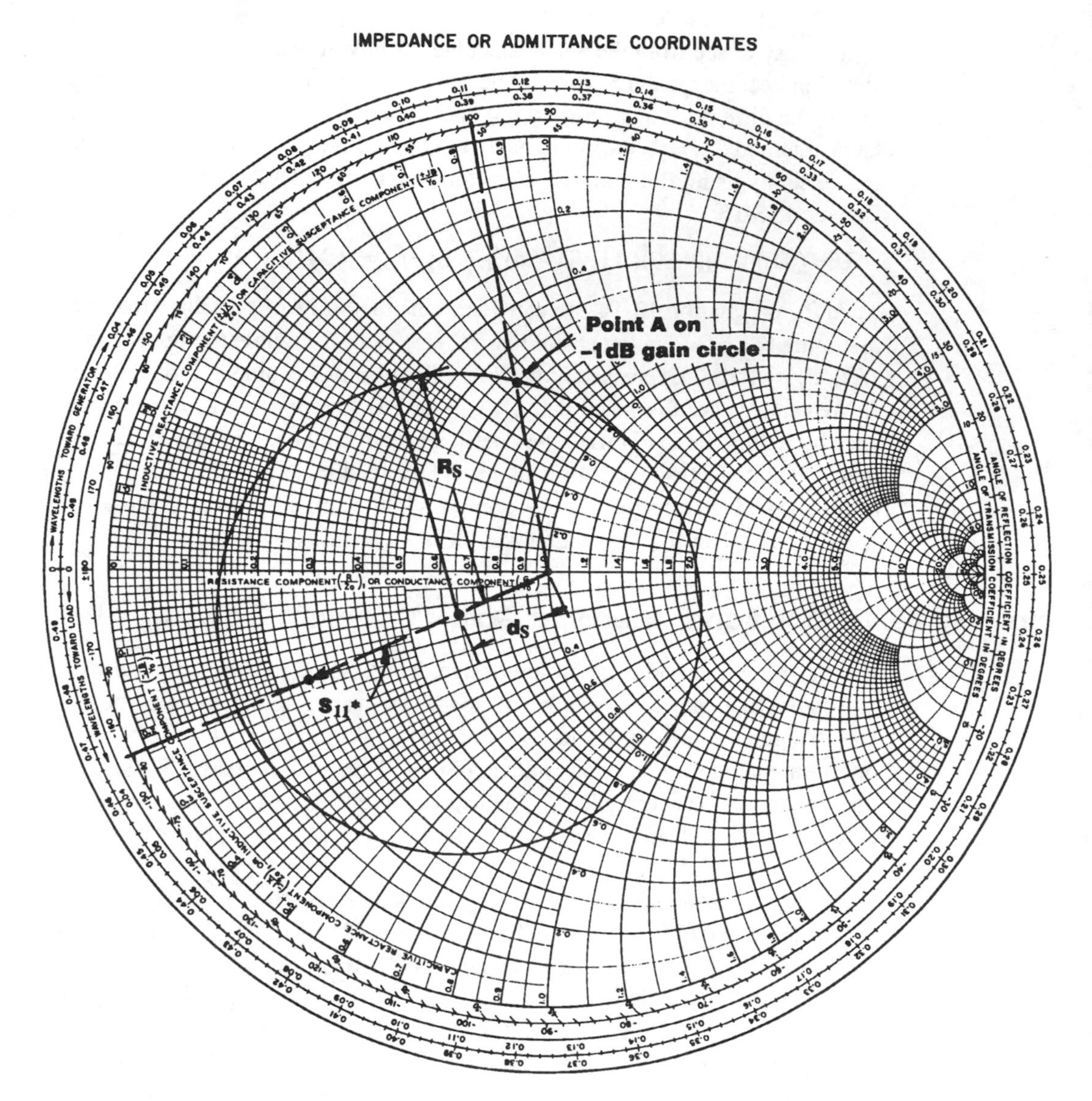

FIGURE 13-8
Smith Chart plot for Example #3.

70.6 or 18.5 dB gain at 500 MHz and 19.3 or 12.9 dB gain at 1000 MHz. Thus our matching circuits must decrease the gain by 4.5 dB at 500 MHz and increase the gain by 1.1 dB at 1 GHz.

We will plan to put all the "gain" or "loss" in the input matching network. As a result of this assumption, it will be easier to draw the gain circles using the previously stated formulas (Equations 13-41 through 13-44) for available gain. One could also use a computer program such as MMICAD (which was the process used to determine the gain circles plotted in Figure 13-9).

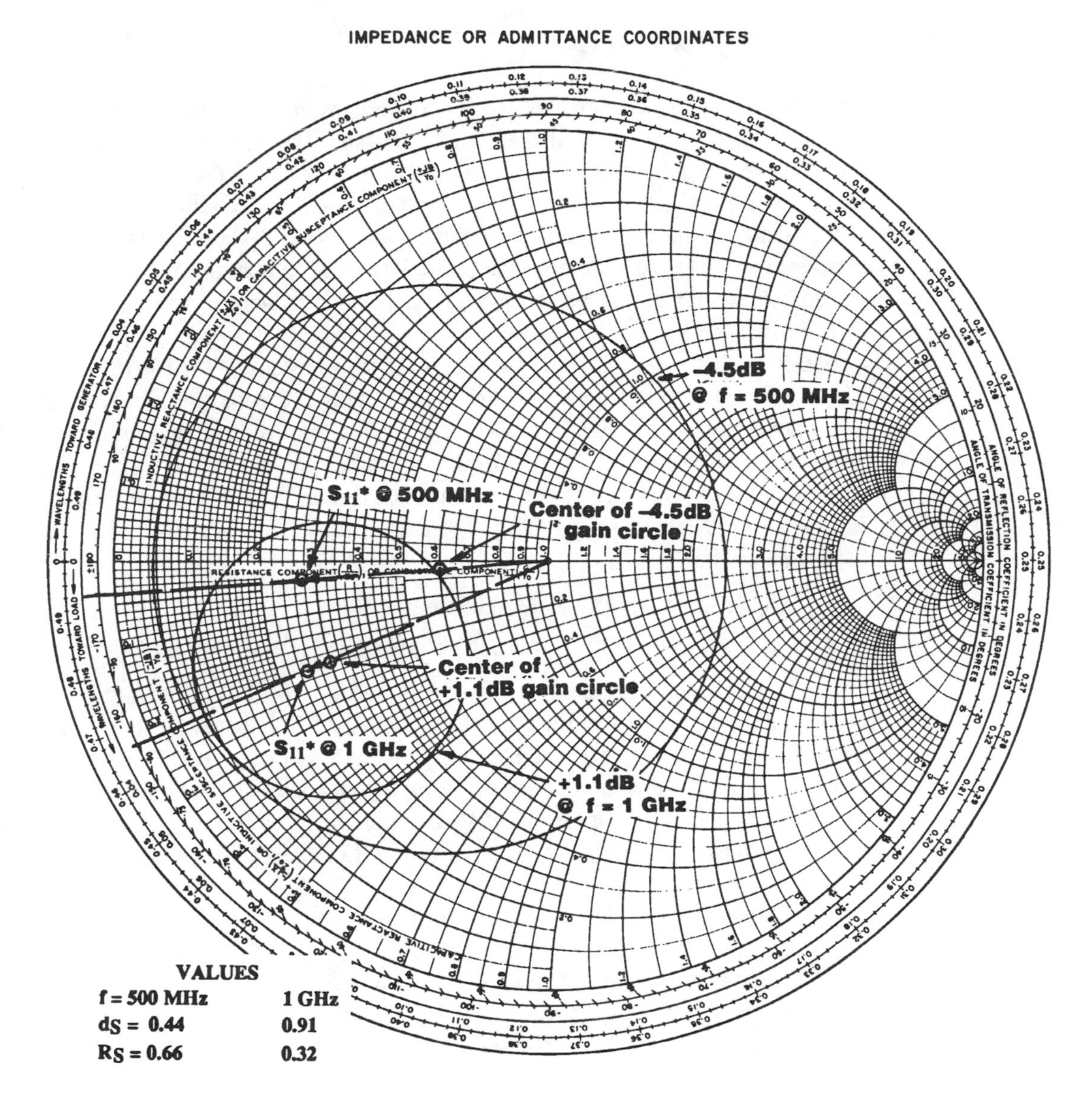

FIGURE 13-9

Smith Chart plot for Example #4. Values for f = 500 MHz: $d_s = 0.44$, $R_s = 0.66$. Values for f = 1 GHz: $d_s = 0.91$, $R_s = 0.32$.

Later, in our final phase of this chapter, we will use these circles to determine matching networks that place the source impedance on the –4.5 dB circle at 500 MHz and at the same time on the +1.1 dB circle at 1 GHz. Then we will calculate the load impedance using Equation 13-13. The process of choosing the proper source reflection coefficient is an iterative process and can be done manually or through the use of a computer optimization program such as MMICAD.

Example #5: Designing for Low Noise

So far we've concerned ourselves only with gain. Many low power amplifiers are required to be low noise also and in this instance we must take into account the effect our matching has on noise figure. Remember, only the source reflection coefficient affects noise figure. It is true that the load reflection coefficient will affect the source reflection coefficient IF $S_{12} \neq 0$. If $S_{12} \neq 0$, then the simplest step to take is to choose the desired value of Γ_s that gives the desired tradeoff between gain and noise figure, then use that value of Γ_s to calculate Γ_L from Equation 13-13.

In this example, we will once again assume the transistor is the MRF571. The frequency is 1 GHz but we will change the bias to 6 volts and 5 mA, which is a more appropriate value for a low noise application. Another reason for these choices is because the MRF571 data sheet has gain and noise figure contours plotted for the conditions stated. Let's further assume that gain must be at least 12 dB and the noise figure not greater than 2 dB. Our new values of S-parameters are:

$S_{11} = 0.61$, angle +178° $S_{12} = 0.09$, angle +37°

$S_{21} = 3.0$, angle +78° $S_{22} = 0.28$, angle −69°

An examination of Figure 13-10 (which is a plot of the 2 dB noise figure circle and the 12 dB gain circle taken from the MRF571 data sheet) shows that the 2 dB noise circle intersects the 12 dB gain circle in two places. Either value of G_s will lead to the result of both the desired gain and noise figure. We can also select a value for G_s on the 12 dB gain circle between these two points that will result in even lower than 2 dB noise figure. Let's select the point "A" shown in Figure 13-10.

The value for Γ_s is estimated to be 0.55 at an angle of 158°. All that's left is to determine the value of Γ_L from the formula $\Gamma_L = [S_{22} + (S_{12}S_{21}\Gamma_s/\{1 - (\Gamma_s S_{11})\}]^*$ It is apparent from the formula that for the case of $S_{12} = 0$, $\Gamma_L = S_{22}^*$. Thus if $S_{12} = 0$, $\Gamma_L = 0.28$ at an angle of 69°. If we take the value of S_{12} into account, $G_L = 0.48$ at an angle of 82°. These values are also shown in Figure 13-10 as points "B" and "C" respectively.

CIRCUIT REALIZATION

For each example given, let's now determine circuits that will realize the desired impedances:

Example #1

The input impedance for Example #1 is plotted in Figure 13-11. The Smith Chart shows $\Gamma_s = 0.6$ at an angle of −156°. This is the value initially determined in Example #1 and plotted in Figure 13-7. The problem is to take the 50 Ω source and make it look like Γ_s. The most common circuit is a low-pass filter

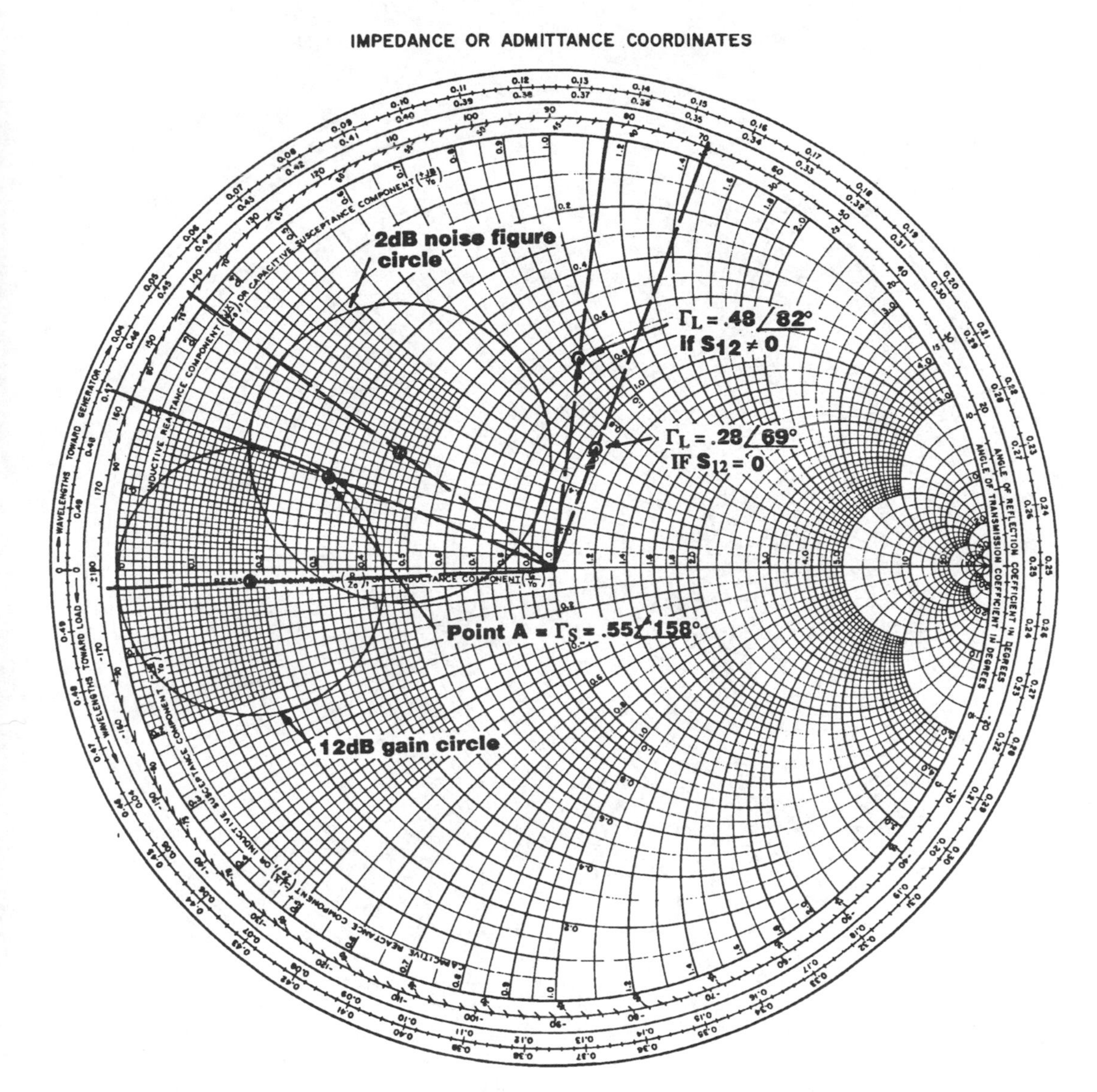

FIGURE 13-10
Smith Chart plot for Example #5.

configuration consisting of a shunt C and a series L. Remember that when using the Smith Chart with shunt elements you use it as an admittance chart, and when using the chart for series elements you use it as an impedance chart. For this reason, Figure 13-11 is plotted on a special Smith Chart graph which shows both normalized impedance and admittance circles simultaneously.

Figure 13-11 shows that a shunt capacitor having a susceptance value of +j1.75 (the arc length from point "O" to point "A") and a series inductance having a positive reactance value of +j0.225 (the arc length from point "A" to point "B") will rotate the 50 Ω source into the normalized value of 0.6 at –156°. It is

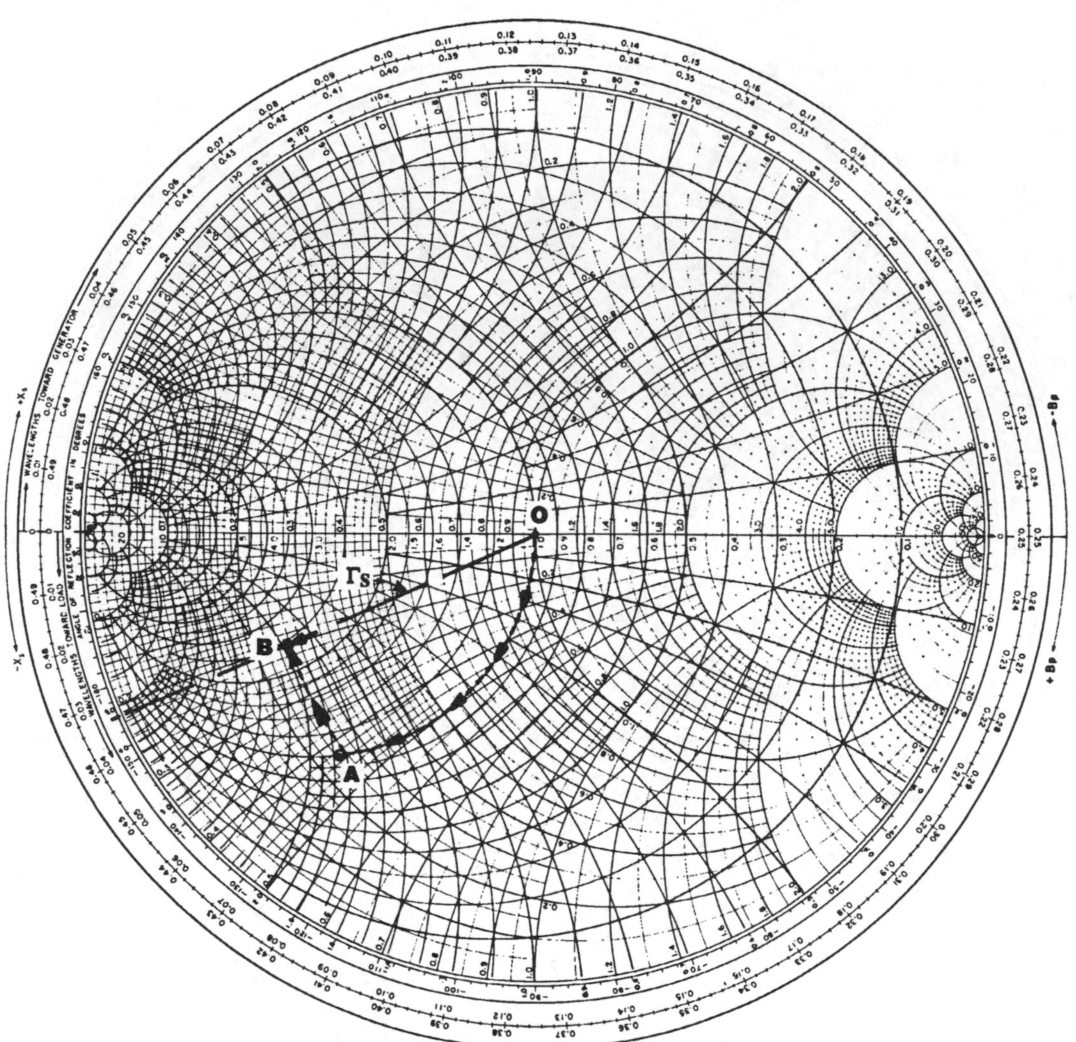

FIGURE 13-11
Input matching for Example #1.

a simple matter to calculate the actual values that give these normalized results at a frequency of 1 GHz. The values turn out to be 5.6 pF and 1.8 nH.

In a similar manner, Figure 13-12 shows the output load impedance also previously plotted in Figure 13-7. Again a network of shunt C and series L will be used to transform the 50 Ω load to the desired load impedance represented by point "D" in Figure 13-12. The arc from point "O" to point "C" calls for a shunt capacitor having a normalized susceptance value of +j 0.5. The arc from point "C" to point "D" requires a series inductance having a normalized value of +j 0.45. Again at a frequency of 1 GHz, these can be realized by a shunt capacitor of approximately 1.6 pF and a series inductance of 3.6 nH. The final circuit configuration is shown in Figure 13-13.

NORMALIZED IMPEDANCE AND ADMITTANCE COORDINATES

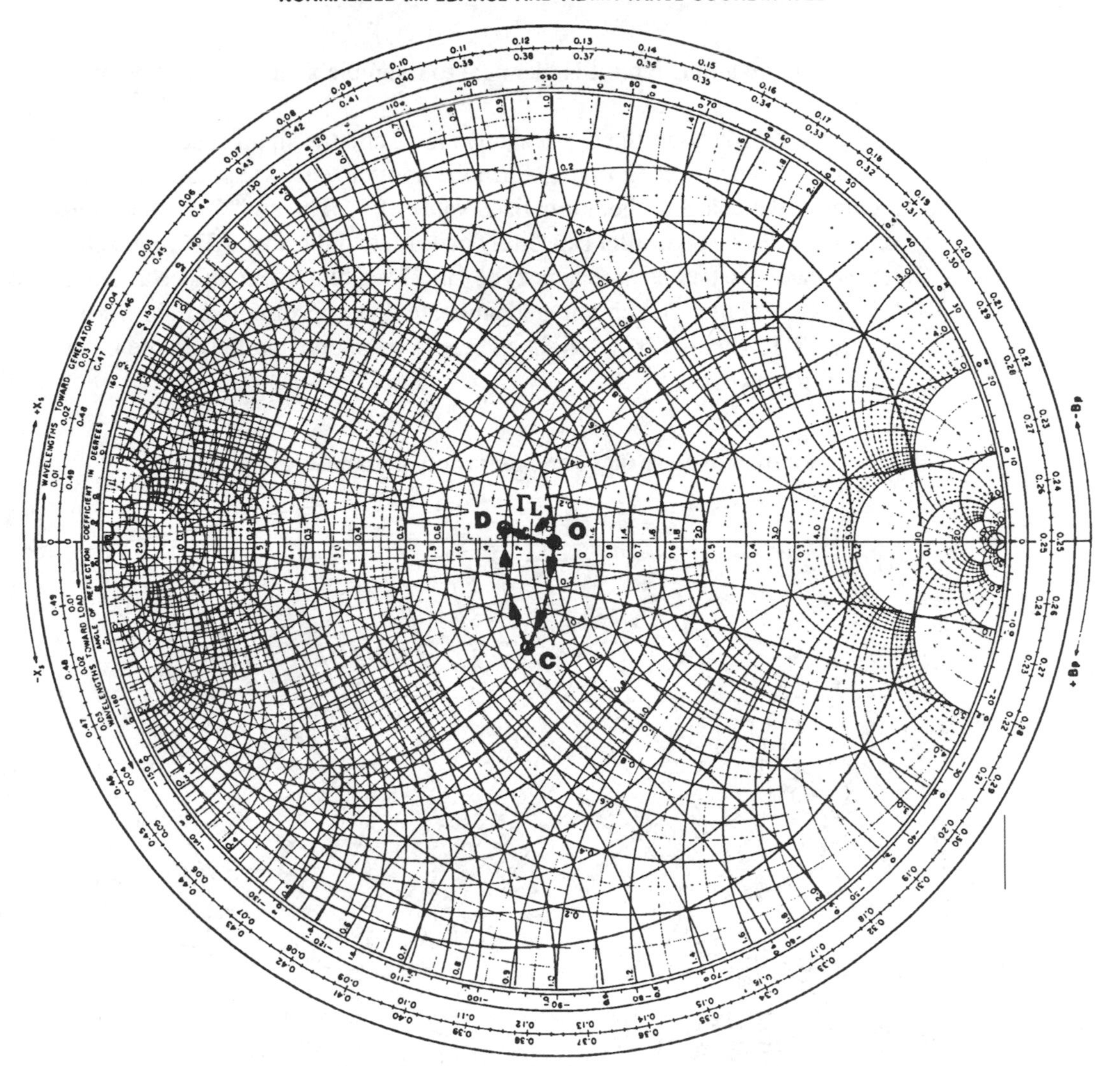

FIGURE 13-12
Output matching for Example #1.

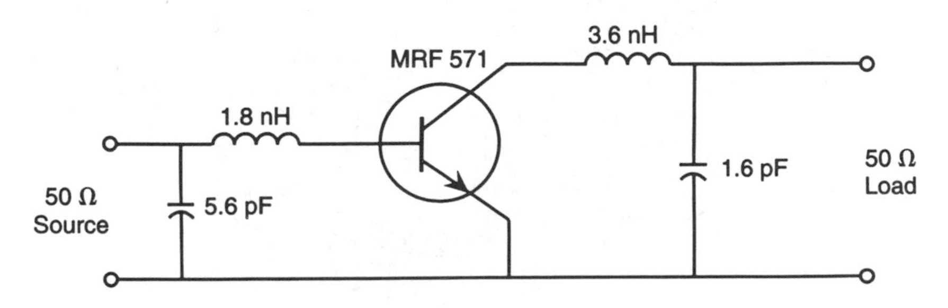

FIGURE 13-13
Narrow band matching for best gain. This circuit can be used to match input and output in Example #1.

Example #2

The output matching circuit is identical to Example #1and will not be repeated here.

Figure 13-14 shows the slight change in Γ_s and the shunt susceptance and series inductance needed to transform the normalized 50 Ω source impedance (center of the chart) to the normalized value of Γ_s. The arc "OA" represents a shunt susceptance (parallel capacitor) of +j1.6 and the arc "AB" represents a series reactance (series inductance) of +j0.275. Component values at 1 GHz are 5.1 pF and 2.2 nH. The circuit is identical to that of Example #1 shown in Figure 13-13 except for the slight change in component values in the input matching network.

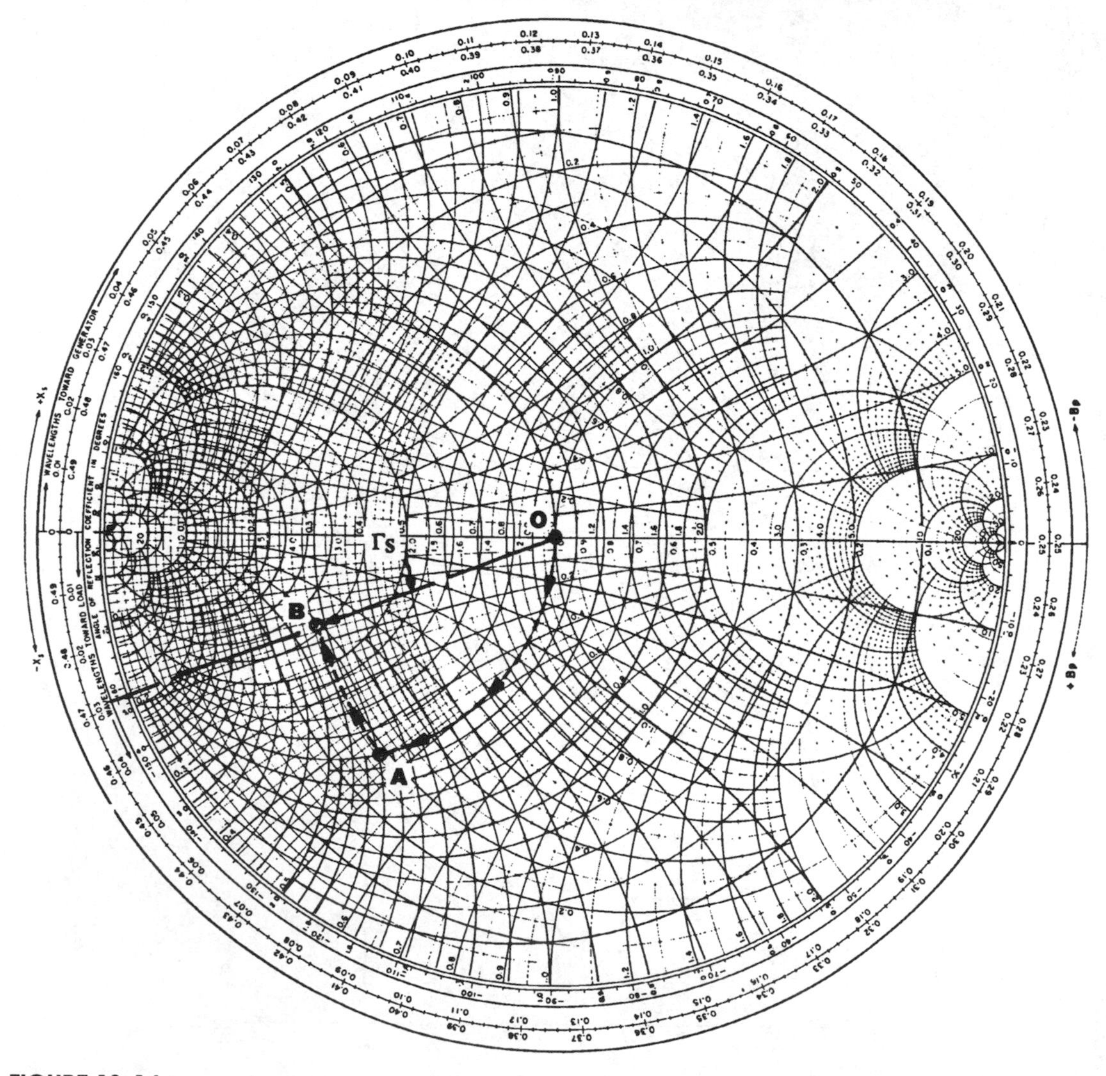

FIGURE 13-14
Input matching for Example #2.

Example #3

The input circuit of Example #3 is shown in Figure 13-15. Point "A" from Figure 13-8 has been re-plotted on a Smith Chart showing both impedance and admittance circles. The value of Γ_s selected was 0.44 at an angle of +100°. The transformation of the 50 Ω source impedance into the desired value represented by Γ_s is again most easily accomplished by using a shunt C and series L as shown in Figure 13-15.

Arc "OC" represents a shunt capacitor having a normalized susceptance of 0.82 while the arc "CA" represents a series inductance having a normalized reactance of approximately 1.15. These values are realized at 1GHz by a capacitance of 2.6 pF and inductance of 9.2 nH. Because the desired output load

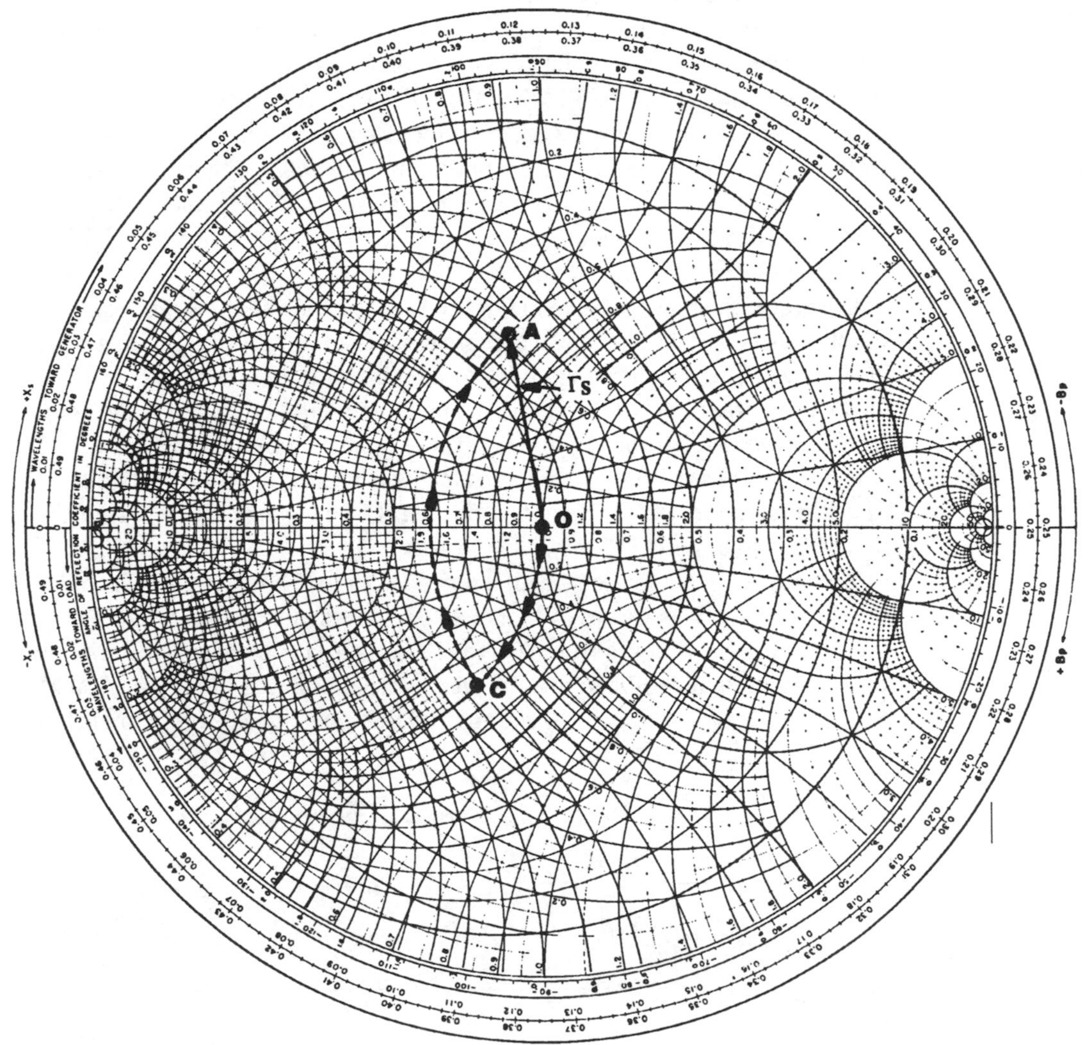

FIGURE 13-15
Input matching for Example #3.

impedance is identical in Example #3 and Example #1, matching the output circuit is also identical to what was shown in Figure 13-12. A final circuit for Example #3 is shown in Figure 13-16.

Example #4

The gain circles shown in Figure 13-9 are re-plotted in Figure 13-17 using the special Smith Chart with normalized impedance and admittance circles. The objective is to devise a matching network that places us on the –4.5 dB gain circle at 500 MHz and at the same time places us on the +1.1 dB gain circle at 1 GHz. In this instance, the MMICAD optimizer program was used to determine an appropriate network which consisted of a shunt C and a combination of series C and L. A similar solution could have been achieved by an iterative process of adjusting the circuit values until a combination was determined which met both objectives simultaneously.

At 500 MHz the shunt susceptance moves us along arc "OA." The same capacitor will move us to point "C" when the frequency increases to 1 GHz. When the frequency is at 500 MHz, the series network must move us along the arc "AB" while the same network must move us from point "C" to point "D" at 1 GHz. Susceptance and reactance values that will achieve these objectives simultaneously are a shunt susceptance of +j0.4 at 500 MHz, increasing to +j0.8 at 1 GHz and an overall series reactance that is –j0.84 at 500 MHz but +j0.09 at 1 GHz.

Component values that give the desired susceptance and reactances are a shunt capacitor or 2.55 pF and a series capacitor of 5.4 pF along with a series inductance of 5.5 nH. This input network is shown in Figure 13-19. Because all the "gain" correction was placed in the input network, the output is the conjugate match or S_{22}^* at each frequency. These points are plotted in Figure 13-18. The problem now is to find a network that will locate simultaneously the matched load (the point "O" in the center of the chart) at point "A" when f = 500 MHz and at point "B" when f = 1 GHz.

A shunt LC network along with a series LC network was chosen such that the shunt network as shown in Figure 13-18 moves along the arc "OC" at 500 MHz but only moves from "O" to "D" at 1 GHz. Likewise, the series network moves the normalized line impedance from point "C" to "A" at 500 MHz and at the same time moves the normalized line impedance from point "D" to

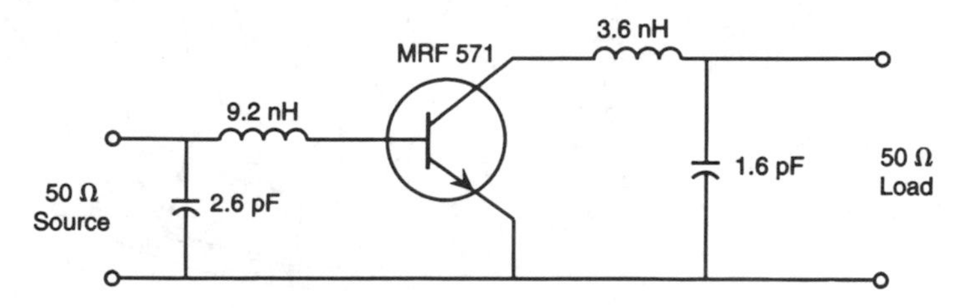

FIGURE 13-16
Circuit realization for Example #3.

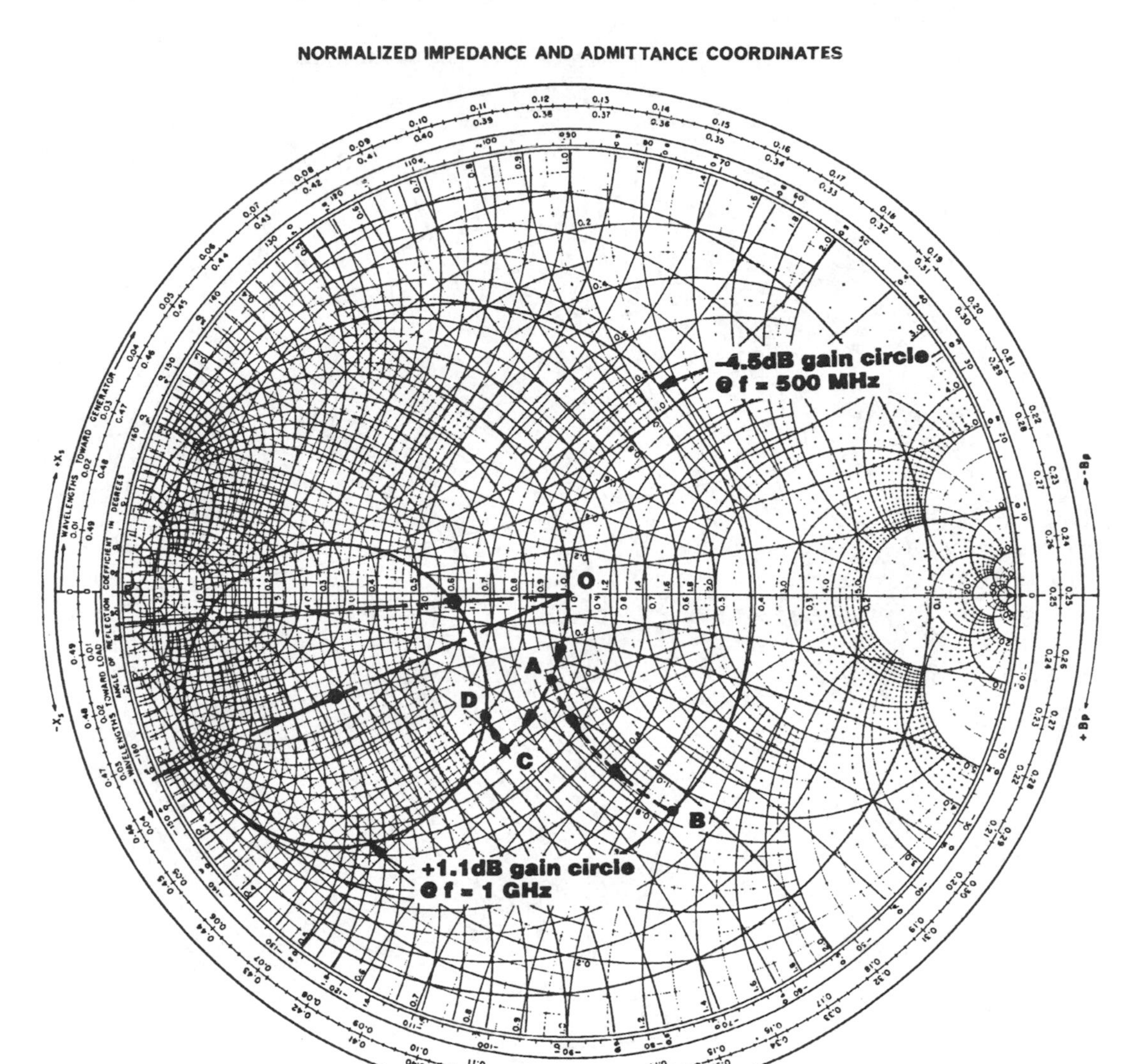

FIGURE 13-17
Input matching for Example #4.

"B" at 1 GHz. The values of L and C that will accomplish these objectives can be determined by solving relatively simple simultaneous equations relating the susceptances and reactances at the two frequencies of interest. Actual values, along with the input network, are shown in Figure 13-19.

Example #5

Figure 13-20 shows the value of Γ_s taken from Figure 13-10 and re-plotted on the special impedance and admittance Smith Charts used to determine graphically the impedance matching networks needed for the particular application.

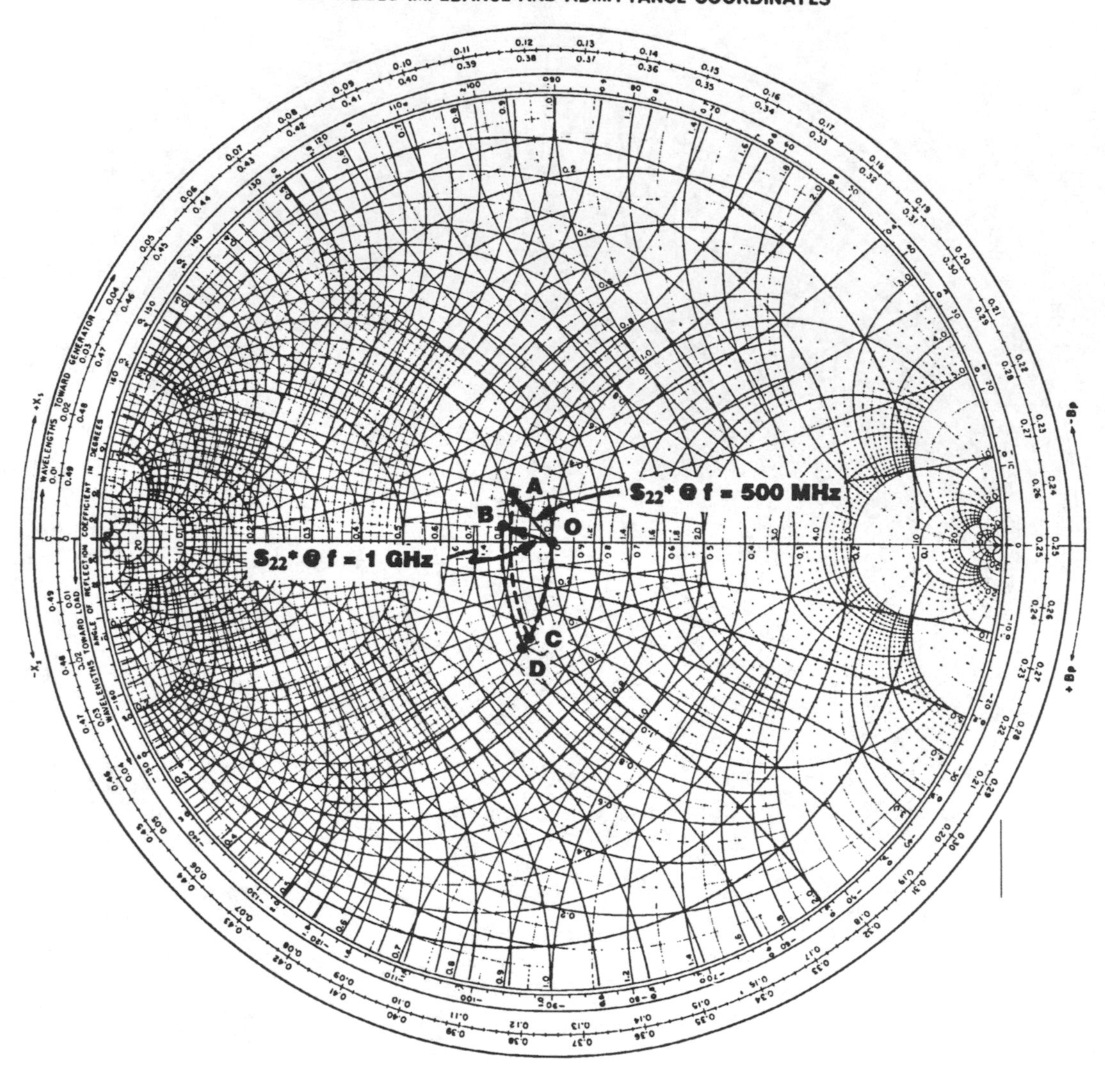

FIGURE 13-18
Output matching for Example #4.

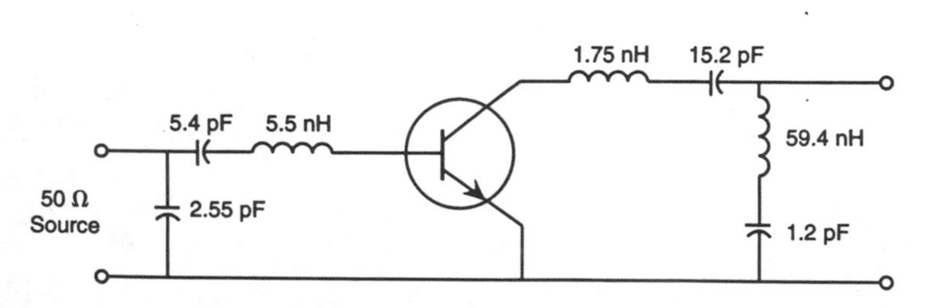

FIGURE 13-19
Circuit realization for Example #4.

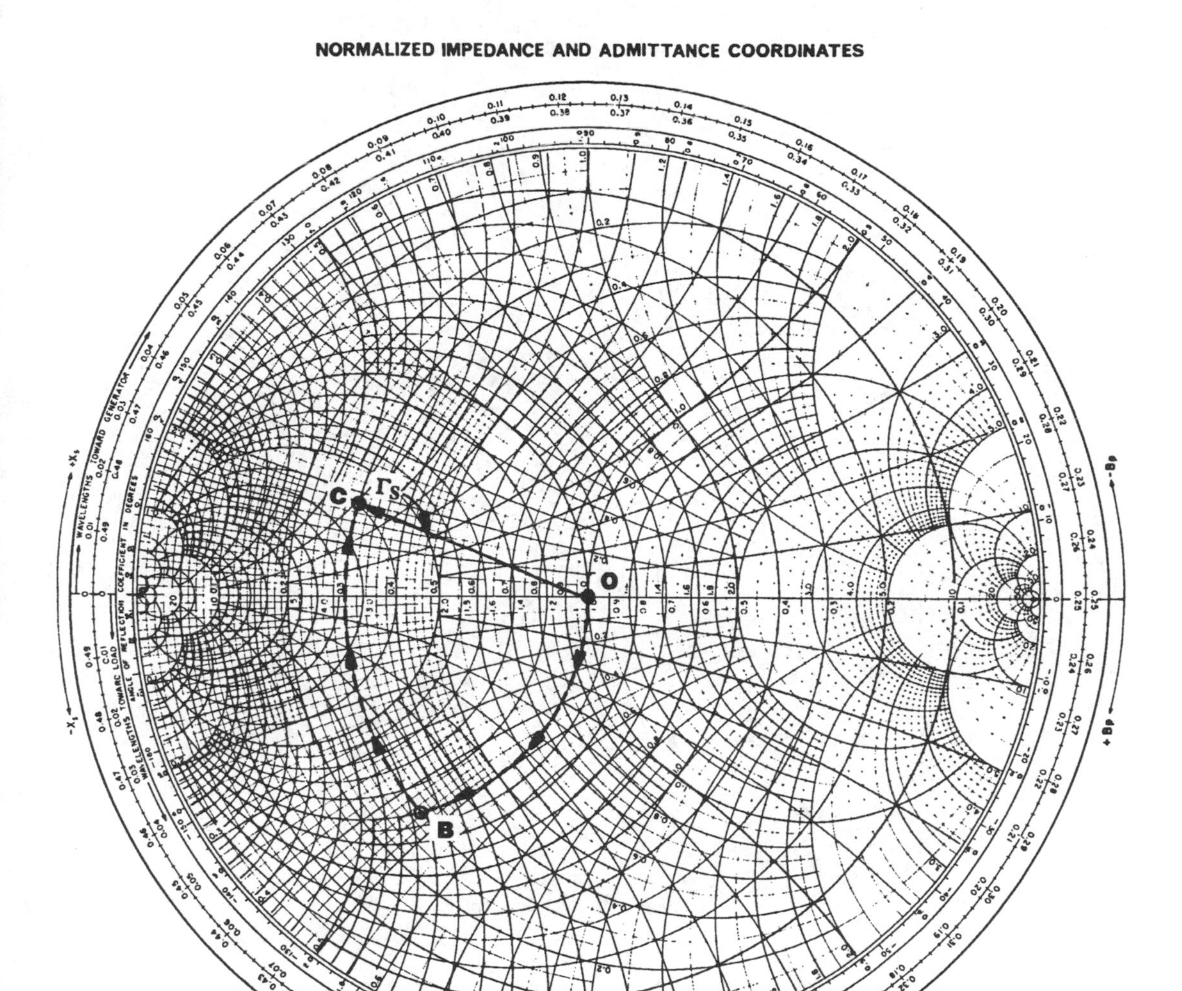

FIGURE 13-20
Input matching for Example #5.

Matching is readily achieved by using a shunt capacitor to transform the 50 Ω source from point "O" to point "B." Then a series inductance finishes the match by transforming the value of point "B" to the desired Γ_s shown as point "C." Values of shunt C and series L that achieve the desired source reflection coefficient are 4.9 pF and 5.1 nH as shown in the circuit configuration of Figure 13-22.

Output impedance matching is shown in Figure 13-21. The matching network if $S_{12} = 0$ is a series capacitance and a shunt inductance that transforms the matched load first to point "A" (series capacitor) and then to point "B" (shunt inductance) or in the case of $S_{12} \neq 0$ to point "C" (series capacitor) and then to point "D" (shunt inductance). Thus for $S_{12} = 0$, arc "OA" is a reactance

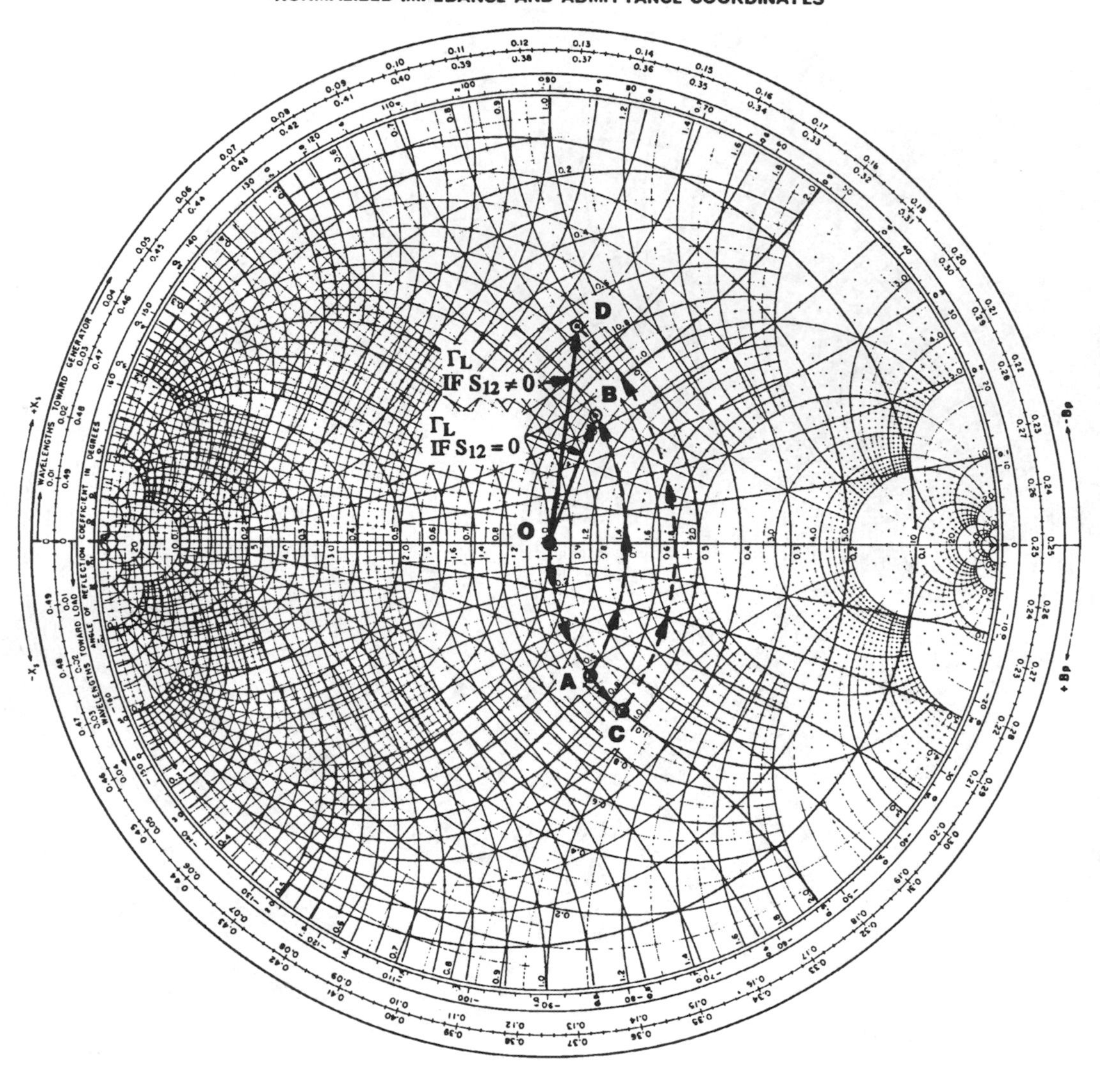

FIGURE 13-21
Output matching for Example #5.

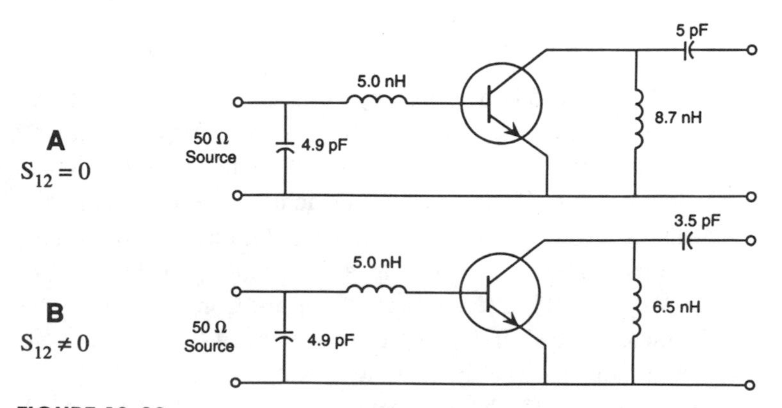

FIGURE 13-22
Circuit configurations for Example #5

of –j0.64 which translates into a series capacitor of 5 pF. Likewise, arc “AB” is a susceptance of +j0.91 which can be realized by a shunt inductance having a value of 8.7 nH as shown in Figure 13-22A.

If the measured value of S_{12} is used, the matching requires a reactance represented by the arc “OC” and a susceptance represented by the arc “CD.” These values are realized by a series capacitor of 3.5 pF and a shunt inductance of 6.5 nH as shown in Figure 13-22B.

References

[1] Chris Bowick, *RF Circuit Design*, Indianapolis: Howard Sams & Co., 1982.

[2] “S-Parameter Design,” Application Note #154, Hewlett-Packard Co., Palo Alto, CA, April, 1972.

[3] Roy Hejhall, “Small Signal Design Using Two-Port Parameters,” Application Note #215A, Motorola Semiconductor Sector, Phoenix, AZ.

[4] Operating Manuals for 8753C and 85046A/B, Hewlett Packard Co., Palo Alto, CA.

[5] J. M. Rollett, “Stability and Power Gain Invariants of Linear Two Ports,” *IRE Transactions—CT*, Volume CT-9, March, 1962, pp. 29-32.

[6] Guillermo Gonzalez, *Microwave Transistor Amplifiers*, Englewood Cliffs, NJ: Prentice-Hall, Inc., 1984.

[7] Dragon Wave 2.0 (for Macintosh PC’s), Nedrud Data Systems, P.O. Box 27020, Las Vegas, NV, 89126.

[8] MMICAD (for IBM PCs), Optotek, 62 Steacie Drive, Kanata, Ontario, Canada, K2K2A9.

[9] Data sheet for MRF571, Motorola Semiconductor Sector, Phoenix, AZ.

Index